04T

Lecture Notes in Control and Information Sciences 301

Editors: M. Thoma · M. Morari

Springer

Berlin
Heidelberg
New York
Hong Kong
London
Milan
Paris
Tokyo

M. de Queiroz · M. Malisoff · P. Wolenski (Eds.)

Optimal Control, Stabilization and Nonsmooth Analysis

With 27 Figures

Springer

Editors

Dr. Marcio S. de Queiroz
Louisiana State University
Department of Mechanical Engineering
Baton Rouge, LA 70803-6413
USA

Dr. Michael Malisoff
Dr. Peter Wolenski
Louisiana State University
Mathematics Department
Baton Rouge, LA 70803-6413
USA

ISSN 0170-8643

ISBN 3-540-21330-9 Springer-Verlag Berlin Heidelberg New York

Library of Congress Control Number: 2004103120

Springer-Verlag is a part of Springer Science+Business Media

springeronline.com

Printed in Germany

Typesetting: Data conversion by the editors.
Final processing by PTP-Berlin Protago-TeX-Production GmbH, Berlin
Cover-Design: design & production GmbH, Heidelberg
Printed on acid-free paper 62/3020Yu - 5 4 3 2 1 0

Preface

Applied mathematics and engineering abound with challenging situations where one wishes to control the behavior of the solution trajectories of dynamical systems. These include optimal control problems, where the goal is to choose a trajectory that optimizes a given cost criterion subject to given constraints, as well as stabilization problems, where one tries to keep the state of the system in a given set or bring the state towards a specified set. Specific issues that subsequently arise include the construction of stabilizing control mechanisms, and the analysis of Hamilton-Jacobi equations whose solutions give the minimum cost in optimal control.

It has become clear over the past two decades that classical mathematical concepts are not wholly adequate to describe and prescribe the behavior of controlled dynamical systems, largely because nonsmoothness arises naturally when min operations are present, or when nonlinear systems develop shocks. The theory of nonsmooth analysis has been developed in part to supply the requisite mathematical tools needed to systematically handle these phenomena. This has led to an explosion of recent results in nonsmooth analysis, accompanied by major breakthroughs in mathematical control theory, optimization, and control engineering.

This volume explores some of these recent developments. The twenty-four papers comprising this volume are based on the contributed and invited talks from the Louisiana Conference on Mathematical Control Theory (MCT'03), which was held at Louisiana State University in Baton Rouge in April 2003. Particular attention is paid in this volume to new results in nonsmooth analysis, and to novel applications of nonsmooth analysis in control and optimization. The contributing authors include major figures and leading young researchers who are at the cutting edge of modern control theory and nonsmooth analysis. While some of the papers collected here are announcements of important new mathematical results, or of novel applications, this volume also contains authoritative surveys by Francis Clarke, Eduardo Sontag, Andrew Teel, and others that will be of broad interest to postgraduates and researchers in systems science, control, optimization, and applied mathematics.

The first section of this book is devoted to optimal control, and to closely related areas, such as Hamilton-Jacobi equation theory and singular perturbations. The section begins with a survey by Andy Teel on receding horizon optimal control, which is an optimization paradigm for stabilizing nonlinear systems. This is followed with a survey by Richard Vinter on feedback strategies in differential games based on the Isaacs equation. Other papers in this section include a survey of new results for systems of Hamilton-Jacobi equations by Daniel Ostrov, and new results by Zvi Artstein on limits of singular perturbations. Also included are new necessary optimality conditions for delay differential inclusions by Boris Mordukhovich and Lianwen Wang, a new value function approach to turnpike theory by Pierre Cartigny and Alain Rapaport, and new results by Gaemus Collins and William McEneaney on min-plus methods for the Hamilton-Jacobi equations associated with H_∞ problems. Finally, Boris Mordukhovich and Ilya Shvartsman discuss issues related to optimization and feedback control of constrained parabolic systems.

The next section of this book is devoted to stabilization, and to related issues such as Lyapunov function theory. The section begins with two surveys co-authored by Eduardo Sontag, the first with David Angeli on monotone systems and their applications in systems biology, and the second with Michael Malisoff on the input-to-state stabilization of asymptotically controllable systems with respect to actuator errors under small observation noise. Also included are two alternative but related approaches to stabilization for systems that lack continuous time-invariant stabilizing feedbacks, the first by Ludovic Rifford on smooth almost global stabilization, and the second by Fabio Ancona and Alberto Bressan on patchy vector field approaches. The section also includes a survey on small gain results for interconnected systems by Zhong-Ping Jiang, an alternative approach to input-to-state stabilization by Lars Grüne called input-to-state dynamical stability that takes the decay of the perturbation term explicitly into account, and a discussion by Andrey Smyshlyaev and Miroslav Krstic on boundary stabilization problems for PDEs. Finally, Bin Xian, Marcio de Queiroz, and Darren Dawson discuss a new approach to stabilizing uncertain systems.

The final section focuses on nonsmooth analysis and its numerous important applications in Hamilton-Jacobi and Lyapunov function theory and optimization. The section begins with a survey by Francis Clarke on the unifying role of nonsmooth analysis in Lyapunov and feedback theory, and in particular, the importance of semiconcave and other nonsmooth control-Laypunov functions for feedback stabilization. This is followed by an extensive two-part analysis of nonsmooth analysis on smooth manifolds by Yuri Ledyaev and Qiji Zhu, duality theory for Hamilton-Jacobi equations by Rafal Goebel and Chadi Nour, and some recent results by Tzanko Donchev, Vinicio Rios, and Peter Wolenski on dynamical systems with non-Lipschitz dynamics. The section closes with a discussion by Boris Mordukhovich and Bingwu Wang on generalized differentials for moving sets and moving mappings.

We would like to express our gratitude to the National Science Foundation and to the Louisiana Board of Regents Enhancement Program for their generous support for our MCT'03 conference (NSF Grant DMS 0300959 and Board of Regents Contract Number LEQSF-2002-04-ENH-TR-26). This support made it possible for international scholars and young researchers without other sources of support to participate in our conference and contribute to this conference volume. We thank Dr. Mark Emmert, Chancellor of Louisiana State University, for delivering the opening remarks at the conference. We also thank the many people who helped us organize MCT'03 and compile this book. In particular, we thank Susan Oncal and Louisiana Travel for coordinating the visits of the international participants, Julie Doucet for handling the conference-related paperwork, and Dr. Thomas Ditzinger and Heather King from Springer-Verlag Heidelberg for publishing this volume.

We especially thank our referees and typesetters. Their careful and prompt work on this volume was very much appreciated. Finally, we thank Zed Pobre and Jeff Sheldon from the Louisiana State University Department of Mathematics for their expert assistance with the computer-related issues that arose during our conference and during the completion of this work.

Baton Rouge, Louisiana
December 2003

Marcio S. de Queiroz,
Michael A. Malisoff,
and Peter R. Wolenski

Contents

Part II Stabilization and Lyapunov Functions

Part III Nonsmooth Analysis and Applications

Part I

Optimal Control, Optimization, and Hamilton-Jacobi-Bellman Equations

Discrete Time Receding Horizon Optimal Control: Is the Stability Robust?

Andrew R. Teel

Department of Electrical and Computer Engineering, University of California, Santa Barbara 93106-9560. teel@ece.ucsb.edu

0 Introduction

Receding horizon optimal control (RHOC), also called model predictive control, is an increasingly popular approach used to synthesize stabilizing feedback control laws for nonlinear systems. For a recent nonlinear RHOC survey, see [31]. In industry, this control method typically is based on discrete-time plant models which makes determining the RHOC feedback law equivalent to solving a finite-dimensional optimization problem. The latter often can be carried out on-line, especially for slow plants like those in the chemical process industry where RHOC has found widespread application [17], [37]. Even for faster plants, RHOC is beginning to be considered as a reasonable feedback control alternative. See the aerospace-related applications in [5], [19].

In addition to the computational advantages that discrete time affords, there are fewer technical issues than in the corresponding continuous-time formulation. (Many such issues are related to topics addressed elsewhere in this volume.) For these reasons, the discrete-time formulation is typically favored by engineers. Interestingly, many of the issues related to robustness of stability in continuous time, unearthed by mathematicians, carry over to the discrete-time setting. These issues have been largely overlooked in the discrete-time RHOC literature. Some exceptions are [10], [39], and [41], where some nongeneric assumptions are made to guarantee robustness.

Herein we discuss the nominal robustness (or lack thereof) of asymptotic stability induced by the standard form of discrete-time RHOC. (We will not address continuous-time RHOC, which has also received considerable attention. See, e.g., [6], [11], [20], [30], [34], [35].) We are motivated by two facts:

1. Asymptotic stability in *discontinuous* discrete-time systems may have no robustness.
2. Feedback control algorithms generated by discrete-time RHOC are often (sometimes necessarily) discontinuous.

M. de Queiroz et al. (Eds.): Optimal Control, Stabilization and Nonsmooth Analysis, LNCIS 301, pp. 3–27, 2004.

The first fact is the same phenomenon reported in [18] for continuous-time systems where the discontinuity comes from implementing a certain optimal control strategy in feedback form. The second fact is also well-known. For discrete-time examples, see [32]. However, despite the warnings that come from [18], little attention has been given to the following question:

Does discrete-time RHOC ever yield closed-loop asymptotic stability with no robustness?

As we will see, the answer is "yes". So a natural follow up question is:

(How) can RHOC be configured, in a computationally tractable way, so that the "zero robustness" phenomenon does not occur?

This paper has two parts. The first part is on asymptotic stability theory for discontinuous discrete-time systems. In this part, we demonstrate the possibility of zero robustness in discontinuous discrete-time systems that are asymptotically stable. Since asymptotic stability in discontinuous systems is not always fragile, it is worthwhile to pursue results that distinguish between the cases of robustness and no robustness. Some results of this type will be stated. This part of the paper is inspired by the breakthrough results in [9] on robustness of stability for continuous-time systems described by differential inclusions or differential equations with discontinuous right-hand side.

The second part of this work is on discrete-time control theory. We will discuss basic notions such as asymptotic controllability, control Lyapunov functions, and robust stabilizability. This part of the paper parallels the seminal continuous-time results in [42]. See also [26], [45]. We will recall the example in [32] showing that discontinuous feedback is sometimes necessary for stabilization. Then we will describe the standard RHOC paradigm (taken from the survey [31]) and show that sometimes it can produce discontinuous feedback (even when it isn't necessary) that yields asymptotic stability with no robustness. We will also emphasize that this is a nonlinear phenomenon: the lack of robustness we demonstrate does not occur when using RHOC to stabilize linear systems with convex (state and input) constraints. Finally, we will discuss reformulations of the standard RHOC algorithm that lead to guaranteed robustness of asymptotic stability. Some of the ideas mentioned are extensions of those in [28].

This paper is a summary of recent results contained in [12], [13], [14], [23], [24], [25], and [33] where many of the technical details may be found.

Part I: Asymptotic Stability Theory

1 Notation and Definitions

First we will consider uncontrolled systems

$$x^+ = f(x) \ , \tag{1}$$

perhaps representing the closed-loop resulting from a feedback control algorithm. The notation in (1) is shorthand for the alternative $x(k+1) = f(x(k))$. We won't necessarily assume that f is continuous. The equation (1) is a special case of the *difference inclusion*

$$x^+ \in F(x) \ , \tag{2}$$

where $F : \mathbb{R}^n \mapsto$ subsets of $\mathbb{R}^n$. The value $x(k+1)$ can be anything in the set $F(x(k))$. Each solution of (2) at the kth iteration will be denoted $\phi(k, x)$. The set of solutions to (2) starting at x will be denoted $\mathcal{S}(x)$. We use $\mathbb{R}_{\geq 0}$ to denote the nonnegative real numbers and $\mathbb{Z}_{\geq 0}$ to denote the nonnegative integers. A function $\beta : \mathbb{R}_{\geq 0} \times \mathbb{Z}_{\geq 0} \to \mathbb{R}_{\geq 0}$ is said to belong to *class-$\mathcal{KL}$* if it is nondecreasing in its first argument, nonincreasing in its second argument, and $\lim_{s \to 0^+} \beta(s, k) = \lim_{k \to \infty} \beta(s, k) = 0$. A function from $\mathbb{R}_{\geq 0}$ to $\mathbb{R}_{\geq 0}$ is said to belong to *class-$\mathcal{G}$* (respectively, *class-$\mathcal{K}_\infty$*) if it is continuous, zero at zero, and nondecreasing (respectively, strictly increasing and unbounded). A function from $\mathbb{R}_{\geq 0}$ to $\mathbb{R}_{\geq 0}$ is said to belong to *class-$\mathcal{PD}$* if it is positive, except possibly at zero, and continuous. For $x \in \mathbb{R}^n$ and $\rho > 0$, the notation $x + \rho\mathcal{B}$ denotes the closed ball of radius ρ centered at x. Let $\mathcal{A}$ denote a closed subset of $\mathbb{R}^n$. Define $|x|_{\mathcal{A}} := \inf_{z \in \mathcal{A}} |x - z|$.

In this section we will consider only global versions of asymptotic stability for the sake of simplicity. This is enough to give the flavor of the results available. For local results, see [25].

Definition 1. *For the system (2) the set $\mathcal{A}$ is said to be* globally asymptotically stable *provided there exists $\beta \in \mathcal{KL}$ such that, for each $x \in \mathbb{R}^n$ and each $\phi \in \mathcal{S}(x)$, $|\phi(k, x)|_{\mathcal{A}} \leq \beta(|x|_{\mathcal{A}}, k)$ for all $k \in \mathbb{Z}_{\geq 0}$.*

Definition 2. *For the system (2) the function $V : \mathbb{R}^n \to \mathbb{R}_{\geq 0}$ is said to be a* global Lyapunov function for $\mathcal{A}$ *provided there exist $\alpha_1, \alpha_2 \in \mathcal{K}_\infty$ and $\rho \in \mathcal{PD}$ such that, for all $x \in \mathbb{R}^n$, $\alpha_1(|x|_{\mathcal{A}}) \leq V(x) \leq \alpha_2(|x|_{\mathcal{A}})$ and*

$$\sup_{w \in F(x)} V(w) \leq V(x) - \rho(|x|_{\mathcal{A}}) \ .$$

We will also be interested in the behavior of perturbed systems

$$x^+ \in F(x + e) + d \ , \tag{3}$$

where the perturbations e and d take small values. We will use $\mathbf{e}$ and $\mathbf{d}$ to denote the perturbation sequences, i.e., $\mathbf{e} = (e(0), e(1), \ldots)$ and $\mathbf{d} = (d(0), d(1), \ldots)$. Also, $\|\mathbf{e}\| := \sup_k |e(k)|$, $\|\mathbf{d}\| := \sup_k |d(k)|$. The set of solutions of (3) starting at x with $\max\{\|\mathbf{e}\|, \|\mathbf{d}\|\} \leq \delta$ will be denoted $\mathcal{S}_\delta(x)$.

Robust asymptotic stability (see Definition 3 below) will correspond to the situation where $\mathcal{S}_\delta(\cdot)$ is qualitatively similar to $\mathcal{S}(\cdot)$, the latter satisfying Definition 1. While it is possible to discuss robustness for noncompact attracting sets, for simplicity in this paper we will consider robustness for *compact* attracting sets. Because of this, sometimes we will restrict statements to the case of compact attracting sets even though more general results are available.

Definition 3. *Let $\mathcal{A}$ be compact. For the system (3) the set $\mathcal{A}$ is said to be globally robustly asymptotically stable if there exists a class-$\mathcal{KL}$ function β and for each $\varepsilon > 0$ and compact set C there exists $\delta > 0$ such that for each $x \in C$ and $\phi_\delta \in \mathcal{S}_\delta(x)$ we have $|\phi_\delta(k,x)|_{\mathcal{A}} \le \beta(|x|_{\mathcal{A}}, k) + \varepsilon$ for all $k \in \mathbb{Z}_{\ge 0}$.*

2 An Example of No Robustness

This example is related to the examples in [18]. A different example can be found in [23]. Consider the system

$$x^+ = f(x)\ , \quad f(x) = \begin{bmatrix} 0 \\ |x| \end{bmatrix} \text{ if } |x_1| \neq 0 \text{ and } f(x) = \begin{bmatrix} 0 \\ 0 \end{bmatrix} \text{ otherwise.} \tag{4}$$

A typical trajectory for the system (4) is illustrated in Figure 1(a).

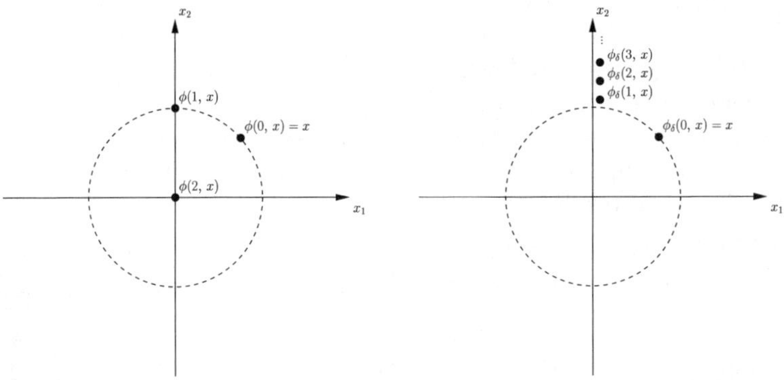

(a) Unperturbed trajectory (b) Perturbed trajectory

Fig. 1. Trajectories for system (4).

The solutions of (4) satisfy

$$|\phi(k,x)| \le |x| \cdot \max\{2-k, 0\} =: \beta(|x|, k) \tag{5}$$

and thus the origin is globally asymptotically stable. However, the asymptotic stability is not robust. In fact, an arbitrarily small additive disturbance with positive components will lead to unbounded trajectories for all initial conditions satisfying $|x_1| \neq 0$. Indeed, in this case the x_1 component of the solution is never zero. So the perturbed solution will satisfy the equation

$$x^+ = \begin{bmatrix} d_1 \\ |x| + d_2 \end{bmatrix} . \tag{6}$$

The solution to (6) is unbounded when $d_1(k) = d_2(k) = \delta > 0$ for all k, no matter how small δ is. A perturbed trajectory is shown in Figure 1(b).

3 Robust Asymptotic Stability

3.1 Regularization

Because of the example in the previous section, is it natural to wonder:

Is the notion of asymptotic stability inadequate for discontinuous discrete-time systems?

or, alternatively,

Is the standard notion of solution inadequate for discontinuous discrete-time systems?

It is well-known that for discontinuous continuous-time systems the standard notion of solution may be inadequate, simply because classical solutions may fail to exist. For such systems, alternative notions of solution have been introduced, such as Filippov or Krasovskii solutions. (For an interesting comparison of Filippov and Krasovskii solutions, see [15].) These solution concepts admit any absolutely continuous function that satisfies a differential inclusion where the set-valued right-hand side is generated from the discontinuous vector field, which is usually assumed to be locally bounded. The differential inclusion is guaranteed to have a solution. It also turns out that if the differential inclusion has an asymptotically stable compact set, the asymptotic stability is robust; see [9] and, for the case of attracting sets, [46].

Unlike for continuous-time systems, solutions of discontinuous discrete-time systems are guaranteed to exist. So there is no obvious need to consider a system with a set-valued right-hand side (a so-called difference inclusion like in (2)). Nevertheless, considering such a generalized notion of solution enables asymptotic stability to become a robust property.

Given a discrete-time system of the form (1), we assume that f is locally bounded and we define the system's regularization to be

$$x^+ \in F(x) := \bigcap_{\delta > 0} \overline{f(x + \delta \mathcal{B})} \ . \tag{7}$$

This regularization is very similar to the regularization of Krasovskii for discontinuous continuous-time systems. The only difference is that the convex hull is not used when creating F. At points x where $f(\cdot)$ is continuous, we have $F(x) = \{f(x)\}$, i.e., $F(x)$ is a singleton. The following statement combines results in [23, 25] and [33].

Theorem 1. *Suppose $\mathcal{A}$ is compact and f is locally bounded. For the system (1) the set $\mathcal{A}$ is globally robustly asymptotically stable if and only if for the regularization (7) of (1) the set $\mathcal{A}$ is globally asymptotically stable.*

Remark 1. In the example of Section 2,

$$F(x) = \left\{ \begin{bmatrix} 0 \\ |x| \end{bmatrix} \right\} \quad |x_1| \neq 0, \quad F(x) = \left\{ \begin{bmatrix} 0 \\ |x| \end{bmatrix} , \begin{bmatrix} 0 \\ 0 \end{bmatrix} \right\} \quad \text{otherwise.} \tag{8}$$

It follows that all points on the nonnegative vertical axis are weak equilibria and thus, for the regularization, the origin is not asymptotically stable.

So one (nonconservative) way to distinguish between asymptotic stability that is robust and asymptotic stability that is not robust is to consider the system's regularization. There is also another way.

3.2 Lyapunov Functions

Again combining results in [23, 25] and [33], we have:

Theorem 2. *Suppose $\mathcal{A}$ is compact and f is locally bounded. For the system (1) the set $\mathcal{A}$ is globally robustly asymptotically stable if and only if the system (1) admits a continuous (in fact, smooth) global Lyapunov function for $\mathcal{A}$.*

Remark 2. Each globally asymptotically stable system, even one for which asymptotic stability is not robust, admits a global Lyapunov function. See, e.g., [36, Lemma 4]. However, according to Theorem 2, for nonrobust systems the Lyapunov function won't be continuous. For the example of Section 2, the (discontinuous) function $V(x) = 2|x|$ if $|x_1| \neq 0$ and $V(x) = |x|$ otherwise, is a global Lyapunov function since $V(f(x)) - V(x) = -|x|$ for all x.

So, another (nonconservative) way to guarantee robustness is to certify asymptotic stability with a continuous Lyapunov function.

Part II: Feedback Control Theory

4 Notation and Definitions

Now we consider control systems

$$x^+ = f(x,u) \ , \quad u \in \mathcal{U} \subset \mathbb{R}^m \ , \tag{9}$$

where $x \in \mathbb{R}^n$ and f is continuous. This notation is equivalent to the alternative notation $x(k+1) = f(x(k), u(k))$. The solution to (9) at the kth iteration will be denoted $\phi(k, x, \mathbf{u})$ where $\mathbf{u} := (u(0), u(1), u(2), \ldots)$. Again for simplicity we will consider making a compact set $\mathcal{A}$ globally attractive.

Definition 4. *Let $\mathcal{A}$ be compact. The system (9) is said to be* globally asymptotically controllable to the set $\mathcal{A}$ *provided there exist $\beta \in \mathcal{KL}$ and for each x there exists $\mathbf{u}$ with $u(i) \in \mathcal{U}$ for all $i \in \mathbb{Z}_{\geq 0}$ such that*

$$|\phi(k, x, \mathbf{u})|_{\mathcal{A}} \leq \beta(|x|_{\mathcal{A}}, k) \ . \tag{10}$$

Definition 5. *A function $V : \mathbb{R}^n \to \mathbb{R}_{\geq 0}$ is said to be a* global control Lyapunov function for (9) with respect to $\mathcal{A}$ *provided there exist α_1, α_2, and $\rho \in \mathcal{PD}$ such that, for all $x \in \mathbb{R}^n$, $\alpha_1(|x|_{\mathcal{A}}) \leq V(x) \leq \alpha_2(|x|_{\mathcal{A}})$ and*

$$\inf_{u \in \mathcal{U}} V(f(x,u)) \leq V(x) - \rho\left(|x|_{\mathcal{A}}\right) \ . \tag{11}$$

In the definition that follows, global robust asymptotic stability under the feedback $u = \kappa(x)$ is as in Definition 3 but for the system

$$x^+ = f(x, \kappa(x + e)) + d \tag{12}$$

instead of the system (3).

Definition 6. *Let $\mathcal{A}$ be compact. For the system (9) the set $\mathcal{A}$ is said to be globally robustly stabilizable by state feedback if there exists a function $\kappa : \mathbb{R}^n \to \mathcal{U}$ such that, for the system (12) the set $\mathcal{A}$ is globally robustly asymptotically stable.*

5 Basic Results

The following result relating the properties defined in the previous section comes from [24] where the case of noncompact sets is also considered.

Theorem 3. *Let $\mathcal{A}$ and $\mathcal{U}$ be compact. The following statements are equivalent:*

1. *The system (9) admits a continuous (in fact, smooth) global control Lyapunov function with respect to $\mathcal{A}$.*
2. *The system (9) is globally asymptotically controllable to $\mathcal{A}$ with locally bounded controls.*
3. *The system (9) is globally robustly stabilizable by state feedback.*

The results in discrete time are somewhat different than those in continuous time. For example, asymptotic controllability in discrete time implies the existence of a *smooth* control Lyapunov function, which is not generic in continuous time. See, for example, [9]. This distinction is not too surprising however, since a negative definite finite difference can readily withstand a smoothing operation. Another distinguishing feature of discrete time is that robust stabilization always can be guaranteed. Again, though, this is not completely unexpected since, in continuous time, robustness to arbitrarily small measurement noise (and disturbances) can be guaranteed by implementing the feedback via sample-and-hold control [8], [44]. This is very similar to solving the problem in discrete time in the first place.

One way to construct a control Lyapunov function is as follows: Given $\beta \in \mathcal{KL}$ that characterizes asymptotic controllability from Definition 4, let $\tilde{\alpha}_1 \in \mathcal{K}_\infty$ satisfy $\tilde{\alpha}_1(\beta(s,t)) \le \tilde{\alpha}_2(s)e^{-t}$ for some $\tilde{\alpha}_2 \in \mathcal{K}_\infty$. (Such functions $\tilde{\alpha}_1, \tilde{\alpha}_2 \in \mathcal{K}_\infty$ exist according to Sontag's Lemma [43, Proposition 7] which extends Massera's classical lemma [29, Section 12].) Then define

$$V(x) = \inf_{\left\{\mathbf{u}: u(i) \in \mathcal{U} \ \forall i \in \mathbb{Z}_{\ge 0}\right\}} \sum_{k=0}^{\infty} \tilde{\alpha}_1(|\phi(k, x, \mathbf{u})|_{\mathcal{A}}) \, . \tag{13}$$

When $\mathcal{U}$ is compact, this function is a continuous control Lyapunov function, and it can be made smooth without destroying the control Lyapunov function property [24].

Robust stabilization is obtained from a control Lyapunov function, as characterized in Definition 5, by picking a feedback function $\kappa : \mathbb{R}^n \to \mathcal{U}$ satisfying, for example,

$$V(f(x, \kappa(x))) \leq V(x) - \frac{1}{2}\rho(|x|_{\mathcal{A}}) \qquad \forall x . \tag{14}$$

Of course, this feedback function is (locally) bounded when $\mathcal{U}$ is compact.

While control Lyapunov functions can be instrumental in solving the feedback stabilization problem, their construction by means of the formula (13) is problematic, in general, because of the infinite computation required. This is the point at which the RHOC formulation enters. In RHOC, a finite horizon optimization problem is used and all that must be computed is the value of the function (which turns out to be a control Lyapunov function under suitable assumptions; see Proposition 1) and an optimizing control sequence at the current state. Also, the quantity $\alpha_1(|\phi(k, x, \mathbf{u})|_{\mathcal{A}})$ in (13) is replaced by a function that is more directly aimed at a given performance objective. But before we get into the details of RHOC, which will be summarized in Section 7, we wish to point out:

6 The Occasional Need for Discontinuous Feedback

There is no reason to expect that continuous feedback stabilizers exist. Consider the system

$$x^+ = x + \begin{bmatrix} u \\ u^3 \end{bmatrix} , \tag{15}$$

which, as shown in [32] or [12], is globally asymptotically controllable to the origin (in four steps) with locally bounded controls. The system (15) is the exact discrete-time model using sample-and-hold control (period $T = 1$) for the continuous-time system $\dot{x}_1 = u$, $\dot{x}_2 = u^3$ for which controllability and stabilizability have been studied extensively.

Any feedback stabilizer for the system (15) must be discontinuous. This is seen by the following argument, which is summarized pictorially in Figure 2. First note that, for any $c > 0$, the set $\mathcal{X}_+(c) := \{x : x_1 \geq c, x_2 \geq c\}$ is invariant under positive controls and, similarly, under negative controls the set $\mathcal{X}_-(c) := \{x : x_1 \leq -c,\ x_2 \leq -c\}$ is invariant. Therefore, a stabilizing feedback must be negative at some point x_+ in $\mathcal{X}_+(c)$ and positive at some point x_- in $\mathcal{X}_-(c)$. But if the stabilizing feedback is continuous then along *any* curve connecting x_+ to x_- there must be some point x_0 where the feedback is equal to zero. This, however, would make x_0 an equilibrium of the closed-loop, meaning that the closed-loop origin would not be asymptotically stable.

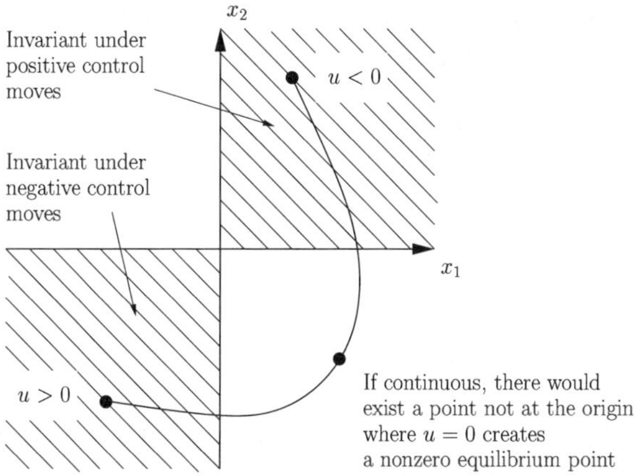

Fig. 2. The occasional need for discontinuous feedback

So we see that, at times, robust stabilization must be accomplished with discontinuous feedback. Let's see if the standard RHOC formulation can be counted on to produce feedbacks that are robustly stabilizing.

7 Receding Horizon Optimal Control

Receding horizon optimal control is an optimization, or performance, based approach to feedback stabilization. The basic idea is to solve an open-loop, finite horizon, optimal control problem on-line using the current state of the system as the initial state. The first value in an optimal control sequence is applied as the feedback control (the rest of the sequence is typically discarded) and the process is repeated at the next time instant. Cf. [27, p. 423]. One of RHOC's biggest selling points is that it is easy to account for control and state constraints in the problem formulation. Using a finite horizon enables practical implementation. The downside is that using a finite horizon doesn't necessarily produce a stabilizing feedback. Because of that, a lot of work over the last few decades, starting with [22] for nonlinear systems, has gone into tailoring the optimization algorithms to guarantee that they are stabilizing. Little attention has been given to computationally tractable means of guaranteeing *robust* stabilization. See the survey [31] and its references for a more thorough discussion.

The standard RHOC formulation, as summarized in [31], is as follows:

Let N be a given positive integer. Define $\mathbf{u}_N := (u(0), u(1), \ldots, u(N-1))$. Let the cost function be

$$J_N(x, \mathbf{u}_N) := g(\phi(N, x, \mathbf{u}_N)) + \sum_{k=0}^{N-1} \ell(\phi(k, x, \mathbf{u}_N), u(k)) \ , \tag{16}$$

where $g : \mathbb{R}^n \mapsto \mathbb{R}_{\geq 0}$ is called the terminal cost and $\ell : \mathbb{R}^n \times \mathbb{R}^m \mapsto \mathbb{R}_{\geq 0}$ is called the stage cost. Consider the following optimization problem:

$$
\begin{aligned}
V_N(x) := \inf_{\mathbf{u}_N} \; & J_N(u, \mathbf{u}_N) \\
\text{subject to} \; & \begin{cases} u(k) \in \mathcal{U}, \; k \in \{0, \ldots, N-1\} \\ \phi(k, x, \mathbf{u}_N) \in \mathcal{X}, \; k \in \{0, \ldots, N-1\} \\ \phi(N, x, \mathbf{u}_N) \in \mathcal{X}_f \;, \end{cases}
\end{aligned}
\tag{17}
$$

where $\mathcal{U} \subseteq \mathbb{R}^m$ is the control constraint set, $\mathcal{X} \subseteq \mathbb{R}^n$ is the state constraint set, and $\mathcal{X}_f \subseteq \mathbb{R}^n$ is the terminal state constraint set. The following assumptions are typical (see [31]):

B1 The set of points for which the constraints can be satisfied, denoted $\mathcal{F}_N$, is nonempty and contains a neighborhood of the origin and, for each $x \in \mathcal{F}_N$, there exists an input sequence $\mathbf{u}_N^*$ such that $V_N(x) = J_N(x, \mathbf{u}_N^*)$.
B2 There exists $\alpha_1 \in \mathcal{K}_\infty$ s.t. $\ell(x, u) \geq \alpha_1(|x|)$ for all $u \in \mathcal{U}$ and all $x \in \mathcal{F}_N$.
B3 There exists $\alpha_2 \in \mathcal{K}_\infty$ such that $V_N(x) \leq \alpha_2(|x|)$ for all $x \in \mathcal{F}_N$.

Remark 3. With the fact that $V_N(x) \geq 0$ for all $x \in \mathcal{F}_N$, Assumption **B3** is equivalent to the condition that $V_N(\cdot)$ is continuous at the origin and vanishes there. This is the case when, for example, $\mathcal{X}_f$ contains the origin in its interior and f, g and ℓ are continuous at the origin and vanish there.

B4 $\mathcal{X}_f \subseteq \mathcal{X}$ and there exists $\kappa_f : \mathcal{X}_f \mapsto \mathcal{U}$ satisfying, for all $x \in \mathcal{X}_f$,
- $f(x, \kappa_f(x)) \in \mathcal{X}_f$, and
- $g(f(x, \kappa_f(x))) - g(x) \leq -\ell(x, \kappa_f(x))$.

Thus, the terminal cost $g(\cdot)$ is a control Lyapunov function on the set $\mathcal{X}_f$ with available decrease related to the stage cost ℓ.

Given $x \in \mathcal{F}_N$, where $\mathcal{F}_N$ comes from **B1**, let $\mathbf{u}_N^*$ be a minimizing control input sequence also coming from **B1**, i.e., $J_N(x, \mathbf{u}_N^*) = V_N(x)$. Next define the RHOC law $\kappa_N : \mathbb{R}^n \mapsto \mathbb{R}^m$ as $\kappa_N(x) := u^*(0)$, the first element of $\mathbf{u}_N^*$.

Since $\mathcal{F}_N$ may not equal $\mathbb{R}^n$, to state the standard RHOC closed-loop stability result we need regional versions of Definitions 1 and 2 for the system

$$x^+ = f(x, \kappa_N(x)) \;. \tag{18}$$

Definition 7. *Let $\mathcal{A}$ be compact and let $\mathcal{R}$ contain a neighborhood of $\mathcal{A}$. For the system (18) the set $\mathcal{A}$ is said to be* asymptotically stable with basin of attraction (containing) $\mathcal{R}$ *provided there exists $\beta \in \mathcal{KL}$ such that for each $x \in \mathcal{R}$ the solution $\phi(\cdot, x)$ satisfies $|\phi(k, x)|_\mathcal{A} \leq \beta(|x|_\mathcal{A}, k)$ for all $k \in \mathbb{Z}_{\geq 0}$.*

Definition 8. *Let $\mathcal{A}$ be compact and let $\mathcal{R}$ contain a neighborhood of $\mathcal{A}$. For the system (18) a function $V : \mathcal{R} \to \mathbb{R}_{\geq 0}$ is said to be a* Lyapunov function for the origin on $\mathcal{R}$ *provided there exist $\alpha_1, \alpha_2 \in \mathcal{K}_\infty$ and $\rho \in \mathcal{PD}$ such that, for all $x \in \mathcal{R}$, $\alpha_1(|x|_\mathcal{A}) \leq V(x) \leq \alpha_2(|x|_\mathcal{A})$ and*

$$V(f(x, \kappa_N(x))) \leq V(x) - \rho(|x|_\mathcal{A}) \;. \tag{19}$$

Proposition 1. *Consider the optimization problem (16)-(17). Under the assumptions* **B1-B4**, *the origin of the closed-loop system* $x^+ = f(x, \kappa_N(x))$ *is asymptotically stable with basin of attraction (containing)* $\mathcal{F}_N$ *and it has the value function* $V_N(\cdot)$ *as a Lyapunov function for the origin on* $\mathcal{F}_N$.

8 Is RHOC's Asymptotic Stability Robust?

8.1 Definition

Because the RHOC algorithm is applicable to the system (15), the RHOC algorithm is capable of generating discontinuous feedback controls. We will see below that it may even generate discontinuous feedback controls when they are not necessary. In light of the discussion in previous sections, robustness of RHOC should thus be a concern.

We are almost at a point where we can address the question posed in the title of this work. However, because the feasible region, $\mathcal{F}_N$, in the RHOC problem may not equal $\mathbb{R}^n$, we still need a precise definition of "robust", like in Definition 3, for the nonglobal case for the system

$$x^+ = f(x, \kappa_N(x+e)) + d \ . \tag{20}$$

Like we have used previously, for each $\delta > 0$, $\mathcal{S}_\delta(x)$ denotes the set of solutions to (20) starting at x with $\max\{\|\mathbf{e}\|, \|\mathbf{d}\|\} \le \delta$. We will consider two robustness notions for (20). First we have the nonglobal version of Definition 3:

Definition 9. *Let* $\mathcal{A}$ *be compact and let* $\mathcal{R}$ *contain a neighborhood of* $\mathcal{A}$. *For the system (20) the set* $\mathcal{A}$ *is said to be* robustly asymptotically stable on $\text{int}(\mathcal{R})$ *if there exists a class-*$\mathcal{KL}$ *function* β *and for each* $\varepsilon > 0$ *and compact set* $\mathcal{C} \subset \text{int}(\mathcal{R})$ *there exists* $\delta > 0$ *such that for each* $x \in \mathcal{C}$ *and* $\phi_\delta \in \mathcal{S}_\delta(x)$ *we have* $|\phi_\delta(k,x)|_\mathcal{A} \le \beta(|x|_\mathcal{A}, k) + \varepsilon$ *for all* $k \in \mathbb{Z}_{\ge 0}$.

Next we add the phrase "when remaining in" and require less of the perturbed trajectories, unless $\mathcal{R} = \mathbb{R}^n$ in which case we recover Definition 9. Definition 10 is the focus of the rest of Section 8 but will not be used in later sections.

Definition 10. *Let* $\mathcal{A}$ *be compact and let* $\mathcal{R}$ *contain a neighborhood of* $\mathcal{A}$. *For the system (20) the set* $\mathcal{A}$ *is said to be* robustly asymptotically stable when remaining in $\text{int}(\mathcal{R})$ *if there exists a class-*$\mathcal{KL}$ *function* β *and for each* $\varepsilon > 0$ *and compact set* $\mathcal{C} \subset \text{int}(\mathcal{R})$ *there exists* $\delta > 0$ *such that: if* $\phi_\delta \in \mathcal{S}_\delta(x)$ *and* $K \in \mathbb{Z}_{\ge 0}$ *satisfy*

$$\phi_\delta(k,x) \in \mathcal{C} \quad \forall k \in \{0, \ldots, K\}, \tag{21}$$

then

$$|\phi_\delta(k,x)|_\mathcal{A} \le \beta(|x|_\mathcal{A}, k) + \varepsilon \quad \forall k \in \{0, \ldots, K\} \ . \tag{22}$$

Now we are ready to ask the question posed in the title of the paper and of this section. We will see below that, using either Definition 9 or 10, the answer is "not necessarily". First, when using Definition 10, it is interesting to note:

8.2 RHOC is Robust for Linear Systems with Convex Constraints

When trying to stabilize a linear system respecting convex constraints [1] the control sequences available to the optimization problem satisfy the condition

$$\mathbf{u}_N \in \mathcal{W}(x) := \{\mathbf{u}_N \in \mathcal{V} : M_1 x + M_2 \mathbf{u}_N \in \mathcal{Y}\} \tag{23}$$

where $\mathcal{V} := \mathcal{U} \times \cdots \times \mathcal{U}$, $\mathcal{Y} := \mathcal{X} \times \cdots \times \mathcal{X} \times \mathcal{X}_f$ and M_1 and M_2 are appropriate matrices. Due to the convexity of the sets $\mathcal{U}$, $\mathcal{X}$ and $\mathcal{X}_f$, the graph of the set-valued map $\mathcal{W}(\cdot)$ is convex. With the assumption that $\mathcal{U}$ is bounded, this implies that the set-valued map $\mathcal{W}(\cdot)$ is locally Lipschitz on the interior of $\mathcal{F}_N$. (See, e.g., [38] or [2, Ch. 1, §3].) With the continuity of the stage and terminal cost, this implies that the value function is continuous on the interior of $\mathcal{F}_N$. (See, e.g., [4, Ch. VI, §3] or [2, pp. 51-54].) Like in Theorem 2, this implies that the asymptotic stability induced by RHOC is robust *when remaining in* the interior of $\mathcal{F}_N$. This result, also stated in [13], is summarized next; cf. [3] and [7], which mention results for the special case of quadratic cost functions.

Proposition 2. *For the system (9), the origin of the RHOC closed-loop produced by the optimization problem (16)-(17) is robustly asymptotically stable when remaining in int*($\mathcal{F}_N$) *if* $f(x,u) = Ax + Bu$, *the sets* $\mathcal{U}$, $\mathcal{X}$ *and* $\mathcal{X}_f$ *are convex,* $\mathcal{U}$ *is bounded, the stage cost* ℓ *and terminal cost* g *are continuous, and* **B1-B4** *hold.*

8.3 No Robustness Due to Terminal State Constraints

We now give a (necessarily nonlinear) example where RHOC is globally asymptotically stabilizing but provides no robustness [13]. The example, which is based on the example in Section 2, is the two dimensional system

$$x^+ = \begin{bmatrix} x_1(1-u) \\ |x|_2 u \end{bmatrix}, \tag{24}$$

where $|\cdot|_2$ denotes the Euclidean norm and where u takes values in the interval $[0,1]$. Thus, the possible values for x^+ are on the line connecting the point $(0, |x|_2)$ to the point $(x_1, 0)$. If the control law $u = 0.5$ is used, the origin is globally exponentially stable, seen using $V(x) = |x|_2^2$ as a Lyapunov function.

Now we show that a stabilizing RHOC algorithm with horizon length $N = 2$ has no robustness. First note that the only way to get to the origin in two steps from any point with $x_1 \neq 0$ is to pick $(u(0), u(1)) = (1, 0)$, and the resulting trajectory is $(x_1, x_2), (0, |x|_2), (0, 0)$. This follows from the fact that the only way to get to the origin in one step is to require $x_1 = 0$ and use $u = 0$ (unless already at the origin in which case any u will do.) Now consider the RHOC approach with $N = 2$, X_f equal to the origin, and ℓ and g continuous,

[1] We also will assume that the control constraint set is bounded, but this can be relaxed.

nonnegative functions that vanish at the origin. In particular, suppose there are class-$\mathcal{K}_\infty$ functions α_1 and α_2 satisfying $\alpha_1(|x|) \le \ell(x,u) \le \alpha_2(|x|)$. Under these conditions, all of the assumptions of standard RHOC are satisfied and the resulting RHOC control is globally asymptotically stabilizing. From the discussion above, it must necessarily be that the RHOC feedback satisfies $\kappa_N(x) = 1$ if $|x_1| \neq 0$. Therefore, the closed-loop system matches the system in Section 2 when $|x_1| \neq 0$. Since arbitrarily small disturbances (or measurement noise) can be used to make (or make it appear that) the x_1 component of the closed-loop is never equal to zero, we can have the system (6), with $|x| = |x|_2$, for all k. Thus the x_2 component of the closed-loop solution can be kept arbitrarily far from the origin. In fact, it can grow without bound in the presence of arbitrarily small disturbances. Thus, the asymptotic stability produced by RHOC is not robust.

The paper [13] includes an additional example where the terminal constraint in the RHOC formulation is the closed unit ball.

8.4 No Robustness Due to Inherent State Constraints

Consider the system

$$\begin{aligned} x_1^+ &= \frac{-(x_1^2+x_2^2)u + x_1}{1+(x_1^2+x_2^2)u^2 - 2x_1 u} \\ x_2^+ &= \frac{x_2}{1+(x_1^2+x_2^2)u^2 - 2x_1 u}\,, \end{aligned} \tag{25}$$

which is the sample-and-hold equivalent, for sampling period $T = 1$, of the continuous-time system, $\dot{x}_1 = (x_1^2 - x_2^2)u$, $\dot{x}_2 = 2x_1 x_2 u$, introduced by Artstein [1]. The key feature of the system (25) is that if the state is initially on the circle $x_1^2 + (x_2 - r)^2 = r^2$ for some $r \in \mathbb{R}$, then any trajectory starting from that initial condition will remain on the same circle, regardless of the input. If $x_2 = 0$ initially, then $x_2(k) = 0$ for all values of k. The circles are shown in Figure 3 where, above the horizontal axis, $u < 0$ generates clockwise movement, $u > 0$ generates counterclockwise movement, and $u = 0$ induces an equilibrium point. The directions are reversed below the horizontal axis. Like for the system (15), since $u = 0$ induces an equilibrium point there cannot exist a continuous feedback law that asymptotically stabilizes the origin.

We will develop an RHOC algorithm to stabilize the origin of the system (25) respecting the input and state constraints $u \in [-1, 1] =: \mathcal{U}$ and $x \in \{x \in \mathbb{R}^2 : x_1 \le c\} =: \mathcal{X}$ where $c \in (0, 1)$. Before this, we point out an important feature of the system when these constraints are in place. Suppose the state of the system is at

$$x_c := \begin{bmatrix} c & c\sqrt{\dfrac{1+c}{1-c}} \end{bmatrix}^T .$$

If we apply the largest input that will drive the state in the clockwise direction, i.e., $u = -1$, we will have

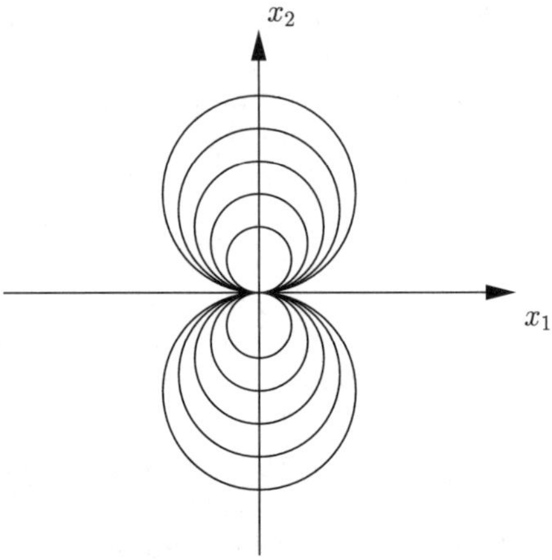

Fig. 3. Artstein's circles .

$$f(x, u) = \left[c \quad c\sqrt{\frac{1-c}{1+c}} \right]^T .$$

Notice that the x_1 component of the state stays on the constraint boundary, $x_1 = c$. This trajectory resides on what we will call the *critical circle* $x_1^2 + (x_2 - r_c)^2 = r_c^2$ where

$$r_c := \frac{c}{\sqrt{1-c^2}}.$$

See Figure 4. Since the trajectories never leave the circle they start on, we can deduce the following property for trajectories with initial conditions in $\mathcal{X}$ and the first quadrant that are outside the critical circle and satisfy

$$x_2 >. c\sqrt{\frac{1+c}{1-c}} :$$

there does not exist a clockwise trajectory converging to the origin satisfying the state and input constraints. It is likely, from points in the first quadrant, that the cost associated with moving in the clockwise direction will be smaller than the cost associated with moving in the counterclockwise direction. We will impose this below. Thus, we will likely find that the RHOC algorithm moves points in the first quadrant in the counterclockwise direction if they are outside the critical circle and satisfy

$$x_2 > c\sqrt{\frac{1+c}{1-c}},$$

while moving them in the clockwise direction otherwise. When c is sufficiently small, e.g., $c = 0.1$, there is room to move on the critical circle without approaching the constraint $x_1 = c$ or the positive vertical axis. Indeed, using $u \in [-1, 0]$ maps the arc (a, b) in Figure 4 to the arc (a, d) in the figure, while using $u \in [0, 1]$ maps the arc (b, d) to the arc (a, d). It follows that, with the likely optimal feedback control strategy, arbitrarily small measurement noise can be used to make the arc (a, d) invariant. If this (compact) arc is contained in the interior of $\mathcal{F}_N$ then the closed-loop will not be robust when remaining in the interior of $\mathcal{F}_N$. We now show that the standard RHOC assumptions can be satisfied while putting the arc (a, d) in the interior of $\mathcal{F}_N$.

First we construct a continuous terminal cost g that is a (global) control Lyapunov function. For each $x \in \mathbb{R}^2$ we define $g(x)$ to be the length of the shortest path from x to the origin while remaining on the invariant circle associated with the state. This function is continuous. It is also even with respect to x_1, i.e., $g((x_1, x_2)) = g((-x_1, x_2))$. Moreover, with the definition

$$\kappa_f(x) := \begin{cases} -\min\{1, k_1(x)\} \ , & x_1 \geq 0 \\ \min\{1, k_2(x)\} \ , & x_1 < 0 \ , \end{cases}$$

where $k_1, k_2 \in \mathcal{PD}$, there exists $\rho \in \mathcal{PD}$ such that

$$g(f(x, \kappa_f(x))) \leq g(x) - \rho(|x|) \qquad \forall x \ , \tag{26}$$

where $f(x, u)$ represents the right-hand side of (25). Next we define $\mathcal{X}_f$ to be a closed disk of radius less than or equal to c centered at the origin. With this definition, $\mathcal{X}_f$ is forward invariant under κ_f. Finally, we pick ℓ to be a continuous function satisfying $\ell(x, \kappa_f(x)) \leq \rho(|x|)$ for all $x \in \mathcal{X}_f$ and, for some $\alpha_1 \in \mathcal{K}_\infty$, $\alpha_1(|x|) \leq \ell(x, u)$ for all x. Moreover, we assume that ℓ is even in x_1, i.e., $\ell((x_1, x_2), u) = \ell((-x_1, x_2), u)$.

With the design parameters set as above, the conditions for Proposition 1 are met and hence the feedback law $\kappa_N(\cdot)$ resulting from the optimization problem (16)-(17) is asymptotically stabilizing for the origin of the system (25) with basin of attraction (containing) $\mathcal{F}_N$. Moreover, $\mathcal{X}_f \subseteq \mathcal{F}_N \subseteq \mathcal{X}$, and $\mathcal{F}_N \to \mathcal{X}$ as $N \to \infty$. In particular, for sufficiently large N, the arc (a, d) in Figure 4 is contained in the interior of $\mathcal{F}_N$. Finally, it can be shown that the RHOC feedback satisfies $\kappa_N(x) \in [-1, 0)$ for all x on the critical circle in the first quadrant while $\kappa_N(x) \in (0, 1]$ for all x outside of the critical circle in the first quadrant with

$$x_2 > c\sqrt{\frac{1+c}{1-c}}.$$

Thus, as indicated above, the closed-loop asymptotic stability is not robust when remaining in the interior of $\mathcal{F}_N$. For more details, see [13].

9 Modified RHOC to Guarantee Nominal Robustness

The possible lack of robustness using the standard RHOC formulation motivates the search for modified, computationally tractable RHOC algorithms

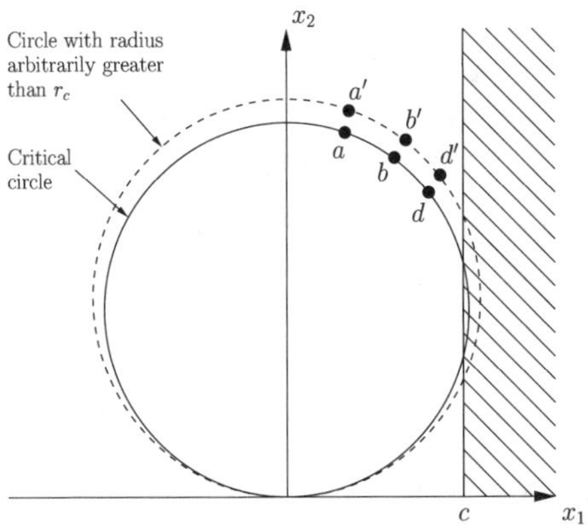

Fig. 4. Perturbed trajectories.

that guarantee robustness. In the process, in order to improve the ease of applicability, it is useful to search for results that don't require finding a local control Lyapunov function to use as the terminal cost in order to provide stability guarantees. Indeed, finding a local control Lyapunov function can be as difficult as finding a stabilizing feedback in the first place. In addition, it is useful to relax the positive definiteness assumption on the stage cost in order to give more flexibility to the design. This is easy to do when the terminal cost is a local control Lyapunov function; it is more challenging otherwise. Another objective to strive for is achieving global asymptotic stability when that is possible. Achieving such a result using the standard formulation is not typical since the terminal constraint set is usually a bounded set and the number of steps required to reach a bounded set often grows without bound as the initial condition grows without bound. Finally, it is useful to extend the basic formulation to the problem of set stabilization.

9.1 No State Constraints

The results in this section are taken from [12]. We again use the cost function (16) and, since we will not be using state or terminal constraints, the value function (17) becomes

$$V_N(x) := \inf_{\mathbf{u}_N} J_N(x, \mathbf{u}_N), \qquad u(k) \in \mathcal{U} \qquad k \in \{0, \ldots, N-1\} \ . \tag{27}$$

We use $\mathcal{F}_N \subseteq \mathbb{R}^n$ to denote the set of points for which the constraints are feasible. When there are no state constraints, $\mathcal{F}_N = \mathbb{R}^n$. With state constraints, as considered in the next section, $\mathcal{F}_N$ is typically a strict subset of $\mathbb{R}^n$.

For simplicity, in what follows we will make global assumptions resulting in global (or nearly global) feedback control results.

Assumption 1 *The functions g and ℓ are continuous.*

Assumption 2 *For each integer N and $x \in \mathcal{F}_N$, there exists a feasible $\mathbf{u}_N^*(x)$ such that $J_N(x, \mathbf{u}_N^*(x)) = V_N(x)$. Moreover, for each compact set $\mathcal{Z} \subset \mathbb{R}^n$ and integer N there exists $\mu > 0$ such that $x \in \mathcal{Z} \bigcap \mathcal{F}_N$ implies $||\mathbf{u}_N^*(x)|| \leq \mu$.*

Assumptions 1 and 2 guarantee that the value function $V_N(\cdot)$ is continuous on $\mathbb{R}^n$. The RHOC law $x \mapsto \kappa_N(x)$ is defined in terms of feasible inputs satisfying Assumption 2 as $\kappa_N(x) = u^*(0)(x)$. Sufficient conditions for the existence of optimal controls, for both finite and infinite horizon constrained optimization problems, can be found in [21] and the references therein.

We use a continuous function $\sigma : \mathbb{R}^n \mapsto \mathbb{R}_{\geq 0}$ as a measure of the state. The feedback control objective will be to drive solutions to the set $\{x : \sigma(x) = 0\}$. Possible choices are $\sigma(x) = |x|^2$ if the target set is the origin, or $\sigma(x) = |x|_{\mathcal{A}}$ if the target set is some set $\mathcal{A}$. For simplicity, we assume:

Assumption 3 *The set $\{x : \sigma(x) = 0\}$ is compact.*

Of course, one would expect some relationship between the target set, expressed in terms of σ, and the components of the cost function g and ℓ. This relationship will come through requirements that correspond to detectability and controllability. We first define the notion of detectability to be used.

Definition 11. *Consider the system (9), a function $\sigma : \mathbb{R}^n \mapsto \mathbb{R}_{\geq 0}$, and a function $\ell : \mathbb{R}^n \times \mathbb{R}^m \mapsto \mathbb{R}_{\geq 0}$. The function σ is said to be* detectable from ℓ with respect to $(\overline{\alpha}_W, \alpha_W, \gamma_W)$ *if $\alpha_W \in \mathcal{K}_\infty$, $\gamma_W \in \mathcal{K}_\infty$, $\overline{\alpha}_W \in \mathcal{G}$ and there exists a continuous function $W : \mathbb{R}^n \mapsto \mathbb{R}_{\geq 0}$ s.t. for all $x \in \mathbb{R}^n$ and $u \in \mathcal{U}$,*

$$
\begin{aligned}
W(x) &\leq \overline{\alpha}_W(\sigma(x)), \\
W(f(x,u)) - W(x) &\leq -\alpha_W(\sigma(x)) + \gamma_W(\ell(x,u)).
\end{aligned}
$$

Assumption 4 *For the system (9), the function σ is detectable from ℓ with respect to some $(\overline{\alpha}_W, \alpha_W, \gamma_W)$.*

Remark 4. If there exists $\underline{\alpha}_\ell \in \mathcal{K}_\infty$ such that $\ell(x,u) \geq \underline{\alpha}_\ell(\sigma(x))$ for all $x \in \mathbb{R}^n$ and $u \in \mathcal{U}$, then σ is detectable from ℓ with respect to $(0 \cdot \mathrm{Id}, \alpha_W, \gamma_W)$ for any pair (α_W, γ_W) satisfying $\gamma_W^{-1} \circ \alpha_W(s) \leq \underline{\alpha}_\ell(s)$ for all $s \in \mathbb{R}_{\geq 0}$. One such pair is $(\alpha_W, \gamma_W) = (\underline{\alpha}_\ell, \mathrm{Id})$.

For upcoming results, it is useful to understand when Assumption 4 is satisfied with linear functions $(\overline{\alpha}_W, \alpha_W, \gamma_W)$. Some examples are:

1. there exists $\varepsilon > 0$ such that $\varepsilon\sigma(x) \leq \ell(x,u)$ for all (x,u). A common case where this bound holds is when $\sigma(x) = |x|^2$ and there exists $\varepsilon > 0$ such that $\varepsilon|x|^2 \leq \ell(x,u)$ for all (x,u). Other situations will be motivated by the controllability assumption, Assumption 5, below.
2. $\sigma(x) = |x|^2$, $\ell(x,u) = x^T C^T C x + u^T D^T D u$, $f(x,u) = Ax + Bu$, and the pair (C, A) is detectable in the usual linear sense. In this case the detectability assumption holds with a quadratic function $W(\cdot)$.

The next assumption is essentially a controllability assumption. It imposes a bound on the value function that is independent of the horizon length.

Assumption 5 *There exists* $\overline{\alpha} \in \mathcal{K}_\infty$ *such that* $V_i(x) \le \overline{\alpha}(\sigma(x))$ *for all* $x \in \mathbb{R}^n$ *and all* $i \in \{1, \ldots, \infty\}$.

For upcoming results, it is useful to understand when Assumption 5 is satisfied with a linear function $\overline{\alpha}$. Some examples are:

1. $x^+ = Ax + Bu$, $\sigma(x) = |x|^2$, $\ell(x,u) = x^T C^T C x + u^T D^T D u$, $g(x) = x^T G^T G x$, $\mathcal{U}$ compact, and all eigenvalues of A have norm less than one.
2. The system (15), $\sigma(x) = |x_1|^3 + |x_2|$, $\ell(x,u) = \sigma(x) + |u|^3$, $g(x) = 15 \cdot \sigma(x)$, and $\mathcal{U} = \mathbb{R}$.
3. The system
$$x^+ = x + \begin{bmatrix} u_1 \\ u_2 \\ u_1 x_2 - u_2 x_1 \end{bmatrix}$$
(the exact discrete-time model of the continuous-time nonholonomic integrator using sample-and-hold control, period $T = 1$), $\sigma(x) = x_1^2 + x_2^2 + 10 \cdot |x_3|$, $\ell(x,u) = \sigma(x)$, $g(x) = 4 \cdot \sigma(x)$, and $\mathcal{U} = \mathbb{R}^2$.

Remark 5. The constants that appear in the latter two examples are not significant for obtaining a linear function $\overline{\alpha}$ in Assumption 5. They are specified here in preparation for Remark 6 which is upcoming. Also for the latter two examples, Assumption 5 with a linear $\overline{\alpha}$ is explained in [12] using discrete-time homogeneity theory; see, e.g., [16]. Note also that $\sigma(x) \le \ell(x,u)$ for both of those examples so that, according to the first item below Remark 4, Assumption 4 also holds with linear functions.

At this point, the following can be asserted [12]:

Theorem 4. *Consider the system (9) and the optimization problem (16),(27) under Assumptions 1-5. Moreover, suppose the functions* $(\overline{\alpha}_W, \alpha_W, \gamma_W)$ *from Assumption 4 and the function* $\overline{\alpha}$ *from Assumption 5 can be taken to be linear. Under these conditions there exists* N^* *such that, for each* $N \ge N^*$, *for the RHOC closed-loop the set* $\{x : \sigma(x) = 0\}$ *is robustly asymptotically stable on* $\mathbb{R}^n$.

Remark 6. For the latter two examples below Assumption 5, it can be shown (see [12]) that $N^* = 4$ suffices for global robust asymptotic stability.

For the case where the functions $(\overline{\alpha}_W, \alpha_W, \gamma_W)$ and $\overline{\alpha}$ cannot be taken to be linear, there is no guarantee that the result is global. However, the result approaches a global one as the optimization problem's horizon length N grows. Since this notion of robustness is new, we need additional definitions. (We will give regional definitions in preparation for the results of the next section.) The first new type of robustness, with $\mathcal{R} = \mathbb{R}^n$, will result when the functions are linear near zero. The second, with $\mathcal{R} = \mathbb{R}^n$, will result without extra requirements on the functions.

Definition 12. *Let $\mathcal{A}$ be compact and let $\mathcal{R}$ contain a neighborhood of $\mathcal{A}$. For the system (20) the set $\mathcal{A}$ is said to be* semiglobally (in the parameter N) robustly asymptotically stable on int($\mathcal{R}$) *if there exists $\beta \in \mathcal{KL}$, for each compact set $\mathcal{C}$ there exists $N^* \in \mathbb{Z}_{\geq 0}$, and for each $\varepsilon > 0$ and $N \geq N^*$ there exists $\delta > 0$ (also depending on $\mathcal{C}$) such that, for each $N \geq N^*$, $x \in \mathcal{C}$, and $\phi_\delta \in \mathcal{S}_\delta(x)$, we have $|\phi_\delta(k,x)|_{\mathcal{A}} \leq \beta(|x|_{\mathcal{A}}, k) + \varepsilon$ for all $k \in \mathbb{Z}_{\geq 0}$.*

Definition 13. *Let $\mathcal{A}$ be compact and let $\mathcal{R}$ contain a neighborhood of $\mathcal{A}$. For the system (20) the set $\mathcal{A}$ is said to be* semiglobally practically (in the parameter N) robustly asymptotically stable on int($\mathcal{R}$) *if there exists $\beta \in \mathcal{KL}$, for each compact set $\mathcal{C}$ and $\varepsilon > 0$ there exist $N^* \in \mathbb{Z}_{\geq 0}$, and for each $N \geq N^*$ there exists $\delta > 0$ (also depending on $\mathcal{C}$ and ε) such that, for each $N \geq N^*$, $x \in \mathcal{C}$, and $\phi_\delta \in \mathcal{S}_\delta(x)$, we have $|\phi_\delta(k,x)|_{\mathcal{A}} \leq \beta(|x|_{\mathcal{A}}, k) + \varepsilon$ for all $k \in \mathbb{Z}_{\geq 0}$.*

Remark 7. The distinction between the semiglobal case (Definition 12) and the semiglobal practical case (Definition 13) is that N^* is independent of ε in semiglobal case. In particular, in the semiglobal case a sufficiently large horizon length results in asymptotic stability, and convergence to the set $\mathcal{A}$ is hindered only by the perturbations e and d. In the semiglobal practical case, N^* depends on ε and convergence to the set $\mathcal{A}$ is hindered both by the perturbations and by the fact that the horizon length is not infinite.

Theorem 5. *Consider the system (9) and the optimization problem (16),(27) under Assumptions 1-5. Moreover, suppose the functions $(\overline{\alpha}_W, \alpha_W, \gamma_W)$ from Assumption 4 and the function $\overline{\alpha}$ from Assumption 5 can be taken to be linear near zero. Then for the RHOC closed-loop the set $\{x : \sigma(x) = 0\}$ is semiglobally robustly asymptotically stable on $\mathbb{R}^n$.*

Theorem 6. *Consider the system (9) and the optimization problem (16),(27) under Assumptions 1-5. Then for the RHOC closed-loop the set $\{x : \sigma(x) = 0\}$ is semiglobally practically robustly asymptotically stable on $\mathbb{R}^n$.*

Remark 8. Sufficient formulas for the N^* appearing, explicitly or implicitly, in Theorems 4-6 are given in [12].

9.2 State Constraints

The results in this section are taken from [14]. We again consider the cost function (16) and, in an important feature borrowed from [28], we allow the state constraints to depend on the step in the optimization horizon. In this case the value function (17) becomes

$$
\begin{aligned}
V_N(x) &:= \inf_{\mathbf{u}_N} \; J_N(u, \mathbf{u}_N) \\
&\text{subject to} \begin{cases} u(k) \in \mathcal{U} & k \in \{0, \ldots, N-1\} \\ \phi(k, x, \mathbf{u}_N) \in \mathcal{X}_{k,N} & k \in \{0, \ldots, N\} \; . \end{cases}
\end{aligned} \tag{28}
$$

Once again we will impose Assumptions 1 and 2. Because of the state constraints, these assumptions don't necessarily guarantee that the value function is continuous. However, they again allow us to define a RHOC law and they are important for robustness. We will also use Assumptions 3 and 4. The spirit of Assumption 5 remains the same but, because of the state constraints, it will be modified slightly to refer to some auxiliary optimization problems. For each $N \in \{1, \ldots, \infty)$ and each $M \in \{1, \ldots, N\}$ we define

$$V_{N,M}(x) := \inf_{\mathbf{u}_M} J_M(x, \mathbf{u}_M)$$
$$\text{subject to} \quad \begin{cases} u(k) \in \mathcal{U} & k \in \{0, 1, \ldots, M-1\} \\ \phi(k, x, \mathbf{u}_M) \in \mathcal{X}_{N-M+k,N} & k \in \{0, 1, \ldots, M\}\,. \end{cases} \tag{29}$$

We use $\mathcal{F}_{N,M}$ to denote the set of points for which the constraints are feasible for the auxiliary problem indexed by (N, M).

Assumption 6 *There exists $\overline{\alpha} \in \mathcal{K}_\infty$ such that, for all $N \in \{1, \ldots, \infty)$, $M \in \{1, \ldots, N\}$, and $x \in \mathcal{F}_{N,M}$, we have $V_{N,M}(x) \le \overline{\alpha}(\sigma(x))$.*

Remark 9. Since $V_{N,N}(x) = V_N(x)$, Assumption 6 covers Assumption 5. If $\mathcal{X}_{k,N}$ does not vary with k or N, then Assumption 6 reduces to Assumption 5 since, in this case, we have $V_{N,M}(x) = V_M(x)$. In general, the supplementary value functions capture information about the various tails of the full horizon optimization problem (16),(28).

So far we have not imposed any assumptions that would rule out the examples of no robustness that we encountered earlier. It is our final condition that is the key to robustness.

Definition 14. *For the system (9), the optimization problem (16),(28) is said to be* robust-optimal feasible *with respect to a given pair (N, M) if the set of feasible points, $\mathcal{F}_N$, is such that for each compact subset $\mathcal{C}_N \subseteq \mathcal{F}_N$ there exists $\varepsilon > 0$ such that for each $x \in \mathcal{C}_N$, each $z \in f(x, \kappa_N(x)) + \varepsilon\mathcal{B}$, and each $k \in \{0, 1, \ldots, N-M\}$, the points $\psi_k := \phi(k, z, [\mathbf{u}]_1^{N-M})$ are such that $\psi_k \in \mathcal{X}_k$ and ψ_{N-M} is feasible for $V_{N,M}$ where $\mathbf{u}$ satisfies $u(0) = \kappa_N(x)$ and $V_N(x) = J_N(x, \mathbf{u})$ and we are using the definition $[\mathbf{u}]_1^{N-M} := (u(1), \ldots, u(N-M))$.*

Assumption 7 *There exist positive integers M and $\tilde{N}$ such that, for all integers $N \ge \tilde{N}$, the problem (28) is robust-optimal feasible with respect to (N, M) and for each pair of integers N_1, N_2 satisfying $\tilde{N} \le N_1 \le N_2$ we have $\mathcal{F}_{N_1} \subseteq \mathcal{F}_{N_2}$.*

Remark 10. In Section 9.3 we will illustrate Assumption 7 on the examples of Sections 8.3 and 8.4.

The next result parallels Theorem 4.

Theorem 7. *Consider the system (9) and the optimization problem (16), (28) under Assumptions 1-4, 6-7. Moreover, suppose the functions* $(\overline{\alpha}_W, \alpha_W, \gamma_W)$ *from Assumption 4 and the function* $\overline{\alpha}$ *from Assumption 6 can be taken to be linear. Under these conditions there exists* N^* *such that, for each* $N \geq N^*$*, for the RHOC closed-loop the set* $\{x : \sigma(x) = 0\}$ *is robustly asymptotically stable on* $\text{int}(\mathcal{F}_N)$.

The next results, which use $\mathcal{F}_\infty := \bigcup_N \text{int}(\mathcal{F}_N)$, parallel Theorems 5 and 6.

Theorem 8. *Consider the system (9) and the optimization problem (16), (28) under Assumptions 1-4, 6-7. Moreover, suppose the functions* $(\overline{\alpha}_W, \alpha_W, \gamma_W)$ *from Assumption 4 and the function* $\overline{\alpha}$ *from Assumption 6 can be taken to be linear near zero. Under these conditions, for the RHOC closed-loop the set* $\{x : \sigma(x) = 0\}$ *is semiglobally robustly asymptotically stable on* $\mathcal{F}_\infty$.

Theorem 9. *Consider the system (9) and the optimization problem (16), (28) under Assumptions 1-4, 6-7. Under these conditions, for the RHOC closed-loop the set* $\{x : \sigma(x) = 0\}$ *is semiglobally practically robustly asymptotically stable on* $\mathcal{F}_\infty$.

Remark 11. Sufficient formulas for the N^* appearing, explicitly or implicitly, in Theorems 7-9 are given in [14]. It turns out that N^* has the form $M + \tilde{N}^*$ where $\tilde{N}^*$ is independent of M. Thus, the smaller M used in Assumption 7 the shorter the required horizon for robust stabilization.

Remark 12. In the results of this section a sufficiently large horizon N^* is needed for the result, and the robustness "margin" δ depends on the actual horizon length N used. There is no guarantee that δ is bounded away from zero as $N \to \infty$.

9.3 Application to Previous Examples

The System (24)

For the system (24), we propose to use $g(x) \equiv 0$, $\ell(x,u) = |x|^2$, $\mathcal{X}_{k,N} = \mathrm{I\!R}^2$ for $k = \{0, \dots, N-1\}$ and $\mathcal{X}_{N,N}$ equals the origin in $\mathrm{I\!R}^2$. So Assumption 1 is satisfied. Since $\mathcal{U} = [0,1]$, Assumption 2 is satisfied. See [21]. We are trying to stabilize the origin, so we take $\sigma(x) = |x|^2$. It follows that Assumption 3 is satisfied. Moreover, Assumption 4 is satisfied by taking $W(x) \equiv 0$, $\overline{\alpha}_W \equiv 0$, $\alpha_W = \gamma_W = \text{Id}$. From the discussion below (24) it follows that $\mathcal{F}_N = \mathrm{I\!R}^2$ for all $N \geq 2$ and Assumption 6 holds with $\overline{\alpha}(s) = 2s$ for all s. As far as Assumption 7 is concerned, we will make use of the following general result [14]:

Proposition 3. *Suppose* $\mathcal{X}_{k,N} = \mathrm{I\!R}^n$ *for all* $k \in \{0, \dots, N-1\}$ *and* $\mathcal{X}_{N,N} = \mathcal{X}_f$*. Suppose the set* $\mathcal{X}_f$ *is reachable from* $\mathrm{I\!R}^n$ *in* M *steps using controls taking values in* $\mathcal{U}$*. Then Assumption 7 holds with* M *and* $\tilde{N} = M$*.*

For the system (24) with the constraint sets mentioned above, the conditions of Proposition 3 hold with $M = 2$. Therefore, all of the conditions of Theorem 7 hold and thus, for sufficiently large horizon length, the RHOC resulting from (16),(28) produces a globally robustly asymptotically stable closed-loop. Using the formulas in [14], it follows that a horizon length $N \geq 7$ suffices for this result.

Artstein's Circles

For the discrete-time model of Artstein's circles (25), we consider general functions g, ℓ and σ, so that Assumptions 1, 3, 4, and 6 are satisfied. In addition, we pick σ so that $\{x : \sigma(x) = 0\}$ is the origin. Assumption 2 will be satisfied, since $\mathcal{U} = [-1, 1]$, as long as the sets $\mathcal{X}_{k,N}$ are closed. See [21]. To facilitate Assumption 7, we use the (closed) sets $\mathcal{X}_{k,N} = \{x : x_1 \leq c_{k,N}\}$ where $c_{0,N} = c$, $c_{k+1,N} = c_{k,N} - N^{-1}\nu$ with $\nu > 0$ sufficiently small. This will guarantee that $\mathcal{F}_N = \mathcal{X}_{0,N}$ for all N. It will also guarantee the conditions of the following general proposition [14]:

Proposition 4. *Suppose for each N there exists $\varepsilon_N > 0$ such that*

$$\mathcal{X}_{k+1,N} + \varepsilon_N \mathcal{B} \subseteq \mathcal{X}_{k,N} \qquad \forall k \in \{0, \ldots, N-1\}$$

and for each $x \in \mathcal{X}_{N,N} + \varepsilon_N \mathcal{B}$ there exists $u \in \mathcal{U}$ such that $f(x,u) \in \mathcal{X}_{N,N}$. Then Assumption 7 holds with $M = 1$ and $\tilde{N} = 1$.

For the system (25) with the constraint sets mentioned above, the conditions of Proposition 4 hold. Therefore, all of the conditions of Theorem 9 hold and thus, for the closed-loop using RHOC with (28), the origin is semiglobally practically robustly asymptotically stable closed-loop on $\{x : x_1 < c\}$. Extra conditions on g, ℓ and σ can be imposed to guarantee the conclusions of Theorems 7 or 8.

10 Conclusion

We have addressed robustness of the asymptotic stability produced by discrete-time receding horizon optimal control (RHOC). The results were covered in two parts. The first part presented discrete-time asymptotic stability theory, especially for systems with discontinuous right-hand sides. Signatures of robustness of asymptotic stability were discussed. Control theory for discrete-time systems, especially the general RHOC method, comprised the second part. We showed by examples that RHOC, when using terminal constraints with short horizons or state constraints, may produce asymptotically stable closed-loops with no robustness. Finally, we illustrated that the standard assumptions of RHOC can be relaxed while simultaneously guaranteeing robust asymptotic stability as long as a sufficiently large optimization horizon is used.

Notably absent from this paper is a discussion of RHOC approaches that attempt to guarantee a prescribed level of robustness. This problem has been addressed in the literature. For references and discussion, see [31, Section 4]. Also, we have ignored completely the computational issues associated with efficiently computing an optimizing control sequence. Interesting results in this direction for constrained linear and piecewise linear systems have begun to appear, e.g., in [3]. Some stability results using suboptimal control sequences in the RHOC framework are discussed in [40].

The RHOC approach to stabilizing nonlinear systems is popular and gaining momentum. However, there are still several fundamental, not very well understood, issues in the area that warrant further investigation. In part, this study is needed to prevent undesirable surprises in future applications.

Acknowledgement. I credit my students, Gene Grimm, Chris Kellett, Michael Messina and Emre Tuna, for their important contributions to the results discussed in this paper. I also thank Rafal Goebel for several useful discussions on graph convexity as it relates to the result in Proposition 2. This work was supported in part by AFOSR under grants F49620-00-1-0106 and F49620-03-1-0203 and NSF under grants ECS-9988813 and ECS-0324679.

References

1. Artstein Z (1983) Stabilization with relaxed controls, Nonlinear Analysis 7:1163–1173
2. Aubin J, Cellina A (1984) Differential Inclusions. Springer-Verlag New York
3. Bemporad A, Morari M, Dua V, Pistikopoulos E (2002) The explicit linear quadratic regulator for constrained systems, Automatica 38:3–20
4. Berge C (1963) Topological Spaces. The Macmillian Company New York
5. Bhattacharya R, Balas G, Kaya M, Packard A (2002) Nonlinear receding horizon control of an F-16 aircraft, J Guidance Control and Dynamics 25:924–931
6. Chen C, Allgöwer F (1998) A quasi-infinite horizon nonlinear model predictive control scheme with guaranteed stability, Automatica 34:1205–1217
7. Chmielewski D, Manousiouthakis V (1996) On constrained infinite-time linear quadratic optimal control, Systems Control Letters 29:121–129
8. Clarke F, Ledyaev Y, Rifford L, Stern R (2000) Feedback stabilization and Lyapunov functions, SIAM J Control Optim 39:25—48
9. Clarke F, Ledyaev Y, Stern R (1998) Asymptotic stability and smooth Lyapunov functions, Journal of Differential Equations 149:69–114
10. De Nicolao G, Magni L, Scattolini R (1996) On the robustness of receding horizon control with terminal constraints, IEEE Transactions on Automatic Control 41:451–453
11. Fontes F (2001) A general framework to design stabilizing nonlinear model predictive controllers, Systems Control Letters 42:127–143
12. Grimm G, Messina M, Teel A, Tuna S (2003) Model predictive control when a local control Lyapunov function is not available. In: Proc American Control Conf Denver CO 4125–4130

13. Grimm G, Messina M, Tuna S, Teel A (2003) Examples of zero robust- ness in constrained model predictive control. In: Proc 42nd IEEE Conference Decision Control Maui HI to appear
14. Grimm G, Messina M, Tuna S, Teel A (2003) Nominally robust model predictive control with state constraints. In: Proc 42nd IEEE Conference Decision Control Maui HI to appear
15. Hájek O (1979) Discontinuous differential equations I-II, J Differential Equations 32:149–185
16. Hammouri H, Benamor S (1999) Global stabilization of discrete-time homogeneous systems, Systems Control Letters 38:5–11
17. Henson M (1998) Nonlinear model predictive control: current status and future directions, Computers Chemical Engineering 23:187–202
18. Hermes H (1967) Discontinuous vector fields and feedback control. In: Jadbabaie A, Hauser J (eds) Differential Equations and Dynamic Systems. Academic Press 155–165
19. Jadbabaie A, Hauser J (2002) Control of a thrust-vectored flying wing a receding horizon-LPV approach, Int J Robust Nonlinear Contr 12:869–896
20. Jadbabaie A, Yu J, Hauser J (2001) Unconstrained receding-horizon control of nonlinear systems, IEEE Trans Automat Control 46:776–783
21. Keerthi S, Gilbert E (1985) An existence theorem for discrete-time infinite-horizon optimal control problems, IEEE Trans Automatic Control AC-30:907–909
22. Keerthi S, Gilbert E (1988) Optimal, infinite horizon feedback laws for a general class of constrained discrete time systems: Stability and moving-horizon approximations, J of Optimization Theory and Applications 57:265–293
23. Kellett C, Teel A (2002) On robustness of stability and Lyapunov functions for discontinuous difference equations. In: Proceedings of the 41st IEEE Conference on Decision and Control 4282–4287
24. Kellett C, Teel A (2003) Further results on discrete-time control-Lyapunov functions. In: Proc 42nd IEEE CDC Maui HI to appear
25. Kellett C, Teel A (2003) Results on converse Lyapunov theorems for difference inclusions. In: Proc 42nd IEEE CDC Maui HI to appear
26. Ledyaev Y, Sontag E (1999) A Lyapunov characterization of robust stabization, Nonlinear Analysis 37:813–840
27. Lee E, Markus L (1986) Foundations of Optimal Control Theory. Robert E. Krieger Publishing Company
28. Limón Marreudo D, Álamo T, Camacho E (2002) Input-to-state stable MPC for constrained discrete-time nonlinear systems with bounded additive uncertainties. In: Proceedings of the 41st IEEE Conference on Decision and Control Las Vegas NV 4619–4624
29. Massera J (1949) On Liapounoff's conditions of stability, Annals of Mathematics 50:705–721
30. Mayne D, Michalska H (1990) Receding horizon control of nonlinear systems, IEEE Trans Automat Control 35:814–824
31. Mayne D, Rawlings J, Rao C, Scokaert P (2000) Constrained model predictive control stability and optimality, Automatica 36:789–814
32. Meadows E, Henson M, Eaton J, Rawlings J (1995) Receding horizon control and discontinuous state feedback stabilization, International Journal of Control 62:1217–1229

33. Messina M, Tuna S, Teel A (2003) Output feedback stabilization by certainty equivalence model predictive control, Automatica submitted
34. Michalska H, Mayne D (1993) Robust receding horizon control of constrained nonlinear systems, IEEE Trans on Automatic Control 38:1623–1633
35. Michalska H, Vinter R (1994) Nonlinear stabilization using discontinuous moving-horizon control, IMA J Math Control Inform 11:321–340
36. Nešić D, Teel A, Kokotović P (1999) Sufficient conditions for stabilization of sampled-data nonlinear systems via discrete-time approximations, Systems Control Lett 38:259–270
37. Qin S, Badgwell T (1997) An overview of industrial model predictive control technology. In: Chemical Process Control V CACHE AIChE Austin TX 232-256
38. Robinson S (1976) Regularity and stability for convex multivalued functions, Mathematics of Operations Research 1:130–143
39. Santos L, Biegler L (1999) A tool to analyze robust stability for model predictive controllers, Journal of Process Control 9:233–246
40. Scokaert P, Mayne D, Rawlings J (1999) Suboptimal model predictive control (feasibility implies stability), IEEE Transactions on Automatic Control 44:648–654
41. Scokaert P, Rawlings J, Meadows E (1997) Discrete-time stability with perturbations: Applications to model predictive control, Automatica 33:463–470
42. Sontag E (1983) A Lyapunov-like characterization of asymptotic controllability, SIAM J Control Optim 21:462–471
43. Sontag E (1998) Comments on integral variants of ISS, Systems Control Lett 34:93–100
44. Sontag E (1999) Clocks and insensitivity to small measurement errors, ESAIM Control Optim Calc Var 4:537–557
45. Sontag E (1999) Stability and stabilization: discontinuities and the effect of disturbances. In: Nonlinear analysis, differential equations and control Montreal QC 1998. Kluwer Acad Publ Dordrecht 551-598
46. Teel A, Praly L (2000) A smooth Lyapunov function from a class-KL estimate involving two positive semidefinite functions, ESAIM Control Optim Calc Var 5:313–367

On the Interpretation of Non-anticipative Control Strategies in Differential Games and Applications to Flow Control

J.M.C. Clark, and R.B. Vinter

EEE Dept. Imperial College London, Exhibition Road, London SW7 2BT UK.
[j.m.c.clark,r.vinter]@imperial.ac.uk

Summary. This paper concerns the notion of feedback strategies in differential games. Classical notions of feedback strategies, based on state feedback control laws for which the corresponding closed loop dynamics uniquely define a state trajectory, are too restrictive for many problems, owing to the absence of minimizing classical feedback strategies or because consideration of classical feedback strategies fails to define, in a useful way, the value of the game. A number of feedback strategy concepts have been proposed to overcome this difficulty. That of Elliot and Kalton, according to which a feedback strategy is a non-anticipative mapping between control functions for the two players, has been widely taken up because it provides a value of the game which connects, via the HJI equation, with other fields of systems science. The non-anticipative mapping approach leaves unanswered, nonetheless, some fundamental questions regarding the representation of optimal feedback strategies. Attempts to solve specific games problems by the method of characteristics, or other techniques, often lead to consideration of optimal feedback strategies in the form of discontinuous state feedback control laws. Such control laws cannot be regarded as classical feedback control strategies because the associated state trajectories are not unique. We provide full details of the proofs of recently announced general theorems that interpret discontinuous state feedback laws as non-anticipative feedback strategies. An application to flow control illustrates the relevance of the theory to process systems engineering.

Keywords: Differential Games, Differential Inclusions, Feedback Control.

M. de Queiroz et al. (Eds.): Optimal Control, Stabilization and Nonsmooth Analysis, LNCIS 301, pp. 29–47, 2004.

1 Introduction

Consider the differential game

$$P(x_0)\begin{cases} \sup_u \inf_v \int_0^{\tau(x)} L(t,x(t),u(t),v(t))dt + g(\tau(x),x(\tau(x))) \\ \text{over feedback controls } u \text{ and open loop controls } v \text{ satisfying} \\ \dot{x}(t) = f(t,x(t),u(t),v(t)) \quad \text{a.e. } t \in [0,\infty), \\ x(0) = x_0, \\ u(t) \in \Omega_1(t),\ v(t) \in \Omega_2(t) \quad \text{a.e. } t \in [0,\infty)\,, \end{cases}$$

in which $\tau(x)$ denotes the first exit time from an open set A of the state trajectory x:

$$\tau(x) \ := \ \inf\{t : x(t) \notin A\}\,.$$

(We define $\tau(x) = \infty$ if $x \in A$ for all $t \geq 0$.)

The data for $P(x_0)$ comprises functions $L : [0,\infty) \times \mathbb{R}^n \times \mathbb{R}^{m_1} \times \mathbb{R}^{m_2} \to \mathbb{R}$, $f : [0,\infty) \times \mathbb{R}^n \times \mathbb{R}^{m_1} \times \mathbb{R}^{m_2} \to \mathbb{R}^n$ and $g : [0,\infty) \times \mathbb{R}^n \to \mathbb{R}$, subsets $\Omega_1 \subset [0,\infty) \times \mathbb{R}^{m_1}$, $\Omega_2 \subset [0,\infty) \times \mathbb{R}^{m_2}$ and $A \subset \mathbb{R}^n$, an n-vector $x_0 \in A$ and numbers $p_1, p_2 \in [1,\infty) \cup \{\infty\}$.

Here, and elsewhere, $D(t) \subset \mathbb{R}^n$ denotes the section $\{x : (t,x) \in D\}$ of a set $D \subset [0,\infty) \times \mathbb{R}^n$.

Define the spaces of open loop control policies for the u and v-players respectively as

$$U := \{u \in L^{p_1}([0,\infty);\mathbb{R}^{m_1}) \ : \ u(t) \in \Omega_1(t) \text{ a.e.}\}\,,$$
$$V := \{v \in L^{p_2}([0,\infty);\mathbb{R}^{m_2}) \ : \ v(t) \in \Omega_2(t) \text{ a.e.}\}\,.$$

Throughout this section we shall assume that, for each $(u,v) \in U \times V$, the differential equation

$$\begin{cases} \dot{x}(t) \ = \ f(t,x(t),u(t),v(t)) \quad \text{a.e. } t \in [0,\infty) \\ x(0) \in x_0 \end{cases}$$

has a unique solution $x \in AC_{loc}([0,\infty);\mathbb{R}^n)$ ($AC_{loc}([0,\infty);\mathbb{R}^n)$, sometimes abbreviated as AC_{loc}, denotes the space of locally absolutely continuous $\mathbb{R}^n$ valued functions on $[0,\infty)$) and $t \to L(t,x(t),u(t),v(t))$ is integrable. To emphasize the dependence, we write the solution $x(u,v)$. For given open loop controls $(u,v) \in U \times V$ then, the cost

$$J(u,v) \ := \ \begin{cases} \int_0^{\tau(x)} L(t,x(t),u(t),v(t))dt + g(\tau(x),x(\tau(x))) & \textit{if } \tau(x) < \infty \\ \infty & \textit{if } \tau(x) = \infty\ , \end{cases}$$

in which $x(t) = x(u,v)(t)$, can be evaluated.

The upper game is that in which the u-player seeks the maximum over feedback strategies u (appropriately defined) of

$$\inf_{v \in V} J(u,v) \tag{1}$$

Note that the infimum is taken over V, the space of open loop controls for the v-player.

The supremum is termed the *upper value of the game.* A similar procedure, but one in which the infimum is taken over closed loop strategies for the v-player and the supremum is taken over open loop strategies for the u-player, yields the *lower value of the game.* Should the upper and lower values coincide, they are said to be the *value of the game.*

Interest in differential games over the past decade results, in no small part, from its relationship with robust non-linear control design [17] and with the asymptotic analysis of optimal stochastic control problems as the driving noise signals become vanishingly small [9]. The link is the Hamilton-Jacobi-Isaacs (HJI) equation. In an appropriate analytical framework, value functions encountered in robust non-linear controller design and small-noise stochastic optimal control, and the upper value in differential games, are all characterized as the unique generalized solution to the same HJI equation, for suitably chosen pay-offs, dynamic models, etc.

The manner in which feedback strategies are defined is a crucial issue in differential games. This is because it can affect the upper value of the game and determine whether, in particular, it can be related to the HJI equation. The classical approach to defining a feedback strategy is to specify a state feedback control law

$$u(t) \;=\; \chi(t, x(t))\,,$$

expressed in terms of a function $\chi : [0,\infty) \times \mathbb{R}^n \to \mathbb{R}^{m_1}$. If the state feedback function is sufficiently regular, then the closed loop system

$$\begin{cases} \dot{x}(t) \;=\; f(t, x(t), \chi(t, x(t)), v(t)) & \text{a.e. } t \in [0,\infty) \\ x(0) \in x_0\,, \end{cases} \tag{2}$$

obtained by using the state feedback control law to eliminate the u-variable, has a unique solution x in $AC_{loc}([0,\infty); \mathbb{R}^n)$. We can therefore speak unambiguously of x as the state trajectory corresponding to v and $u(t) = \chi(t, x(t))$ and evaluate the pay-off.

The major weakness of the classical approach is that, for many problems of interest, we need to enlarge the class of feedback strategies beyond those defined by regular state feedback laws, to investigate the upper value via the HJI equation, etc. For an important class of problems, a simple example of which we study in Section 4, heuristic analysis suggests an optimal feedback strategy in the form of a discontinuous (switching) state feedback control function. But such a feedback control function makes no sense in a classical framework, because it gives rise to an ill-posed closed loop dynamic system – one which fails unambiguously to determine the state trajectory for a given open loop strategy of the v-player, or which fails to determine a state trajectory altogether.

In [10], Elliot and Kalton proposed a radically different approach to defining feedback strategies, in order to circumvent the serious difficulties associated with badly behaved state feedback control laws. It was to work instead with feedback strategies $\mathcal{U}$ for the u-player, interpreted as non-anticipative mappings from V to U:

$$\{G : V \to U \,:\, \forall\, v, v' \in V,\, T \geq 0\,,\\ `v|_{[0,T]} = v'|_{[0,T]}\text{' implies '}G(v)|_{[0,T]} = G(v')|_{[0,T]}\text{'}\}\,.$$

Here, $v|_{[0,T]} = v'|_{[0,T]}$ means that v and v' coincide on $[0,T]$ a.e.

Elements in $\mathcal{U}$ will henceforth be referred to as *non-anticipative feedback strategies*. Introduction of non-anticipative feedback strategies eliminates at a stroke difficulties over evaluating the pay-off associated with an open loop v-player strategy $v \in V$ and a closed loop u-player strategy $G \subset \mathcal{U}$; it becomes simply a matter of evaluating J at the open loop controls $u \in U$ and $v \in V$, where $u = G(v)$. The fact, too, that, under unrestrictive and directly verifiable hypotheses, the upper value associated with the class of non-anticipative feedback strategies can be characterized as a solution to the HJI equation, in some generalized sense, accounts for the emphasis which is given to this approach in the current research literature. (See [2].)

The class of non-anticipative feedback strategies includes classical state feedback control strategies, since a regular state feedback control function χ defines a mapping from V to the space of state trajectories via (2), and then from V to U, via the state feedback control law. This last mapping is obviously non-anticipative. On the other hand, there are many non-anticipative feedback strategies that cannot be associated with a classical state feedback control law.

For all its advantages, the framework of Elliot and Kalton, introduced in order to provide a consistent definition of the value of a differential game, leaves unresolved certain questions about the interpretation of non-anticipative feedback strategies and their implementation. This is significant since, in applications of differential games to non-linear robust controller design for example, it is the optimal feedback control strategy for the u-player and not the value of the game, that is of primary importance.

The most widely used technique for solving differential games problems is the method of characteristics. For a fixed initial state and state feedback control law $u = \chi(t, x)$ for the u-player, we solve a Hamiltonian system of equations, supplied by the Maximum Principle, to determine an optimal open-loop strategy for the v-player. The feedback function is then adjusted to ensure satisfaction of the HJI equation [11].

Often, the optimal state feedback control function χ obtained in this way is *multi-valued*, owing to the non-uniqueness of optimal state feedback strategies for the u-player, for certain initial conditions. We shall see an example of this phenomenon in Section 4. The optimal state feedback control 'law' takes the form

$$u(t) \in \bar{\chi}(t, x(t), v(t)). \tag{3}$$

The overbar indicates the multi-valued nature of the state feedback control 'function'. We arrange for greater flexibility in the ensuing analysis by allowing $\bar{\chi}$ to depend on $v(t)$ as well as t and $x(t)$. It is by no means obvious that (3) can be interpreted as a non-anticipative feedback strategy.

This brings us to one of the main objectives of this paper. It is to give conditions under which the u-player can choose a non-anticipative feedback strategy $G \subset \mathcal{U}$, compatible with a given multi-valued state feedback control law (3). The important point here is that there may be many functions $u \in U$ and $x \in AC_{loc}([0,\infty);\mathrm{IR}^n)$ satisfying

$$\begin{cases} \dot{x}(t) = f(t, x(t), u(t), v(t)) \quad \text{a.e. } t \in [0,\infty) \\ x(0) = x_0 \\ u(t) \in \bar{\chi}(t, x(t), v(t)) \quad \text{a.e. } t \in [0,\infty)\ , \end{cases} \tag{4}$$

for each $v \in V$; Section 3 provides a selection theorem, asserting that u and x can be chosen *in a non-anticipative manner.*

In many cases that have been studied in the literature [11], the optimal multi-valued state feedback law for the u-player is single-valued on regions of (t,x)-space, and the corresponding optimal open loop controls for the v-player give rise to state trajectories which intersect region boundaries on a time set of measure zero. Often, in these circumstances, there is a unique state trajectory associated with the optimal state feedback control strategy for the u-player and the associated optimal open loop control for the v-player. It might be thought then that there is no need to confront non-uniqueness of solutions $(u,x) \in U \times AC_{loc}$ to (4) because such pathological situations seldom arise. This is to ignore the fact that, in order to make sense of the cost function

$$\inf_{v \in V} J(u,v)$$

for the u-player, we must give meaning to

$$J(u \in \bar{\chi}(t,x,v), v)$$

for *every* $v \in V$, including those v's which compel state trajectories to lie in region boundaries, on a set of positive measure.

It is a common perception that the consideration of non-anticipative feedback strategies 'resolves' the problem of interpreting nonlinear feedback laws in differential games. It does not follow directly from the definitions, however, that the notion of generalized feedback strategy can be used to give precise meaning to solutions to specific differential games problems that take the form of a discontinuous feedback law, such as those studied by Isaacs [11]. We highlight this gap and give precise, verifiable and unrestrictive conditions under which it can be bridged, i.e., under which there exist generalized feedback strategies compatible with a given discontinuous feedback law. In addition, we illustrate the applications of this theory to process systems engineering by studying a controller design problem arising in flow control.

Relevant set theoretic techniques for obtaining such conditions have been developed by Cardaliaguet and Paskacz [3] in connection with related issues in two-player dynamical systems. These concern, among other things, conditions for the existence of generalized feedback strategies for the first player ensuring that the state evolves in a given set, whatever the strategy of the second player. This paper brings out the role of such techniques in reconciling generalized feedback strategies and nonlinear feedback laws encountered in differential games applications. In addition, it provides an illustrative example from flow control.

Finally, we remark that the questions raised in this paper, concerning the relationship between classical discontinuous feedback and 'generalized' feedback strategies, are specific to the non-anticipative feedback strategies of Elliot and Kalton. There are other general procedures for defining feedback strategies in a differential game, that allow for discontinuities and also yield a value of the game satisfying the HJI equation, in some generalized sense. The most widely used is that associated with the name of Krasovskii, and developed by A. I. Subbotin and his co-workers (see [16]). Here the issue of interpeting a nonlinear feedback law as a generalized feedback strategy does not arise, since these procedures give meaning to (1), via suitable time discretization and limit taking schemes, when v is a classical (and possibly discontinuous) feedback law.

2 Existence of Non-anticipative Solutions to Controlled Differential Inclusions

As we shall see, a convenient starting point for studying non-anticipative feedback strategies, compatible with a given discontinuous feedback control law, is to examine 'non-anticipative' solutions to the controlled differential inclusion:

$$\begin{cases} \dot{x}(t) \in F(t, x(t), v(t)) & \text{a.e. } t \in [0, \infty) \\ x(0) = x_0 \,. \end{cases} \tag{5}$$

Here, $F : [0, \infty) \times \mathbb{R}^n \times \mathbb{R}^{m_2} \rightrightarrows \mathbb{R}^n$ is a given multi-function and x_0 is a given n-vector.

Fix $p_2 \in [0, \infty) \cup \{\infty\}$ and a multi-function $\Omega_2 : [0, \infty) \rightrightarrows \mathbb{R}^{m_2}$. As in Section 1, define the set of open-loop trajectories for the v-player:

$$V \;=\; \{v \in L^{p_2}_{loc}([0, \infty); \mathbb{R}^{m_2}) \;:\; v(t) \in \Omega_2(t) \text{ a.e. } \}$$

Consider the differential equation

$$\begin{cases} \dot{x}(t) = f(t, x(t), v(t)) & \text{a.e. } t \in [0, \infty) \\ x(0) = x_0 \,, \end{cases} \tag{6}$$

in which $f : [0, \infty) \times \mathbb{R}^n \times \mathbb{R}^{m_2} \to \mathbb{R}^n$ is a given function and x_0 is a given n-vector. Assume that, for each $v \in V$, (6) has a unique solution $x(v) \in$

$AC_{loc}([0,\infty);\mathbb{R}^n)$ on $[0,\infty)$. The following property is a consequence of the assumed uniqueness of solutions to (6) and of the fact that the concatenation of solutions to the differential equation (for fixed v) is also a solution: for any $v, v' \in V$ and $T \geq 0$,

$$\text{`}v|_{[0,T]} = v'|_{[0,T]}\text{'} \qquad \text{implies} \qquad \text{`}x(v)|_{[0,T]} = x(v')|_{[0,T]}\text{'}.$$

Theorem 1 below generalizes this property to controlled differential inclusions. When the multi-function F replaces the function f, we can expect there will be many state trajectories satisfying the dynamic constraint (5), for given $v \in V$. The theorem asserts that we can, nonetheless, *select* solutions to the differential inclusion (5) which retain the non-anticipativity property.

The following terminology will be helpful. Given $v \in V$ and $T \in [0,\infty) \cup \{\infty\}$, we will say that a locally absolutely continuous function $x : [0,\infty) \to \mathbb{R}^n$ is an $F(.,.,v)$-trajectory on $[0,T)$ provided $x(0) = x_0$ and $\dot{x}(t) \in F(t,x(t),v(t))$ a.e. $t \in [0,T)$.

Theorem 1. *Consider the controlled differential inclusion (5). Assume that*

(a) For each $v \in V$, there exists an $F(.,.,v)$-trajectory on $[0,\infty)$.
(b) For each $v \in V$ and $T \geq 0$, any $F(.,.,v)$-trajectory on $[0,T)$ has an extension as an $F(.,.,v)$-trajectory on $[0,\infty)$.

Let $\mathcal{X} : V \rightrightarrows AC_{loc}([0,\infty);\mathbb{R}^n)$ be the multi-function

$$\mathcal{X}(v) := \{x \in AC_{loc}([0,\infty);\mathbb{R}^n) : x \text{ is an } F(.,.,v)\text{-trajectory on } [0,\infty)\}.$$

Then there exists a function $X : V \to AC_{loc}([0,\infty);\mathbb{R}^n)$ such that

$$X(v) \in \mathcal{X}(v) \quad \text{for all } v \in V$$

and, for any $T \geq 0$ and $v, v' \in V$,

$$\text{`}v|_{[0,T]} = v'|_{[0,T]}\text{' implies `}X(v)|_{[0,T]} = X(v')|_{[0,T]}\text{'}.$$

The theorem is proved in Section 5.

Remarks

(i) Other versions of the theorem may be stated, in which hypotheses (a) and (b) are replaced by sufficient conditions for their validity, of a more directly verifiable kind, drawn from the differential inclusions literature ([8], [1]). We prefer however to axiomatize, through our choice of hypotheses, the essential properties of controlled differential inclusions for the existence of a non-anticipatory family of solutions. This is partly to lend transparency to the proof and, partly, because off-the-shelf sufficient conditions for (a) and (b) are not well-suited to the applications made below.

(ii) Theorem 1 concerns the existence of non-anticipative families of solutions to a controlled differential inclusion over the semi-infinite time interval $[0, \infty)$. Suppose that F is defined only on $[0, \bar{T}] \times \mathbb{R}^n \times \mathbb{R}^{m_2}$, for some fixed $\bar{T}$. A finite interval version of the theorem, in which $[0, \bar{T}]$ replaces $[0, \infty)$, is easily derived, by extending F to all of $[0, \infty) \times \mathbb{R}^n \times \mathbb{R}^{m_2}$, setting

$$F(t, x, v) = \{0\} \quad \text{for } t > \bar{T},$$

applying Theorem 1 and interpreting the asserted properties in terms of the controlled differential inclusion on the original domain. The modified theorem is the same as Theorem 2.1, except that '$[0, \bar{T}]$' replaces $[0, \infty)$ throughout and '$T \geq 0$' is replaced by '$T \in [0, \bar{T}]$'. In the modified condition (a), '$F(.,.,v)$-trajectory on [0,T]' has the obvious meaning.

In abstract terms, Theorem1 is a theorem on 'non-anticipative selections' of non-anticipative multi-functions, appropriately defined. Cardaliaguet and Plaskacz [3] provide an axiomatic framework, centered on the notion of a 'non-expansive' multifunction between ultra-metric spaces, for proving selection theorems of this nature in a general context. The assertions of Theorem 1 can be deduced from [3], Lemma 7.1. In the appendix, we provide a direct proof, specific to present requirements.

3 Existence of Non-anticipative Feedback Strategies Compatible with a Multi-valued State Feedback Control Law

The analytical tools of Section 2 will now be used to resolve questions of existence of non-anticipative feedback strategies compatible with a given multi-valued state feedback control function, posed in Section 1. We restrict attention to the following special case of the dynamic constraint in problem $\mathrm{P}(x_0)$:

$$\dot{x}(t) = d(t, x(t)) + A(t, x(t))u(t) + B(t, x(t))v(t) \quad \text{a.e. } t \in [0, \infty), \quad (7)$$

$$x(0) = x_0, \quad (8)$$

$$u(t) \in \bar{\chi}(t, x(t), v(t)) \quad \text{a.e. } t \in [0, \infty). \quad (9)$$

Here $d : [0, \infty) \times \mathbb{R}^n \to \mathbb{R}^n$, $A : [0, \infty) \times \mathbb{R}^n \to \mathbb{R}^{n \times m_1}$ and $B : [0, \infty) \times \mathbb{R}^n \to \mathbb{R}^{n \times m_2}$ are given functions, $\bar{\chi} : [0, \infty) \times \mathbb{R}^n \times \mathbb{R}^{m_2} \rightrightarrows \mathbb{R}^{m_1}$ is a given multifunction and x_0 is a given n-vector. Take also subsets $\Omega_1 \subset \mathbb{R}^{m_1+1}$ and $\Omega_2 \subset \mathbb{R}^{m_2+1}$.

We assume that

$$\bar{\chi}(t, x, v) \in \Omega_1(t) \quad \text{for all } (t, x, v) \in [0, \infty) \times \mathbb{R}^n \times \mathbb{R}^{m_2}.$$

Define

$$U^\infty = \{u \in L^\infty_{loc}([0,\infty);\mathbb{R}^{m_1}) \,:\, u(t) \in \Omega_1(t) \text{ a.e. }\},$$
$$V^1 = \{v \in L^1_{loc}([0,\infty);\mathbb{R}^{m_2}) \,:\, v(t) \in \Omega_2(t) \text{ a.e. }\},$$

i.e., the sets of open loop controls for the u and v-players introduced in Section 1, when $p_1 = \infty$ and $p_2 = 1$.

Theorem 2. *Consider the two player dynamic system (7)– (9). Assume that*

(H_1) d, A and B are continuous functions and there exist

$$k_1 \in L^1_{loc}([0,\infty);\mathbb{R}), \quad k_\infty \geq 0$$

such that

$$|d(t,x)| + |A(t,x)| \leq k_1(t)(1+|x|) \quad \textit{for all } (t,x) \in [0,\infty) \times \mathbb{R}^n,$$
$$|B(t,x)| \leq k_\infty(1+|x|) \quad \textit{for all } (t,x) \in [0,\infty) \times \mathbb{R}^n.$$

(H_2) $\bar{\chi}$ has a closed graph, $\bar{\chi}(t,x,v)$ is convex for each $(t,x,v) \in [0,\infty) \times \mathbb{R}^n \times \mathbb{R}^{m_2}$ and the multi-function $t \rightrightarrows \Omega_1(t)$ is bounded on bounded sets.

Define the multi-function $\mathcal{Z} : V^1 \rightrightarrows U^\infty$ by

$$\mathcal{Z}(v) := \{u \in U^\infty \,:\, \exists\, x \in AC_{loc}([0,\infty);\mathbb{R}^n) \;\; s.t. \;\; (7)-(9) \;\; \text{are satisfied}\}.$$

Then there exists a mapping $Z : V^1 \to U^\infty$ such that

$$Z(v) \in \mathcal{Z}(v) \quad \textit{for all } v \in V^1$$

and, given any $v, v' \in V^1$ and $T \geq 0$ such that

$$v|_{[0,T]} = v'|_{[0,T]},$$

we have

$$u|_{[0,T]} = u'|_{[0,T]},$$

where $u = Z(v)$ and $u' = Z(v')$.

Proof. Define the multi-function $\tilde{F} : [0,\infty) \times \mathbb{R}^{n+m_1} \times \mathbb{R}^{m_2} \rightrightarrows \mathbb{R}^{n+m_1}$ by

$$\tilde{F}(t,(x,y),v) =$$

$$\left\{ \begin{pmatrix} d(t,x) + A(t,x)u + B(t,x)v \\ u \end{pmatrix} \in \mathbb{R}^{n+m_1} \,:\, u \in \bar{\chi}(t,x,v) \right\},$$

in which we regard the partitioned vector $(x,y) \in \mathbb{R}^{n+m_1}$ as the dependent variable. We shall show in Section 5 that the controlled differential inclusion

$$(\dot{x},\dot{y}) \in \tilde{F}(t,(x,y),v) \tag{10}$$
$$(x(0),y(0)) = (x_0,0) \tag{11}$$

satisfies the hypotheses of Theorem 1. Assume, then, this to be the case. We are assured of the existence of a mapping $X : V^1 \to AC_{loc}([0,\infty); \mathbb{R}^{n+m_1})$ such that

$$X(v) \in \{(x,y) \in AC_{loc}([0,\infty); \mathbb{R}^{n+m_1}) : (x,y) \text{ satisfies (10) and (11)}\}$$

and, for all $v, v' \in V^1$ and $T \geq 0$ such that

$$v|_{[0,T]} = v'|_{[0,T]},$$

we have

$$x|_{[0,T]} = x'|_{[0,T]} \quad \text{and} \quad y|_{[0,T]} = y'|_{[0,T]},$$

where $(x,y) = X(v)$. Now define $Z : V^1 \to U^\infty$ to be

$$Z(v)(t) = \dot{y}(t) \quad \text{a.e. } t \in [0,\infty),$$

where $y \in AC_{loc}$ is the second component of $X(v)$.

It is easy to derive from the above stated properties of the mapping X that, for any $v, v' \in V^1$ and $T \geq 0$,

$$v|_{[0,T]} = v'|_{[0,T]} \quad \text{implies} \quad Z(v)|_{[0,T]} = Z(v')|_{[0,T]}.$$

Thus, the mapping Z has the properties asserted in the theorem statement.

Remarks

(i) The space of open loop controls for the v-player V^1 is taken to be a subset of $L^1_{loc}([0,\infty); \mathbb{R}^{m_2})$. This is the largest L^p_{loc} space, and provides the most comprehensive theorem statement (the non-anticipativity property is asserted to hold over the largest set of v-player controls). U is taken to be a subset of $L^\infty_{loc}([0,\infty); \mathbb{R}^{m_1})$. This is a natural choice of function space in the present context, since $t \to \Omega_1(t)$ is bounded on bounded sets, and the control law (9) compels u to be bounded on bounded sets.

(ii) Note that, under the stated hypotheses, there is no guarantee that the dynamic equations (7)-(9) have a unique solution x for each $(u,v) \in U^\infty \times V^1$. (The Lipschitz continuity conditions usually imposed for this purpose are absent from the hypotheses.) Thus, for the specified dynamic equations, the proof of Theorem 2 gives an affirmative answer to a broader question than that posed in Section 1, namely: can a *state trajectory* as well as a u-player control be matched to each v-player control in a non-anticipative manner?

4 An Application to Flow Control

The following controller design problem arises in flow control. The volume of water y in a tank is related to the rate of change of inflow v and the rate of change of outflow u according to the following differential equation.

$$d^2y/dt^2 = -u + v\,.$$

Here, v is regarded as a disturbance and u as a control variable. Normalizing the variable y, we can represent the constraints that the tank neither empties nor overflows as

$$-1 \le y \le +1\,.$$

There is also a constraint on the maximum rate of change of outflow. Normalizing the variable u, this can be express as

$$-1 \le u \le +1\,.$$

Fix $k \ge 0$ and the initial values of y and dy/dt. We seek a feedback control policy

$$u = \chi(y, dy/dt)$$

to maximize the minimum over all disturbance histories of

$$\min \left\{ \int_0^{\tau(x)} |v(t)|^2 dt + k\tau(x) \;:\; v \in L^2_{loc} \right\},$$

where x is the state trajectory corresponding to u and v, and $\tau(x)$ is the first time that the tank either empties or overflows.

The purpose here is to offer some protection against the tank either emptying or overflowing. According to the above formulation, this goal is achieved by maximizing the minimum disturbance energy (plus a term consisting of the scaled exit time) required to bring about such an event.

The design problem fits the framework of Section 1, when we make the following identifications:

$$\begin{array}{l} n = 2,\ m_1 = 1,\ m_2 = 1,\ p_1 = \infty,\ p_2 = 2\,, \\ L(t,x,u,v) \;=\; v^2, \quad g(\tau, x) = k\tau\,, \\ f(t,x,u,v) \;=\; (x_2, u - v)\,, \\ \Omega_1 = [-1,+1],\ \Omega_2 = \mathbb{R},\ A = (-1,+1) \times \mathbb{R}\,. \end{array}$$

The following information about the solution to the flow problem is established in a forthcoming paper:

Proposition 1. *Consider the differential games problem of Section 1, with the above identifications. Take any $k \ge 0$ and $x_0 \in A$. Then there exists a C^∞, monotone decreasing function $s_k : [-1,+1] \to \mathbb{R}$ with the following properties. Define*

$$\bar{\chi}(x = (x_1, x_2)) \;=\; \begin{cases} \{+1\} & \text{if } \; x_2 > s_k(x_1), \\ [-1,+1] & \text{if } \; x_2 = s_k(x_1), \\ \{-1\} & \text{if } \; x_2 < s_k(x_1)\,. \end{cases}$$

Take any $x_0 \times \mathbb{R}$ and non-anticipative feedback strategy $G^* : V \to U$ such that

$$u(t) \in \chi(x(t)) \quad a.e. \ ,$$

where $u = G^*(v)$ and x is the state trajectory corresponding to u and v for initial value x_0. Then

$$\inf_{v \in V} J(G^*(v), v) \geq \sup_{G \in \mathcal{U}} \inf_{v \in V} J(G(v), v) \, .$$

In other words, G^* achieves the upper value of the differential game.

The preceding proposition leaves open the question of whether there exists a non-anticipative feedback strategy compatible with the specified set-valued state feedback. The existence of such a strategy is confirmed, however, by Theorem 2.

Theorem 3. *There exists a non-anticipative feedback strategy $G^* : V \to U$ such that, for all $v \in V$,*

$$u(t) \in \bar{\chi}(x(t)) \quad a.e. \ ,$$

where $u = G^*(v)$ and x is the state trajectory corresponding to u and v, i.e., x is the unique $AC_{loc}([0,\infty); \mathbb{R})$ function satisfying

$$\begin{cases} \dot{x} = (x_2, u - v) & a.e. \ t \in [0, \infty) \\ x(0) = x_0 \, . \end{cases}$$

Any such non-anticipative feedback strategy is an optimal non-anticipative feedback strategy for the u-player.

Proof. The extra information in this theorem is the existence of a non-anticipative feedback strategy with the stated properties. Existence of such a strategy follows directly from Theorem 2, the hypotheses in which are satisfied when, as here, the dynamics are linear and time invariant, Ω_1 is a constant multi-function, taking value a closed, convex, bounded set and $\bar{\chi}$ has the properties asserted in the proposition.

Comments

1. Traditionally, PID (proportional + integral + derivative) controllers have been used in surge tank control. McDonald and McAvoy [14] identified the Maximum Rate of Change of Outflow (MROC) as a key performance indicator. Indeed it is the *rate of change* of outflow that has adverse effects on downstream process units (in the form of disturbed sediments, turbulence, etc.). McDonald and McAvoy determined the controller, in feedback form, that minimizes MROC, for step changes in input flow rate. The MacDonald/McAvoy controller was significant for introducing modern controller design methods into this branch of process control. It suffers, nonetheless, from a number of defects. It takes account of only a restricted set of

disturbance signals (step changes in inflow or, equivalently, impulsive disturbances). It is designed in such a way that, following a disturbance, the surge tank either fills or empties in finite time. Finally, it ignores controller saturation.
Kantor [12] examined variants on the MacDonald controller and also proposed sliding-mode type controllers that do not cause the tank to fill or empty. A radical departure from either methods, first described in [5] (see also [6]), was to adopt the differential games formulation described above. The most significant feature of this approach was the manner in which it takes account of MROC. MROC features most naturally as a *constraint* describing the limits on rate of change of outflow, if adverse downstream effects are to be avoided. Accordingly, MROC is merely constrained rather than minimized in the differential games formulation and the 'extra degrees of freedom' resulting from this can be exploited to protect the surge tank from overflowing or emptying. We note also that the differential games controller both takes account of controller constraints and also is 'optimal' with respect to a much richer class of disturbance signals than MacDonald's.

2. Figure 1 on the next page illustrates the switching curves $s_k(.)$ for a number of values of the parameter k. Note that all curves pass through the origin, and the absolute value of the slope increases as k increases. In the case $k = 0$, the switching curve is given by the following parametric formula

$$x_1(\alpha) = \frac{(1 + 3\alpha - 2\alpha^2 + 2\alpha^3 - 3\alpha^4 - \alpha^5)}{(1 + 3\alpha - 2\alpha^2 - 2\alpha^3 + 3\alpha^4 + \alpha^5)}$$

$$x_2(\alpha) = \frac{\sqrt{12}\alpha(\alpha - 1)}{(1 + \alpha)^{1/2}(1 + 3\alpha - 2\alpha^2 - 2\alpha^3 + 3\alpha^4 + \alpha^5)^{1/2}}$$

(the denominators in these formulae are positive for all non-negative α's), in which α ranges of over the interval

$$\frac{4}{3 + \sqrt{17}} \leq \alpha \leq \frac{3 + \sqrt{17}}{4} .$$

The relationships describing the switching curves in the case $k > 0$, which are much more complicated, will be described in a forthcoming paper.

5 Proofs

Proof of Theorem 1. We can assume, without loss of generality, that the set V of open loop controls for the v-player is non-empty since, otherwise, the assertions of the theorem are true vacuously. Write

$$S := \{(v, x) \in V \times AC_{loc}([0, \infty); \mathbb{R}^n) \, : \, (5) \text{ is satisfied } \}$$

and

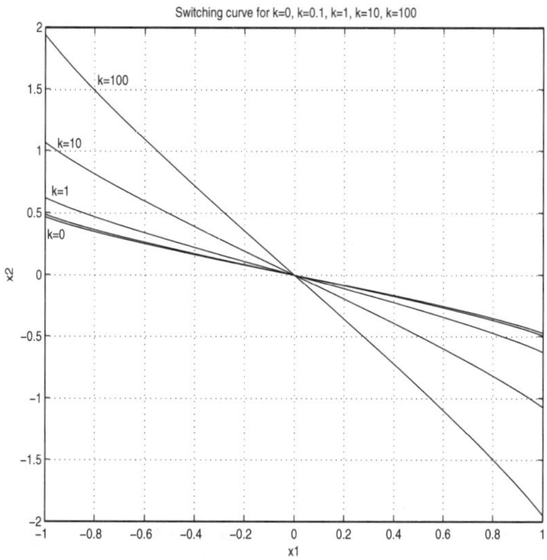

Fig. 1. Switching Curves for the Surge Tank Controller

$$Q := \{D \subset S : (v,x),(v',x') \in D \text{ and } `v|_{[0,T]} = v'|_{[0,T]} \text{ for some } T \geq 0' \text{ implies } `x|_{[0,T]} = x'|_{[0,T]}'\}\,.$$

We emphasize that Q is a family of *subsets* of S. Note that, for any $D \in Q$ and $v \in V$, there exists at most one element x such that $(v,x) \in Q$; this fact follows directly from the defining properties of Q. Thus, elements in Q are subgraphs, i.e., graphs of functions mapping subsets of V into AC_{loc}. Furthermore, Q is non-empty, since it contains point sets $\{(v,x)\}$, where v is an arbitrary element $v \in V$ and x is an $F(.,.,v)$-trajectory on $[0,\infty)$. We verify presently the following claim:

Claim: There exists $D^* \in Q$ with the property: for each $v \in V$ there is a unique $x \in AC_{loc}([0,\infty);\mathbb{R}^n)$ such that $(v,x) \in D^*$. That is, D^* is a graph over V.

The assertions of Theorem 1 will follow immediately. Indeed, the set D^* is, according to the claim, the graph of a mapping $G : V \to AC_{loc}$. Since $D^* \in Q$, the mapping has the requisite non-anticipativity properties for it to qualify as a member of $\mathcal{U}$.

It remains therefore to verify the claim. To this end, we introduce a partial ordering on Q:

$$D << D' \quad \text{if and only if} \quad D \subset D'\,.$$

Let $\mathcal{C} = \{D_\alpha : \alpha \in \mathcal{A}\}$ be any chain in Q, i.e., a subclass on which the partial ordering induces a total ordering.

Define

$$M = \cup_{\alpha \in \mathcal{A}} D_\alpha .$$

Notice that $M \in Q$. To see this, take any $T \geq 0$ and $(v,x),(v',x') \in M$ such that $v|_{[0,T]} = v'|_{[0,T]}$. Then $(v,x) \in D_\alpha$ and $(v',x') \in D_{\alpha'}$, for some $\alpha, \alpha' \in \mathcal{A}$. Since $\mathcal{C}$ is totally ordered by inclusion, $(v,x),(v',x') \in D_\beta$, where either $\beta = \alpha$ or $\beta = \alpha'$. In either case $x|_{[0,T]} = x'|_{[0,T]}$, since $D_\beta \in Q$. We have shown that $M \subset Q$. Since

$$D_\alpha \subset M \quad \text{for all} \quad \alpha \in \mathcal{A},$$

it follows that

$$D_\alpha << M \quad \text{for all } \alpha \in \mathcal{A}\,,$$

i.e., M is a maximal element in $\mathcal{C}$. We have shown that every chain in Q has a maximal element in Q. Zorn's Lemma [13] now tells us that Q itself has a maximal element, which we write D^*.

We show that D^*, chosen in this way, has the properties asserted in the claim.

Suppose the claim is false. As D^* is a subgraph, this implies that there exists $\bar{v}$ such that

$$\{x \in AC_{loc}([0,\infty); \mathbb{R}^n) \,:\, (x,\bar{v}) \in D^*\} = \emptyset\,. \tag{12}$$

Let

$$T^* := \sup\{T \geq 0 : \exists\,(v,x) \in D^* \text{ s.t. } \bar{v}|_{[0,T]} = v|_{[0,T]}\,.\}$$

Case 1: $T^* = +\infty$. In this case, there exists a sequence of times $T_i \uparrow +\infty$ and a sequence $\{(v_i, x_i)\}$ in D^* such that

$$v_i|_{[0,T_i]} = \bar{v}|_{[0,T_i]} \quad \text{for each } i\,.$$

Since $D^* \in Q$,

$$x_i|_{[0,T_i]} = x_j|_{[0,T_i]} \text{for all } j \geq i\,.$$

We can therefore define a function $\bar{x} \in AC_{loc}([0,\infty); \mathbb{R}^n)$ according to

$$\bar{x}(t) = x_i(t)\,,$$

where i is any index value such that $t < T_i$. Since $\dot{\bar{x}}(t) \in F(t, \bar{x}(t), \bar{v}(t))$ a.e. and $x(0) = x_0$, we have $(\bar{v}, \bar{x}) \in S$. But, by (12), $(\bar{v}, \bar{x}) \notin D^*$. Define

$$D' := D^* \cup \{(\bar{v}, \bar{x})\}\,.$$

We show that $D' \subset Q$. Indeed, take any $T > 0$ and distinct elements $(v,x),(v',x') \in D'$ such that

$$v|_{[0,T]} = v'|_{[0,T]}\,.$$

If $(\bar{v}, \bar{x})$ does not coincide with either (v, x) or (v', x') then $x|_{[0,T]} = x'|_{[0,T]}$, since $D^* \in Q$. If, on the other hand, $(v, x) = (\bar{v}, \bar{x})$ and $(v', x') \in D^*$, say, then

$$v_i|_{[0,T]} = \bar{v}|_{[0,T]} \quad \text{and} \quad x_i|_{[0,T]} = \bar{x}|_{[0,T]}$$

for any i such that $T_i > T$. But then, since $(v_i, x_i) \in D^*$, we have $x|_{[0,T]} = x'|_{[0,T]}$. It has been verified that $D' \in Q$. But $D' \neq D^*$ and $D^* << D'$. This contradicts the fact that D^* is a maximal element in Q. So (12) is false and, in this case, the claim is true.

Case 2: $0 < T^* < +\infty$. Take $T_i \uparrow T^*$ and a sequence $\{(v_i, x_i)\}$ in D^* such that $v_i|_{[0,T_i]} = \bar{v}|_{[0,T_i]}$, for each i. Define $\bar{x} \in AC_{loc}([0,\infty); \mathbb{R}^n)$ according to

$$\bar{x}(t) = x_i(t),$$

where i is any index value such that $T_i > t$. Now $\bar{x}$ is an $AC_{loc}([0,\infty); \mathbb{R}^n)$ function on $[0, T^*)$ such that $\bar{x}(0) = x_0$ and $\dot{\bar{x}} = F(t, \bar{x}, \bar{v})$ a.e. Under the hypotheses of Theorem 2.1, $\bar{x}$ admits an extension as a *loc*ally absolutely continuous function on $[0, \infty)$ satisfying $\bar{x}(0) = x_0$ and $\dot{\bar{x}} = F(t, \bar{x}, \bar{v})$ a.e. It follows that $(\bar{v}, \bar{x}) \in S$.

By (12), $(\bar{v}, \bar{x}) \notin D^*$. Define

$$D' := D^* \cup \{(\bar{v}, \bar{x})\}.$$

We show that $D' \in Q$. To this end, take any $T > 0$ and distinct elements $(v, x), (v'x') \in D'$ such that $v|_{[0,T]} = v'|_{[0,T]}$. If $(\bar{v}, \bar{x})$ coincides with neither (v, x) nor (v', x'), then $x|_{[0,T]} = x'|_{[0,T]}$, since $D^* \in Q$. We can assume then that $(v, x) = (\bar{v}, \bar{x})$ and $(v', x') \in D^*$. Notice that the case '$T > T^*$' cannot arise, since, by definition of T^*, there does not exist $(v', x') \in D^*$ such that $v'|_{[0,T]} = \bar{v}|_{[0,T]}$. We can assume then that $T \leq T^*$. Fix j. For any $i \geq j$, we have

$$x_i|_{[0,T_j \wedge T]} = x'|_{[0,T_j \wedge T]} .$$

But $x_i|_{[0,T]} = x|_{[0,T]}$. Since $T_j \uparrow T^*$, we deduce that

$$x|_{[0,T]} = x'|_{[0,T]} .$$

We have shown that $D' \in Q$. But this is not possible, since $D^* << D'$ and $D^* \neq D'$, yet D^* is a maximal element in Q. So (12) is false and, in this case too, the claim is confirmed.

Case 3: $T^* = 0$. In this case, there is no $(u, v) \in D^*$ and $T > 0$ such that $v|_{[0,T]} = \bar{v}|_{[0,T]}$. Under the hypotheses of Theorem 2.1, there exists $\bar{x} \in AC_{loc}([0,\infty); \mathbb{R}^n)$ such that $(\bar{v}, \bar{x}) \in S$. By (12), $(\bar{v}, \bar{x}) \notin D^*$. Define

$$D' := D^* \cup \{(\bar{v}, \bar{x})\}.$$

Take any distinct elements $(v, x), (v', x') \in D'$ and $T > 0$ such that $v|_{[0,T]} = v'|_{[0,T]}$. If neither (v, x) nor (v', x') coincides with $(\bar{v}, \bar{x})$, then $x|_{[0,T]} = x'|_{[0,T]}$, since $(v,x),(v',x') \in D^*$. The remaining case to consider is that when $(v,x) = (\bar{v},\bar{x})$ and $(v',x') \in D^*$, say. But this cannot arise, since $T^* = 0$ and therefore there are no elements $(v',x') \in D^*$ with $v'|_{[0,T]} = \bar{v}|_{[0,T]}$. We have shown that $D^* \;<<\; D'$ and $D^* \neq D'$. This contradicts the fact that D^* is maximal element in Q, so (12) is false. We have confirmed the claim in the remaining case to be considered. Proof of the theorem is complete.

Completion of the Proof of Theorem 2. Recall that it remains to check that the multi-function $\tilde{F}$:

$$\tilde{F}(t,(x,y),v)$$

$$= \left\{ \begin{pmatrix} d(t,x) + A(t,x)u + B(t,x)v \\ u \end{pmatrix} \in \mathbb{R}^{n+m_1} \;:\; u \in \bar{\chi}(t,x,v) \right\},$$

satisfies the hypotheses of Theorem 2.1 We now make good this omission.

Fix $v \in L^1_{loc}([0,\infty);\mathbb{R}^{m_2})$, $T \geq 0$ and $z_0 \in \mathbb{R}^{n+m_1}$. Define

$$\begin{aligned} \mathcal{S}_T(z_0) := \{z : [0,T) \to \mathbb{R}^{n+m_1} \;&:\; z \text{ is locally absolutely continuous} \\ & z(0) = z_0 \text{ and } \dot{z} \in \tilde{F}(t,z,v(t)) \text{ a.e.}\}\,. \end{aligned}$$

It suffices to confirm the two conditions:

(a)$'$there exists a locally absolutely continuous solution $z : [T,\infty) \to \mathbb{R}^{n+m_1}$ to

$$\left.\begin{aligned} \dot{z} &\in \tilde{F}(t,z,v(t)) \quad \text{a.e. } t \in [T,\infty) \\ z(T) &= z_0 \end{aligned}\right\}, \tag{13}$$

(b)$'$there exists an integrable function $\beta : [0,T] \to R$ such that

$$|\dot{z}(t)| \leq \beta(t) \quad \text{a.e.},$$

for every $z \in \mathcal{S}_T(z_0)$.

Indeed, setting $T = 0$ and $z_0 = (x_0,0)$, we see that (a)$'$ implies (a). On the other hand, the 'uniform integrability' condition (b)$'$ implies that any locally absolutely continuous solution of (13) on $[0,T)$ can be extended as an absolutely continuous solution to the closed interval $[0,T]$. If (a)$'$ also holds, it can be further extended as a locally absolutely continuous solution to all of $[0,\infty)$. We have shown that (a)$'$ and (b)$'$ imply (b).

The fact that (b)$'$ is satisfied is simply deduced from the linear growth hypotheses on f, A and B, with the help of Gronwall's lemma. So it remains only to show that, under the hypotheses of Theorem 2, condition (a)$'$ is satisfied.

The principal difficulty in verifying (a)$'$ arises from the fact that v is an arbitrary $L^1_{loc}([0,\infty);\mathbb{R}^{m_2})$ function. If v were continuous, (a)$'$ would follow from standard existence theorems for global solutions to differential inclusions, whose right sides are closed graph multi-functions, taking values in the class of convex sets and which have at most linear growth, with respect to the x variable [1], [18]. For discontinuous v, however, a little more analysis is required. One possibility is to approximate v by continuous functions v_i, $i = 1, 2, \ldots$, and extract a solution to the original differential inclusion in the limit, by means of a sequential compactness argument. This is what we do below, but with one extra refinement: it is to introduce a change of independent variable before passage to limit. This ensures satisfaction of the relevant uniform integrability hypotheses which justify the limit taking involved.

Consider the one-to-one mapping

$$\theta(t) = \int_T^t (1 + |v(s)|) ds \,.$$

With respect to the new independent variable $\tau = \theta(t)$, the differential inclusion (13) becomes

$$\begin{cases} d\bar{z}(\tau)/d\tau \in (1 + |\bar{v}(\tau)|)^{-1} \tilde{F}(\theta^{-1}(\tau), \bar{z}(\tau), \bar{v}(\tau)) & \text{a.e. } \tau \geq \theta(T) \\ \bar{z}(\theta(T)) = z_0 \,, \end{cases}$$

in which $\bar{v}(\tau) = v(\theta^{-1}(\tau))$.

By Lusin's Theorem (Thm 2.23 in [15]), there exists a sequence of continuous functions $\bar{v}_i : [0, \infty) \to \mathbb{R}^{m_2}$ such that

$$\mathcal{L}\{t \in [T, \infty) : \bar{v}_i \neq \bar{v}(t)\} \to 0 \quad \text{as} \quad i \to \infty \,.$$

($\mathcal{L}$ denotes Lebesgue measure.)

For each i, there is a locally absolutely continuous arc $\bar{z}_i$ on $[0, \infty)$ satisfying

$$\begin{cases} d\bar{z}_i(\tau)/d\tau \in (1 + |\bar{v}_i(\tau)|)^{-1} \tilde{F}(\theta^{-1}(\tau), \bar{z}_i(\tau), \bar{v}_i(\tau)) & \text{a.e. } \tau \geq \theta(T) \\ \bar{z}_i(\theta(T)) = z_0 \,, \end{cases} \tag{14}$$

by well-known global existence theorems for existence of solutions to differential inclusions [1], [18]. (Apply Theorem 12.2.1 of [18], for example, treating t as a state variable.) Now making use of standard sequential compactness theorems [7], [18] (see, for example, Theorem 2.5.3 of [18]) with reference to an increasing sequence of time-intervals converging to $[0, \infty)$ and extracting diagonal subsequences, we can justify extracting a subsequence (we do not re-label) such that, as $i \to \infty$,

$$\bar{z}_i \to \bar{z} \quad \text{uniformly on bounded intervals,}$$

for some locally absolutely continuous function $\bar{z}$ on $[0, \infty)$ satisfying (14). Making an inverse change of variables $t = \theta^{-1}(\tau)$, we obtain a locally absolutely continuous arc $z(t) = \bar{z}(\theta^{-1}(t))$ to (13) (in which $T = 0$). Property (b)$'$ has been confirmed and the proof is complete.

References

1. Aubin J-P (1991) Viability theory. Birkhäuser Boston
2. Bardi M, Capuzzo-Dolcetta I (1997) Optimal control and viscosity solutions of Hamilton-Jacobi equations. Birkhäuser Boston
3. Cardaliaguet M, Plaskacz S (2000) Invariant solutions of differential games and Hamilton-Jacobi-Isaacs equations for time-measurable Hamiltonians, SIAM J Control and Optim 38:1501-1520
4. Clark J, James M, Vinter R (2004) The interpretation of discontinuous feedback strategies as non-anticipative feedback strategies in differential games, IEEE Trans Aut Control to appear
5. Clark J, Vinter R (2000) A differential games problem arising in surge tank control. In: Proc UKACC Int Conf on Control 2000 Session 3 Optimal Control Cambridge
6. Clark J, Vinter R (2003) A differential games approach to flow control. In: Proc 43rd IEEE Conference on Decision and Control Hawaii to appear
7. Clarke F (1983) Optimization and nonsmooth analysis. Wiley-InterScience New York
8. Deimling K (1992) Multivalued differential equations. de Gruyter Berlin
9. Dupuis P, McEneaney W (1997) Risk-sensitive and robust escape criteria, SIAM J Control Optim 35: 2021-2049
10. Elliot R, Kalton N (1972) The existence of value in differential games. AMS Mem Amer Math Soc Providence
11. Isaacs R (1965) Differential games. Wiley New York
12. Kantor J (1989) Non-linear sliding mode controller and objective function for surge tanks, Int J Control 50: 2025-2047
13. Kingman J, Taylor S (1966) Introduction to measure and probability. Cambridge Univ Press Cambridge
14. McDonald K, McAvoy T (1986) Optimal Averaging Level Control, AIChE J 32: 75-86
15. Rudin W (1966) Real and complex analysis. McGraw Hill New York
16. Subbotin A (1995) Generalized solutions of first-order PDEs the dynamical optimization perspective. Birkhäuser Boston
17. Van de Shaft A (2000) Nonlinear state space H^∞ control theory. Birkhäuser Boston
18. Vinter R (2000) Optimal control. Birkhäuser Boston

Nonuniqueness in Systems of Hamilton-Jacobi Equations

Daniel N. Ostrov

Department of Mathematics and Computer Science, Santa Clara University, Santa Clara, CA 95053. dostrov@scu.edu

1 Introduction

Control theory provides a thorough description of the unique viscosity solution to the scalar Hamilton-Jacobi equation

$$\begin{aligned} u_t + H(u_x) &= 0 \qquad x \in (-\infty, \infty) \quad t \in [0, \infty) \\ u(x,0) &= g(x), \end{aligned} \tag{1}$$

where the Hamiltonian function, $H(p)$, is both convex and grows superlinearly in p. It is natural to consider how far these control theory methods can be extended to describe the solution of systems of Hamilton-Jacobi equations

$$\begin{aligned} \mathbf{u}_t + \mathbf{H}(\mathbf{u}_x) &= \mathbf{0} \qquad x \in (-\infty, \infty) \quad t \in [0, \infty) \\ \mathbf{u}(x,0) &= \mathbf{g}(x), \end{aligned} \tag{2}$$

where $\mathbf{u}$ and $\mathbf{g}$ are n-dimensional vectors and the Hamiltonian, $\mathbf{H} : \mathbb{R}^n \to \mathbb{R}^n$, has appropriate convexity and growth assumptions. In this paper we will see that there are some important uniqueness issues that can arise even when $n = 2$; that is, in systems of the form

$$\begin{aligned} u_t + H_1(v_x, u_x) &= 0 \qquad x \in (-\infty, \infty) \quad t \in [0, \infty) \\ v_t + H_2(u_x, v_x) &= 0 \\ u(x,0) &= g_1(x) \\ v(x,0) &= g_2(x) \end{aligned} \tag{3}$$

where both H_1 and H_2 grow superlinearly and are convex in their second variables when their first variables are fixed.

To describe these uniqueness issues for system (3), we begin in section 2 by reviewing the properties of the viscosity solution to the scalar equation

M. de Queiroz et al. (Eds.): Optimal Control, Stabilization and Nonsmooth Analysis, LNCIS 301, pp. 49–59, 2004.

$$u_t + H\left(f(x,t), u_x\right) = 0 \qquad (4)$$
$$u(x,0) = g(x)$$

where $f, H,$ and g are continuous. Just as u_x is typically discontinuous in the solution for equation (4), both u_x and v_x will typically be discontinuous in the solution for system (3). Therefore, if we want to make a connection between (4) and each of the two equations in (3), we must consider how to define a solution to (4) when f is a discontinuous function, which we do in section 3. In section 4, we consider how to generalize this solution to the solution for system (3), but will only be able to determine necessary conditions for the control theory representation of the solution. To achieve sufficient conditions (and uniqueness of the solution), it will become clear that our only hope is to look at the limit of solutions to stochastic control problems. Finally, in section 5, we will see that there are examples of systems where the limit of the solutions to these stochastic control problems is not unique; that is, there are cases where no unique solution can be defined for (4).

2 Scalar Hamilton-Jacobi Equations with Dependence on Continuous Functions

We begin by reviewing the properties of the control theory form of the viscosity solution to

$$u_t + H\left(f(x,t), u_x\right) = 0 \qquad (5)$$
$$u(x,0) = g(x)$$

where f, H, and g are continuous and, for any fixed value of f, $H(f,p)$ is convex in p and grows superlinearly in p

$$\left(\text{i.e., } \lim_{|p|\to\infty} \frac{H(f,p)}{|p|} \to \infty\right).$$

We know that the unique viscosity solution to this problem has the following control theory representation:

$$u(x_0,t_0) = \min_{x(\cdot)\in AC} \left[\int_0^{t_0} L\left(f\left(x(t),t\right), \dot{x}(t)\right) dt + g\left(x(0)\right)\right] \qquad (6)$$

where AC represents the set of absolutely continuous functions defined on $[0,t_0]$ such that $x(t_0) = x_0$ and the Lagrangian function, L, is defined by $L(f,p) = \max_a \left[ap - H(f,a)\right]$.

A curve $(x(t),t)$ in the (x,t) plane where $x(t)$ is a function that minimizes the cost in (6) is called a "characteristic path." The set of points (x_0,t_0) where u_x for the solution defined in (6) is not continuous will generally form curves in the (x,t) plane. These curves where u_x is discontinuous are called "shocks". Characteristic paths can "enter" shocks, which means that as a characteristic path evolves from the initial condition $t = 0$, it may intersect (enter) a shock at a later time. (More specifically, for most points on the shock, there is one characteristic path that enters the point on the shock by coming in from the left of the shock on the (x, t) plane and one characteristic path that enters the point on the shock by coming in from the right of the shock.) However, characteristic paths cannot "exit" shocks, which means that once a characteristic path enters a shock, it never will leave the shock at a later time.

The property that characteristics cannot exit shocks implies that the cost of the Lagrangian along shocks is never really needed since, in the case of a point not on a shock, the unique characteristic path between the initial condition and this point cannot travel (or even intersect) a shock and, in the case of a point on a shock, if there is a characteristic path between the initial condition and this point which travels along the shock, there must also be at least one alternative characteristic path between the initial condition and this point which does not travel along (or even intersect) a shock.

While this inability of characteristic paths to exit from shocks is an automatic property of (6), in other contexts it is considered to be the property that yields uniqueness of the solution to (5) . For example, it corresponds to the classical entropy condition in the field of conservation laws. It also corresponds to the unique viscous limit; that is, the unique solution, u, in (6) can be defined as the pointwise limit as $\varepsilon \to 0^+$ of the (unique) solutions, u^ε, to

$$\begin{aligned} u_t^\varepsilon + H\left(f(x,t), u_x^\varepsilon\right) &= \varepsilon u_{xx}^\varepsilon \\ u^\varepsilon(x,0) &= g(x). \end{aligned} \tag{7}$$

Here $\varepsilon > 0$ is called the "viscosity" due to equations for fluid flow of this form in which ε represents the fluid's viscosity. The term "viscosity solution", as coined by Crandall and Lions, has the same origin; that is, for equation (5), the unique viscosity solution corresponds to this viscous limit (see [3] and [4]). The form that the u^ε take is also of interest; specifically, the u^ε are smoothed versions of u looking much like u being convolved with a mollifier of strength ε, so the u_x^ε are continuous functions that converge to the discontinuous function u_x as $\varepsilon \to 0^+$.

3 Scalar Hamilton-Jacobi Equations with Dependence on Discontinuous Functions

Now we consider what happens when the function f in (5) is not continuous along a single smooth curve in the (x, t) plane. (Note, however, that the results we present are actually extendable to f being discontinuous along a finite number of piecewise smooth, possibly intersecting curves in the plane.) We define $\gamma(t)$ so that this curve of discontinuity for f in the (x, t) plane is given by $(\gamma(t), t)$ where the domain of t can be any subset of $[0, \infty)$. For shorthand, in the rest of this paper, when we use γ (without any functional dependence), we are referring to this curve in the (x,t) plane as opposed to when we use $\gamma(t)$ (with the functional dependence), which is the function needed to define the γ curve in the plane.

The γ curve of discontinuity potentially causes both existence and uniqueness issues. If we still insist on our previous condition that characteristics cannot exit shocks, we typically have no solutions, so our previous condition must be loosened. On the other hand, if we loosen this condition too much, we are certain to lose uniqueness. To resolve these problems, we first recall that for continuous f, we obtained uniqueness by requiring that u be stable with respect to adding viscosity to the equation, which caused u_x^ε to be a continuous approximation (like a convolution of u_x) that converged to the discontinuous function u_x. Similarly, we should require any solution where f is discontinuous to be stable in a similar manner. So let us define $f^\varepsilon(x,t)$ to be the convolution with respect to the x variable of $f(x,t)$ with a mollifier of strength ε. Since the f^ε are continuous functions, we can define $u^\varepsilon(x,t)$ to be the unique viscosity solution to

$$u_t^\varepsilon + H\left(f^\varepsilon(x,t), u_x^\varepsilon\right) = 0 \qquad (8)$$
$$u^\varepsilon(x,0) = g(x),$$

which also has a control theory representation via equation (6). Of course, we now wish to define the solution u to the Hamilton-Jacobi equation where f is discontinuous to be the pointwise limit of the u^ε as $\varepsilon \to 0^+$. It turns out (see [8]) that this works; that is, the limit exists and so the solution is defined. Even better, this solution has a control theory representation:

$$u(x_0,t_0) = \min_{x(\cdot)\in AC}\left[\int_0^{t_0} L\left(\hat{f}\left(x(t),t\right), \dot{x}(t)\right) dt + g\left(x(0)\right)\right] \qquad (9)$$

$$\text{where}\hat{f}(x,t) = \begin{cases} f(x,t) \text{ at any } (x,t) \text{ not on } \gamma \\ \underset{f\in\left[\liminf\limits_{y\to x} f(y,t), \limsup\limits_{y\to x} f(y,t)\right]}{\operatorname{argmin}} L\left(f, \dot{\gamma}(t)\right) \text{ at any } (x,t) \text{ on } \gamma. \end{cases}$$

Unlike in the case where f is continuous and we don't need the value of the Lagrangian along shocks, here it is crucial since the discontinuity in f generally induces a shock throughout the entire γ curve, but characteristics can now — and, in fact, often will — travel along γ (meaning the characteristics can now exit shocks). Therefore, the key part of (9) is the second line in the definition of $\hat{f}$ since this shows how to define the cost of the Lagrangian along a shock occurring on γ. Specifically, it says is that if we look at any point on γ, we find the limit of f as we approach this point both from the left of γ and the right of γ and then we choose the value between these two limits that minimizes the Lagrangian assuming that we are considering a path that continues movement along γ.

The most interesting cases arising from this notion of a stable solution occur when the value of $\hat{f}$ on γ is strictly between the limits of f from the left and from the right of γ. These are interesting because even though the solution is defined as the limit of viscosity solutions, in this case the solution itself is not a viscosity solution in any sense. There are many notions of how "viscosity solutions" should be defined when f is discontinuous (both when the Hamilton-Jacobi equation is time dependent as we have here, e.g., [6], [7], and when the equation does not depend on time, e.g., [2], [5], [9]) but the stable notion of the solution defined in (9) actually agrees with none of these definitions (if they apply) when $\hat{f}$ falls strictly between its limit from the left and its limit from the right.

As an example, consider the Hamilton-Jacobi equation

$$\begin{aligned} u_t + (u_x)^2 - [f(x)]^2 &= 0 \\ u(x,0) &= 0 \\ \text{where } f(x) &= \begin{cases} -1 \text{ if } x < 0 \\ \ \ 1 \text{ if } x \geq 0. \end{cases} \end{aligned} \tag{10}$$

Since the Hamiltonian for this case is $H(f,p) = p^2 - f^2$, our corresponding Lagrangian is

$$L(f,p) = \frac{p^2}{4} + f^2,$$

and, since $u(x,0) = 0$, we have that $g(x) = 0$. For $x \neq 0$, we have that $\hat{f} = f$, but to determine $\hat{f}$ on the line $x = 0$, which is the γ curve of discontinuity in this case, we set $p = 0$ (since $\dot{\gamma}(t) = 0$), and determine the value of f between the limit of f from the left of γ (which is -1) and the limit of f from the right (which is 1) that minimizes $L(f,0) = f^2$. Clearly, this Lagrangian is minimized when $f = 0$, so we have that

$$\hat{f}(x) = \begin{cases} -1 \text{ if } x < 0 \\ \ \ 0 \text{ if } x = 0 \\ \ \ 1 \text{ if } x > 0. \end{cases}$$

Note that if $x \neq 0$, $\hat{f}^2 = 1$ and therefore

$$L(\hat{f},p) = \frac{p^2}{4} + 1,$$

whereas if $x = 0$, $\hat{f}^2 = 0$ and therefore

$$L(\hat{f},p) = \frac{p^2}{4}.$$

In other words, traveling along the γ line, $x = 0$, is particularly cheap, so we expect this line to attract many characteristic paths, which is exactly what happens. When we think about stability, this makes sense. By convolving f, we have access near $x = 0$ to all of the values of f between -1 and 1 and so it is logical that in each of these convolved cases, the minimum cost paths will be attracted toward the cheap locations near $x = 0$ where $f = 0$ no matter how small the convolution.

Now we determine the minimum cost path (i.e., the characteristic path) using (9). If $|x_0| \le t_0$, then the path starts at the origin and moves up the cheap line $x = 0$ until $t = t_0 - |x_o|/2$ and then it moves in a straight line directly from $(0, t_0 - |x_o|/2)$ to (x_0, t_0) (which corresponds to the speed $|\dot{x}(t)| = 2$.) On the other hand, if $|x_0| > t_0$, then we are so far away from the cheap path at $x = 0$ that it is too expensive to travel to it, and so instead we stay on the straight line $x = x_0$ from the initial condition, $(x_0, 0)$, to (x_0, t_0) which at least cheapens the cost by making $p = 0$. This leads to the final stable solution being

$$u(x,t) = \min\left[t, |x|\right].$$

The fact that characteristics exit from both sides of the γ line prevents this solution from fitting with any of the other notions of viscosity solutions. In fact, for all of these notions, one must pick either the limit from the left or the limit from the right for the value of f on γ, which leads to $u(x,t) = t$.

4 Control Theory Representations for Hyperbolic Systems of Equations

We are now ready to look at our system of equations (3) :

$$\begin{aligned} u_t + H_1(v_x, u_x) &= 0 \qquad x \in (-\infty, \infty) \quad t \in [0, \infty) \\ v_t + H_2(u_x, v_x) &= 0 \\ u(x,0) &= g_1(x) \\ v(x,0) &= g_2(x) \end{aligned}$$

where for any fixed f, both $H_1(f,p)$ and $H_2(f,p)$ are C^1 functions that are convex in p and grow superlinearly in p. If possible, we would like to solve this system of equations by simultaneously solving two corresponding control problems. This implies having two characteristic curves, which we know from partial differential equations theory only occurs in hyperbolic systems of PDE. Therefore, we limit ourselves to hyperbolic systems; that is systems where, for all (u_x, v_x) pairs, the matrix

$$\begin{bmatrix} \frac{\partial H_1}{\partial u_x} & \frac{\partial H_1}{\partial v_x} \\ \\ \frac{\partial H_2}{\partial u_x} & \frac{\partial H_2}{\partial v_x} \end{bmatrix} \tag{11}$$

has two real eigenvalues.

As before, we would like to define our solution (u, v) to be the unique viscous limit; i.e.,

$$(u, v) = \lim_{\underline{\varepsilon} \to (0^+, 0^+)} (u^{\underline{\varepsilon}}, v^{\underline{\varepsilon}}),$$

where $\underline{\varepsilon} = (\varepsilon_1, \varepsilon_2)$ and $(u^{\underline{\varepsilon}}, v^{\underline{\varepsilon}})$ is the solution to

$$\begin{aligned} u_t^{\underline{\varepsilon}} + H_1\left(v_x^{\underline{\varepsilon}}, u_x^{\underline{\varepsilon}}\right) &= \varepsilon_1 u_{xx}^{\underline{\varepsilon}} \qquad x \in (-\infty, \infty) \quad t \in [0, \infty) \\ v_t^{\underline{\varepsilon}} + H_2\left(u_x^{\underline{\varepsilon}}, v_x^{\underline{\varepsilon}}\right) &= \varepsilon_2 v_{xx}^{\underline{\varepsilon}} \\ u^{\underline{\varepsilon}}(x, 0) &= g_1(x) \\ v^{\underline{\varepsilon}}(x, 0) &= g_2(x). \end{aligned} \tag{12}$$

This leads to a key question in the field of hyperbolic conservation laws: does this limit exist? There are numerous results in the field; one of the most general is a recent result by Bressan and Bianchini [1], which states that this limit exists provided that (1) $\varepsilon_1 = \varepsilon_2$, (2) $|g_1(x)|$ and $|g_2(x)|$ stay sufficiently small, and (3) the system is strictly hyperbolic, which means that, for all (u_x, v_x) pairs, the two eigenvalues of the matrix (11) are never equal.

Again, for simplicity, we restrict ourselves to systems where there is a single shock curve in the (x, t) plane (i.e., a single curve where potentially u_x and v_x are discontinuous) and we denote this curve by γ. It is tempting to try to find the solution by generalizing the formula in (9) in a straightforward manner such as

$$\begin{aligned} u(x_0, t_0) &= \min_{x_1(\cdot) \in AC} \left[\int_0^{t_0} L_1\left(\hat{v}_x\left(x_1(t), t\right), \dot{x}_1(t)\right) dt + g\left(x(0)\right) \right] \\ v(x_0, t_0) &= \min_{x_2(\cdot) \in AC} \left[\int_0^{t_0} L_2\left(\hat{u}_x\left(x_2(t), t\right), \dot{x}_2(t)\right) dt + g\left(x(0)\right) \right] \end{aligned} \tag{13}$$

where

$$\hat{v}_x(x, t) = \begin{cases} v_x(x, t) \text{ at any } (x, t) \text{ not on } \gamma \\ \operatorname*{argmin}_{v_x \in \left[\liminf_{y \to x} v_x(y, t), \limsup_{y \to x} v_x(y, t)\right]} L_1\left(v_x, \dot{\gamma}(t)\right) \text{ at any } (x, t) \text{ on } \gamma \end{cases}$$

and $$\hat{u}_x(x, t) = \begin{cases} u_x(x, t) \text{ at any } (x, t) \text{ not on } \gamma \\ \operatorname*{argmin}_{u_x \in \left[\liminf_{y \to x} u_x(y, t), \limsup_{y \to x} u_x(y, t)\right]} L_2\left(u_x, \dot{\gamma}(t)\right) \text{ at any } (x, t) \text{ on } \gamma, \end{cases}$$

but this is wrong as these two control problems may be inconsistent with each other and lead to no solution. Essentially, the presence of the second ε factor causes blurring which makes it impossible for the control path to always be able to take advantage of the argmin of the Lagrangian. So, as before, the key question is how do we determine the correct value for $\hat{u}_x$ and $\hat{v}_x$ on the shock, γ?

Our answer comes from the fact that both u and v are required to be continuous at the shock. This means that

$$\dot{\gamma}(t) = \frac{H_1(v_x^+, u_x^+) - H_1(v_x^-, u_x^-)}{u_x^+ - u_x^-} = \frac{H_2(u_x^+, v_x^+) - H_2(u_x^-, v_x^-)}{v_x^+ - v_x^-} \tag{14}$$

where

$$u_x^+ = \lim_{x \to \gamma(t)^+} u_x(x,t),$$

$$u_x^- = \lim_{x \to \gamma(t)^-} u_x(x,t),$$

$$v_x^+ = \lim_{x \to \gamma(t)^+} v_x(x,t),$$

and

$$v_x^- = \lim_{x \to \gamma(t)^-} v_x(x,t).$$

Condition (14) is referred to as the "Rankine-Hugoniot jump condition" in the field of conservation laws. This condition yields the proper values for $\hat{u}_x$ and $\hat{v}_x$ on the shock: specifically, if we are on the shock, we must choose $\hat{v}_x$ so that

$$L_1(\hat{v}_x, \dot{\gamma}(t)) = \frac{H_1(v_x^+, u_x^+)u_x^- - H_1(v_x^-, u_x^-)u_x^+}{u_x^+ - u_x^-}$$

and $\hat{u}_x$ so that

$$L_2(\hat{u}_x, \dot{\gamma}(t)) = \frac{H_2(u_x^+, v_x^+)v_x^- - H_2(u_x^-, v_x^-)v_x^+}{v_x^+ - v_x^-}.$$

Experimentally, we find that with this definition $\hat{v}_x$ stays between v_x^- and v_x^+ and $\hat{u}_x$ stays between u_x^- and u_x^+ as we would expect.

Unfortunately, these are only necessary conditions for the solution; they do not provide for a unique solution necessarily. This is because we still have not applied the viscous limit condition, so, in fact, our only hope for uniqueness is to look at the limit as $\underline{\varepsilon} \to (0^+, 0^+)$ of the stochastic control representations for the solutions to (12). Interestingly, this does not always work as we will see in the next section where we present a case where $\lim_{\underline{\varepsilon} \to (0^+,0^+)}(u^{\underline{\varepsilon}}, v^{\underline{\varepsilon}})$ does not exist; specifically, we show that

$$\lim_{\varepsilon_1 \to 0^+} \lim_{\varepsilon_2 \to 0^+} (u^{\underline{\varepsilon}}, v^{\underline{\varepsilon}}) \neq \lim_{\varepsilon_2 \to 0^+} \lim_{\varepsilon_1 \to 0^+} (u^{\underline{\varepsilon}}, v^{\underline{\varepsilon}}).$$

5 Example of Nonuniqueness for the Viscous Limit

We consider the following decoupled system of equations, which is an example of the system in (12):

$$\begin{aligned} u_t + u_x^2 &= \varepsilon_1 u_{xx} \qquad x \in (-\infty, \infty) \quad t \in [0, \infty) \\ v_t + v_x^2 - u_x^2 &= \varepsilon_2 v_{xx} \\ u(x,0) &= -|x| \\ v(x,0) &= 0. \end{aligned} \tag{15}$$

Note that for simplicity we have now dropped the $\underline{\varepsilon}$ superscript in (12) for both u and v. System (15) is hyperbolic but not strictly hyperbolic since, if $u_x = v_x$, the two eigenvalues of the matrix (11) are equal.

We begin with the first equation in (15), which is easier to analyze since it does not depend upon v. In fact, this is just the viscous Burgers' equation, which we can analyze by first using the Hopf-Cole transformation $u = -\varepsilon_1 \ln(w)$, which transforms the first equation into the linear heat equation $w_t = \varepsilon_1 w_{xx}$. We then solve the linear heat equation and transform back into the u form of the solution where we see that as ε_1 gets smaller, u_x converges to $-\tanh(x/\varepsilon_1)$ more and more quickly as time progresses. So, essentially, we can substitute

$$u_x^2 = \tanh^2\left(\frac{x}{\varepsilon_1}\right)$$

into the second equation in (15).

With this substitution, the stochastic control representation for the solution to the second equation in (15) is

$$v(x_0, t_0) = \min_{b(x(\cdot),\cdot)\in M} E\left[\int_0^{t_0} \tanh^2\left(\frac{x(t)}{\varepsilon_1}\right) + \frac{b^2\left(x(t),t\right)}{4} dt\right] \tag{16}$$

subject to the dynamics

$$dx(t) = b(x(t),t)dt + \sqrt{2\varepsilon_2}dB, \qquad x(t_0) = x_0 \tag{17}$$

where B is a Brownian motion and M is the set of all functions, b, of the form $b(x(t),t)$ that are measurable in t. Note that when $x(t) << \varepsilon_1$,

$$\tanh^2\left(\frac{x(t)}{\varepsilon_1}\right) \approx 0$$

so we have a cheaper path for (16), whereas when $x(t) >> \varepsilon_1$,

$$\tanh^2\left(\frac{x(t)}{\varepsilon_1}\right) \approx 1,$$

which is more expensive.

Now we consider what happens as $(\varepsilon_1, \varepsilon_2) \to (0^+, 0^+)$. If we take the limit as $\varepsilon_1 \to 0^+$ first, we have that

$$\lim_{\varepsilon_1 \to 0^+} \tanh^2\left(\frac{x(t)}{\varepsilon_1}\right) = 1$$

except at $x(t) = 0$ where

$$\lim_{\varepsilon_1 \to 0^+} \tanh^2\left(\frac{x(t)}{\varepsilon_1}\right) = 0.$$

In other words, the path for (16) is expensive unless we can stay on the line $x(t) = 0$, but there is zero probability of staying on this line for a subset of time that has positive measure due to the stochastic noise generated by the positive number ε_2. Therefore, we can replace $\tanh^2(x(t)/\varepsilon_1)$ by 1 in (16) if we take the limit as $\varepsilon_1 \to 0^+$ first. If we then take the limit as $\varepsilon_2 \to 0^+$, the stochastic noise goes away, causing the b functions that minimize (16) to converge to 0. Therefore, by the dynamics in (17), we have the minimizing paths converging to the straight lines $x(t) = x_0$ and our solution converges to $v(x_0, t_0) = t_0$. In other words

$$\lim_{\varepsilon_2 \to 0^+} \lim_{\varepsilon_1 \to 0^+} v(x,t) = t.$$

On the other hand, if, instead of letting $\varepsilon_1 \to 0^+$ first, we instead let $\varepsilon_2 \to 0^+$ first, we lose the stochastic noise, which makes paths where $x(t) << \varepsilon_1$ possible. This means that it is possible for the (now deterministic) paths to stay on the line $x(t) = 0$ where $\tanh^2(x(t)/\varepsilon_1) = 0$. As we then let $\varepsilon_1 \to 0^+$, the $\tanh^2$ term converges to 1 everywhere except on the line $x(t) = 0$ where the $\tanh^2$ term converges to 0. Since we are back to deterministic paths, we see that we have exactly the same situation as we had for the example in equation (10); that is, the line $x(t) = 0$ attracts characteristic paths because it is cheap and available, so we have the same solution as in our previous example, $v(x_0, t_0) = \min[t_0, |x_0|]$. Therefore,

$$\lim_{\varepsilon_1 \to 0^+} \lim_{\varepsilon_2 \to 0^+} v(x,t) = \min[t, |x|],$$

and we see that the viscous limit is not unique as it depends upon the way in which we let the two viscosities vanish.

References

1. Bianchini S, Bressan A (2003) Vanishing viscosity solutions of nonlinear hyperbolic systems, Annals of Mathematics to appear
2. Camilli F, Siconolfi A (2003) Hamilton-Jacobi equations with measurable dependence on the state variable, preprint
3. Crandall M, Evans L, Lions P-L (1984) Some properties of viscosity solutions of Hamilton-Jacobi equations, Trans Amer Math Soc 282:487–502
4. Crandall M, Lions P-L (1983) Viscosity solutions of Hamilton-Jacobi equations, Trans Amer Math Soc 277:1–42
5. Deckelnick K, Elliott C (2003) Uniqueness and error analysis for Hamilton-Jacobi equations with discontinuities, preprint

6. Dupuis P (1992) A numerical method for a calculus of variations problem with discontinuous integrand. In: Applied Stochastic Analysis New Brunswick NJ 1991 (Lecture Notes in Control and Inform Sci 177). Springer Berlin 90–107
7. Ishii H (1985) Hamilton-Jacobi equations with discontinuous Hamiltonians on arbitrary open subsets, Bull Fac Sci Engrg Chuo Univ 28:33–77
8. Ostrov D (2002) Solutions of Hamilton-Jacobi equations and scalar conservation laws with discontinuous space-time dependence, Journal of Differential Equations 182:51–77
9. Soravia P (2002) Boundary value problems for Hamilton-Jacobi equations with discontinuous Lagrangian, Indiana Univ Math J 51:451–477
10. Tourin T (1992) A Comparison theorem for a piecewise Lipschitz continuous Hamiltonian and application to shape-from-shading problems, Numer Math 62:75–85

On Impulses Induced by Singular Perturbations

Zvi Artstein *

Department of Mathematics, The Weizmann Institute of Science, Rehovot 76100, Israel. zvi.artstein@weizmann.ac.il

Summary. The possibility that a singularly perturbed controlled dynamics generates impulses in a slow dynamics coupled to it, is investigated. A scheme for the analysis and generation of such impulses is offered in the case where the control affects the perturbed dynamics only. Examples are displayed along with general comments on the underlying conditions.

1 Introduction

It is desirable, at times, that a state trajectory in a controlled system exhibits a discontinuity. A simple toy example where the state and the control variables are scalars is

$$\begin{aligned} &\text{minimize} && \int_0^1 |x(t) - 1|dt \\ &\text{subject to} && \tfrac{dx}{dt} = u \\ & && x(0) = 0 \\ & && u \in (-\infty, \infty). \end{aligned} \tag{1.1}$$

A direct inspection reveals that the best performance would be achieved with an instantaneous jump of the state from $x(0) = 0$ to $x(0+) = 1$ followed by the trajectory $x(t) = 1$ for $t \in (0, 1]$. Such a jump cannot be a result of an ordinary control $u(\cdot)$. It can be approximated, however, by admissible controls with large values around $t = 0$. For instance, the larger the m the better the approximation of optimality if the following scheme is adopted. Define the control to be $u(t) = m$ for $t \in [0, m^{-1}]$ and $u(t) = 0$ otherwise. A general theory of such controlled jumps and of the controls which drive them, is available in the literature. See for instance, McShane [6], Murray [8], Warga [9]. The theory examines, in particular, the relaxed limit of the

*Incumbent of the Hettie H. Heineman Professorial Chair in Mathematics. Research supported by a grant from the Israel Science Foundation.

M. de Queiroz et al. (Eds.): Optimal Control, Stabilization and Nonsmooth Analysis, LNCIS 301, pp. 61–72, 2004.

controls. Notice, however, that the jumps cannot be generated by ordinary controls. Furthermore, a sequence of ordinary controls employed in generating approximations closer and closer to the optimal value of the jump is necessarily unbounded.

In the present paper we examine a situation where, again, the impulse is desirable, but the jump is generated through a singularly perturbed variable, namely a fast state variable, say y, coupled with the slow state x. The following variant of (1.1) demonstrates the situation:

$$\begin{aligned} &\text{minimize} && \int_0^1 |x(t) - 1|dt \\ &\text{subject to} && \tfrac{dx}{dt} = y \\ & && \varepsilon\tfrac{dy}{dt} = u \\ & && x(0) = 0, \quad y(0) = 0 \\ & && u \in [-1, 1]. \end{aligned} \tag{1.2}$$

In particular, the state equations in (1.2) reflect a double integrator with a high amplifier gain; systems of this kind have been examined in the literature in contexts other than the impulses sought after in the present paper.

A direct inspection reveals that an optimal behavior of (1.2) is achieved with a jump in the slow state from $x(0) = 0$ to $x(0+) = 1$ followed by the trajectory $x(t) = 1$ for $t \in (0, 1]$. With an initial condition $x(0+) = 1$ the latter trajectory can be a result of $y(t) = 0$ for $t \in (0, 1]$. The existence of the small parameter multiplying y enables bounded controls on a small interval to yield, for small $\varepsilon > 0$, arbitrarily large state values for the variable y, which in turn may generate the impulse in the slow variable. Hence we may restrict the controls in the singularly perturbed case to be bounded, as is the case in (1.2).

The toy example (1.2) falls within the framework of the celebrated theory of singularly perturbed control systems from Kokotovic, Khalil and O'Reilly [4]; in particular, for examples of a double integrator with a high amplifier gain, see, e.g., [4, Example 2.2, Example 4.3]. As is customary in singularly perturbed systems, one is interested in the case where ε is small, namely, one is interested in the case $\varepsilon \to 0+$. Traditionally, the research and its applications are concerned with slow trajectories which exhibit continuous behavior. The present paper attempts to develop techniques that cope with impulses of the slow state.

In the next section we state the general underlying problem of generating an impulse by a singularly perturbed variable. In the rest of the paper we solve example (1.2) explicitly and provide a scheme for solving, under some conditions, the problem of generating an impulse in a class of problems. Some comments concerning the geometrical structure and limitations of strategies which create state impulses are offered as well and, in particular, a possible connection with the method of graph completion of measure differential inclusions is indicated.

2 Problem Statement

Although motivated by optimization problems like (1.2) and similar problems related to design and stability, we concentrate in this paper on the structure of the impulse rather than on the whole problem. Thus, a singularly perturbed system is given as follows:

$$\begin{aligned} \frac{dx}{dt} &= f(x, y, u) \\ \varepsilon \frac{dy}{dt} &= g(x, y, u) \\ u &\in U, \end{aligned} \tag{2.1}$$

where $x \in \mathbb{R}^n$, $y \in \mathbb{R}^m$ and $U \subseteq \mathbb{R}^k$. Following the standard practice of handling singularly perturbed control systems (see Kokotovic et al. [4]) we are interested in the limit behavior of solutions of (2.1) as $\varepsilon \to 0+$.

For the specific problem of generating an impulse, we assume that a desirable continuous trajectory is defined on a time interval preceding, say, $t = 0$ and is the limit as $\varepsilon \to 0+$ of admissible trajectories; its value at the end point, say $(x(0), y(0))$, is given. We assume that the desirable trajectory for the interval past $t = 0$, say $(0, T]$, is prescribed and can be generated as the limit as $\varepsilon \to 0+$ of admissible trajectories; its limit as $t \to 0+$ is also given, say $(x(0+), y(0+))$. Part of the assumption is that if an initial point, say (x_0, y_0), satisfies the properties that x_0 is close enough to $x(0+)$ and y_0 is within a prescribed bounded open set containing $y(0+)$, then an appropriate control can be applied to generate a good approximation of the desired limit trajectory. The difference in the assumptions, namely, that the y-variable need not get near the target point $y(0+)$, stems from the possibility to generate a boundary layer which at negligible time and cost would drive y_0 to $y(0+)$; namely, the property is rooted in the applications. See, e.g., [4, Sections 6.2, 7.2]. When $x(0) \neq x(0+)$ a jump has to be generated at $t = 0$, namely, the jump is a limit as $\varepsilon \to 0+$ of trajectories generated by admissible controls on small intervals around $t = 0$. These considerations lead us to the following formulation of the problem:

Problem Statement 2.1. The values $(x(0), y(0))$ and $(x(0+), y(0+))$ and a bounded open set B around $y(0+)$ are prescribed. Determine time intervals $[0, \tau(\varepsilon)]$ where $\tau(\varepsilon) \to 0$ as $\varepsilon \to 0$, and determine control strategies $u_\varepsilon(\cdot)$ defined, respectively, on $[0, \tau(\varepsilon)]$ such that when the controls are plugged in (2.1), the solutions, say $(x_\varepsilon(\cdot), y_\varepsilon(\cdot))$, satisfy the following stable boundary value property: For any $\beta_1 > 0$ there exists a $\beta_0 > 0$ such that for ε small enough if $(x_\varepsilon(0), y_\varepsilon(0))$ is within a β_0-neighborhood of $(x(0), y(0))$ then $x_\varepsilon(\tau(\varepsilon))$ is within a β_1-neighborhood of $x(0+)$ and $y_\varepsilon(\tau(\varepsilon))$ is in B. It is preferable that the control be in a feedback form independent of ε if possible.

It is clear how problem (1.2) fits into the framework of the preceding formulation. Indeed, a solution to (1.2) within Problem Statement 2.1 would be a generation of a jump from $(x(0), y(0)) = (0, 0)$ to $(x(0+), y(0+)) = (1, 0)$ at the time $t = 0$, and then maintaining the trajectory in an obvious way close to $(x(t), y(t)) \equiv (1, 0)$. In the present paper we concentrate on the generation of the impulse.

3 Solving (1.2)

We provide here a direct solution to problem (1.2) within the framework of Problem Statement 2.1 (see the closing paragraph of the preceding section). The procedure will be commented on and generalized in the sequel.

Inspired by the graph completion technique in measure differential equations (we elaborate on the connection later on) we consider the change of time variables given by

$$s = \varepsilon^{-1/2} t. \tag{3.1}$$

When applied to the state equations in (1.2) it results in

$$\begin{aligned} \frac{dx}{ds} &= \varepsilon^{1/2} y \\ \frac{dy}{ds} &= \varepsilon^{-1/2} u. \end{aligned} \tag{3.2}$$

A further manipulation yields a second-order equation for the slow variable x as follows.

$$\frac{d^2 x}{ds^2} = u. \tag{3.3}$$

Notice that (3.3) is ε-free. At this point we may wish to consider (3.3) along with the boundary values induced by the desired jump, namely

$$x(0) = 0, \quad x(\sigma) = 1 \tag{3.4}$$

for some constant $\sigma > 0$. A solution to this problem can be easily expressed as a feedback of the slow variable. Indeed, plugging

$$u(x) = -x + 1/2 \tag{3.5}$$

into (3.3) when $x(0) = y(0) = 0$ yields a trajectory satisfying (3.4) with $\sigma = \pi$. It is also easy to see that the suggested control is feasible. Furthermore, the feedback is stable in the sense that an initial condition near 0 reaches within $[0, \pi]$ a terminal condition near 1. Reversing the change of time variables yields a near-jump of the slow state as required in the problem statement, for the ε-dependent time intervals determined by $\tau(\varepsilon) = \varepsilon^{1/2}\pi$, namely,

$$x(0) = 0, \quad x(\varepsilon^{1/2}\pi) = 1. \tag{3.6}$$

Since, as noted earlier, the transition between the boundary values is stable, a jump in the slow variable has been established. A solution to the problem would be complete if in addition to (3.6) one gets

$$y(0) = 0, \quad y(\varepsilon^{1/2}\pi) = 0 \tag{3.7}$$

with a similar stability property (irrespective of the prescribed open set B). A direct inspection reveals that these requirements are satisfied. Indeed, the definition $v = \varepsilon^{1/2}y$ transforms (3.2) into a harmonic oscillator.

In conclusion we may state that the feedback (3.5) generates the desired jump for the singularly perturbed problem (1.2).

4 A Comment on the Time Consumed by the Jump

The solution to problem (1.2) as provided in the preceding section employs the time intervals $[0, \varepsilon^{1/2}\pi]$, namely, the time consumed by the jump is $O(\varepsilon^{1/2})$ (that is, big Oh of $\varepsilon^{1/2}$). A common change of time scales in the singular perturbations trait (see Kokotovic et al. [4]) is $t = \varepsilon s$. Such a change transforms the fast flow y to an ordinary flow and transforms the relatively slow variable x to a slow variable. We wish to point out that under standard conditions a time interval of $O(\varepsilon)$ will not suffice in the generation of a jump, as follows:

Observation 4.1. Suppose that in system (2.1) the constraint set U is compact and the functions $f(x, y, u)$ and $g(x, y, u)$ grow at most linearly with (x, y). Then an impulse according to Problem Statement 2.1 with $x(0) \neq x(0+)$ cannot be achieved when $\tau(\varepsilon) = O(\varepsilon)$.

Proof. The argument is based on simple calculus of inequalities (see, e.g., Hartman [3, Chapter III]). Suppose that $|g(x, y, u)| \leq c(|y| + 1)$ when x is bounded, say, by b where $b > |x(0)|$. Denote by $(x_\varepsilon(\cdot), y_\varepsilon(\cdot))$ a solution of (2.1) satisfying $(x_\varepsilon(0), y_\varepsilon(0)) = (x(0), y(0))$. Then when $|x| < b$

$$|y_\varepsilon(t)| \leq 1 + |y(0)|e^{\frac{c}{\varepsilon}t}. \tag{4.1}$$

If the previous estimate holds for the slow flow as well, namely, $|f(x, y, u)| \leq c(|y| + 1)$ when $|x| \leq b$, then as long as x satisfies the restriction we have

$$|x_\varepsilon(\tau) - x(0)| \leq \int_0^\tau (1 + |y(0)|e^{\frac{c}{\varepsilon}t})dt. \tag{4.2}$$

Plugging in (4.2) the expression $\tau = \tau(\varepsilon) = O(\varepsilon)$ implies that $|x_\varepsilon(\tau) - x(0)| \to 0$ as $\varepsilon \to 0$. This verifies in particular that for ε small the x-variable stays within the bound b and that an impulse in the x-state cannot occur within the interval $[0, \tau(\varepsilon)]$. This completes the proof.

It is clear from estimate (4.2) that when $\tau(\varepsilon)\varepsilon^{-1} \to \infty$ as $\varepsilon \to 0$, a jump may occur as is the case with the derivations presented in the previous section.

5 A General Case

Motivated by the example solved in Section 3 and the comment in the previous section, we display a general solution scheme for a nonlinear double integrator problem with a high amplifier gain. The change of the time variable in such problems yields an ε-free second-order equation for the slow variable. An ε-dependent generalization is worked out in the next section. The system we analyze is given by

$$\begin{aligned} \tfrac{dx}{dt} &= f(y) \\ \varepsilon \tfrac{dy}{dt} &= g(x,y,u) \\ u &\in U, \end{aligned} \tag{5.1}$$

namely, a special case of (2.1). We assume that the vector fields are continuously differentiable. We employ the change of time variables introduced in Section 3, namely

$$s = \varepsilon^{-1/2} t. \tag{5.2}$$

With (5.2) the system (5.1) takes the form

$$\begin{aligned} \tfrac{dx}{ds} &= \varepsilon^{1/2} f(y) \\ \tfrac{dy}{ds} &= \varepsilon^{-1/2} g(x,y,u). \end{aligned} \tag{5.3}$$

Since $f(\cdot)$ is assumed continuously differentiable it follows that $x(\cdot)$ solves a second-order equation as follows (here $f'(\cdot)$ is the derivative of $f(\cdot)$):

$$\frac{d^2x}{ds^2} = f'(y)g(x,y,u). \tag{5.4}$$

Notice that, as was the case in Section 3, the second-order equation is ε-free.

Solution Scheme 5.1. Consider the system (5.1) along with a desired impulse determined by $(x(0), y(0)), (x(0+), y(0+))$ and B as described in Problem Statement 2.1. Look for a feedback $u(x,y)$ which when applied to both (5.4) and the y-equation in (5.3) implies that for ε-small the initial condition $x(0)$ is steered on an interval $[0,\sigma]$ to a small neighborhood of $x(0+)$ and the steering is stable, and at the same time for ε small, a small neighborhood of $y(0)$ is steered to B. Such a feedback obviously generates the desired impulse.

Remark 5.2. The advantage of the preceding solution scheme is that the small parameter does not explicitly appear in (5.4). (It enters via the fast variable y.) In particular, our handling of the slow state is inspired by the technique developed in Bressan and Rampazzo [1]-[2] and Motta and Rampazzo [7] for measure differential equations. In the present case no such equation is explicitly given but the outcome of carrying out the solution scheme should be the same as approximating the solution given by a graph completion as developed in the references cited. We do not carry out such a limit procedure in the present paper but wish to point out that when the limit arguments are carried out, the convergence considerations of Liu and Sussmann [5] and the sampling technique offered by Wolenski [10] would be very useful.

We conclude this section with three examples. The first two demonstrate how the solution scheme can be used. The third indicates that one should not be deceived by the simplicity of the reduction.

Example 5.3. Consider the following version of the high gain amplifier with a nonlinear block presented in Kokotovic et al. [4, Example 2.2]:

$$\begin{aligned} \tfrac{dx}{dt} &= y \\ \varepsilon \tfrac{dy}{dt} &= -x - N(y) + u \\ x(0) &= 0, \quad y(0) = 0 \\ u &\in (-\infty, \infty), \end{aligned} \tag{5.5}$$

with $N(y)$ a prescribed bounded continuous function. System (5.5) is a simple variant of (5.1). Suppose that an impulse has to be generated at $t = 0$ reaching the boundary condition $(x(0+), y(0+)) = (1, 0)$. Applying the previous consideration yields the following ε-free second-order equation for the x-variable.

$$\frac{d^2x}{ds^2} = -x - N(y) + u. \tag{5.6}$$

The resulting equation for the y-dynamics is

$$\frac{dy}{ds} = \varepsilon^{-1/2}(-x - N(y) + u). \tag{5.7}$$

Now we can employ a control feedback of the form $u(y) = N(y) + 1/2$ and get a feedback loop identical to the solution of (1.2). The stability arguments for both the slow and the fast variables are the same in the present equation as in the solution in Section 3. Thus the given feedback, which results in a bounded control function, generates the desired impulse.

Example 5.4. Consider a version of the high gain amplifier where the slow state is driven by a nonlinear version of the fast variable as follows.

$$\begin{aligned} \tfrac{dx}{dt} &= y^2 \\ \varepsilon \tfrac{dy}{dt} &= u \\ x(0) &= 0, \quad y(0) = 0 \\ u &\in [-1, 1]. \end{aligned} \tag{5.8}$$

Suppose that an impulse has to be generated at $t = 0$ reaching the boundary condition $(x(0+), y(0+)) = (1, 0)$. Applying the suggested substitution results in the following second-order equation for the x-variable:

$$\frac{d^2x}{ds^2} = 2yu. \tag{5.9}$$

The resulting equation for the y-dynamics is

$$\frac{dy}{ds} = \varepsilon^{-1/2}u. \tag{5.10}$$

Since in (5.9) the y variable multiplies the control variable, one cannot employ a feedback which would transform the equation into the harmonic oscillator as was done in the analysis of the previous example. Consider however, the following feedback rule.

$$u(x,y) = \begin{cases} 1 & \text{when } |y| \le 2 \\ \frac{1}{2y}(-x + 1/2) & \text{otherwise.} \end{cases} \tag{5.11}$$

It is easy to see that the control (5.11) satisfies the constraints. With this control feedback, the y-part of (5.10) implies that starting near $y = 0$ and when ε is small, the fast variable reaches the value 2 on a small portion of the interval, say, $[0, \pi]$. At this point a discontinuity in the control occurs which becomes small as ε gets smaller (but which does not disturb the solution, and can be gotten rid of at the cost of complicating the formulation). This means that except for a small portion of the interval $[0, \pi]$, the right hand side of (5.9) is equal to the expression (3.5), namely the equation is equivalent to the harmonic oscillator around $(1/2, 0)$. On the small portion where the equation is not identical to the harmonic oscillator the right hand side of the equation is bounded. This implies that the impulse of the slow variable with the desired stability is obtained. One has to check, however, the behavior of the fast variable. It follows from plugging the suggested control in (5.10) that the solution of the latter exists on, say, $[0, \pi]$. It follows from (5.8) that $y^2 = \frac{dx}{dt}$. Hence $y^2(\pi)$ is also close to zero as required. This verifies that the feedback (5.11) provides a solution.

Example 5.5. We show that checking the behavior of the fast variable cannot be skipped even when the second-order equation seems to produce the desired impulse. Consider the following system with x and y scalars.

$$\begin{aligned} \frac{dx}{dt} &= 1 - e^{-|y|} \\ \varepsilon \frac{dy}{dt} &= u \\ x(0) = 0, &\quad y(0) = 0 \\ u &\in (-\infty, \infty). \end{aligned} \tag{5.12}$$

One can see right away that although the control may be unbounded, an impulse of the x-variable in (5.12) cannot be generated. Indeed, the right hand side of the x-equation is bounded. However, when the derivations offered in the present section are carried out, one arrives at the following second-order equation for the x variable.

$$\frac{d^2x}{ds^2} = e^{-|y|}u. \tag{5.13}$$

Now, setting $u(y) = e^{|y|}(-x + 1/2)$ yields the equation

$$\frac{d^2x}{ds^2} = -x + 1/2, \tag{5.14}$$

namely, identical to the second-order system which generated the impulse in (1.2) and in Example 5.3. The explanation is that the second-order equation (5.4), as manifested, say, in (5.14) does not stand alone. When one inspects (as one should) how the fast variable y behaves in this example, one arrives at an equation for which it is easy to conclude that the solution blows up in finite time, and for ε small the blow up time is small even on the s scale.

6 A Further Generalization

The second-order controlled equation for the slow variable employed in the previous section is ε-free. The reason is that the slow variable in (5.1) is driven by a function of the fast variable only. We show in this section that under some reasonable conditions a variant of the method can be applied under a more general structure. We maintain however, the property that the controls enter only through the fast equation. (Without this property, one faces a much more intricate problem.) Thus, the system is given by

$$\begin{aligned} \tfrac{dx}{dt} &= f(x, y) \\ \varepsilon \tfrac{dy}{dt} &= g(x, y, u) \\ u &\in U, \end{aligned} \tag{6.1}$$

which is again a special case of (2.1). We assume that the vector fields are continuously differentiable.

Employing the change of time variables given in (3.1) or (5.2) the system (6.1) takes the form

$$\begin{aligned} \tfrac{dx}{ds} &= \varepsilon^{1/2} f(x, y) \\ \tfrac{dy}{ds} &= \varepsilon^{-1/2} g(x, y, u). \end{aligned} \tag{6.2}$$

Since $f(\cdot, \cdot)$ is assumed continuously differentiable, it follows that $x(\cdot)$ solves the second-order equation

$$\frac{d^2x}{ds^2} = \frac{\partial f}{\partial y}(x, y)\ g(x, y, u) + \varepsilon \frac{\partial f}{\partial x}(x, y)\ f(x, y). \tag{6.3}$$

Here $\frac{\partial f}{\partial y}(x, y)$ and $\frac{\partial f}{\partial x}(x, y)$ are the obvious matrices of partial derivatives at the given states. Notice that unlike the case in the previous section, the second-order equation (6.3) is not ε-free.

Solution Scheme 6.1. Consider the system (6.1) along with a desired impulse determined by $(x(0), y(0)), (x(0+), y(0+))$ and B as described in Problem Statement 2.1. Under a condition which guarantees that the perturbation term in (6.3) tends to zero as $\varepsilon \to 0+$ (see next two paragraphs) find a feedback $u(x, y)$ which when applied to the zero order part of (6.3), namely to

$$\frac{d^2x}{ds^2} = \frac{\partial f}{\partial y}(x, y)\ g(x, y, u), \tag{6.4}$$

the initial condition $x(0)$ is steered on an interval $[0, \sigma]$ to a small neighborhood of $x(0+)$, the outcome trajectory is bounded and the steering is stable; at the same time, check the y-equation in (6.2) to make sure that for ε small, a small neighborhood of $y(0)$ is steered to B.

Condition 6.2. We provide a condition which guarantees that the perturbation term in (6.3) is small and that Solution Scheme 6.1 can be applied. Suppose that for some positive $\delta > 0$ and a constant $C(x)$ which is bounded for bounded x, the estimate

$$\left| \frac{\partial f}{\partial x}(x, y)\ f(x, y) \right| \le C(x)(1 + |y|^{2-\delta}) \tag{6.5}$$

holds, namely, the growth of the indicated function is less than quadratic. If, in addition, the feedback $u(x, y)$ makes $g(x, y, u(x, y))$ a bounded function then the perturbation tends to 0 as $\varepsilon \to 0$. Indeed, $g(x, y, u(x, y))$ bounded, say by c, together with the y-equation in (6.2) imply that $|y_\varepsilon(s)| \le c\varepsilon^{-1/2}$. Inequality (6.5) implies in turn that the perturbation term in (6.3) is less than $\varepsilon C(x)(1 + \varepsilon^{-1+1/2\delta})$, which implies that when x stays bounded, the perturbation tends to zero as $\varepsilon \to 0+$. Then a simple continuous dependence argument would reveal that the feedback which generates the impulse for the ε-free system (6.4) also generates the same impulse in the presence of the perturbation.

Condition 6.3. We provide an alternative to Condition 6.2. Suppose that

$$\left| \frac{\partial f}{\partial x}(x, y)\ f(x, y) \right| \le C(x) \log^k(1 + |y|) \tag{6.6}$$

for some integer k. If, in addition, when the feedback $u(x, y)$ is applied the inequality $|g(x, y, u(x, y))| \le c(x)(1+|y|)$ holds, then the desired perturbation tends to 0 as $\varepsilon \to 0$. Indeed, Gronwall inequality (see e.g., [3]) then implies that when x is bounded, an inequality of the type $|y_\varepsilon(s)| \le \varepsilon^{-1/2}e^{cs}$ holds. Plugging this estimate in the perturbation term which for bounded x is bounded by $\varepsilon C \log^k(1 + |y|)$ results in a bound of order $\varepsilon(\log \varepsilon)^k$ which tends to 0 as

$\varepsilon \to 0$. Once the perturbation term tends to 0 it is clear that the feedback which generates the impulse for the ε-free system (6.4) generates the same impulse in the presence of the perturbation.

We conclude with an illustration of the preceding scheme.

Example 6.4. Consider the system

$$\begin{aligned} \tfrac{dx}{dt} &= ((x-1/2)^2+1)\log(|y|+1) \\ \varepsilon\tfrac{dy}{dt} &= u \\ x(0) &= 0, \quad y(0)=0 \\ u &\in [-1,1]. \end{aligned} \tag{6.7}$$

Suppose that an impulse has to be generated at $t=0$ reaching the boundary condition $(x(0+), y(0+)) = (1,0)$. The derived second-order equation (for simplicity we display it when $y \geq 0$) is

$$\frac{d^2x}{ds^2} = \frac{(x-1/2)^2+1}{y+1}u + \varepsilon(2x-1)\log^2(y+1). \tag{6.8}$$

For the unperturbed version of (6.8) the desired impulse can be generated by the feedback $u(x,y) = (-x+1/2)(y+1)((x-1/2)^2+1)^{-1}$ (for x and y nonnegative). It is easy to see that both estimates pointed out in Condition 6.3 are satisfied. The solution to the unperturbed closed loop system on $[0,\pi]$ is $x(s) = 1/2(1-\cos s)$ and (as pointed out in Condition 6.3) the perturbation is of order $\varepsilon(\log\varepsilon)^k$. This implies that the right hand side of the fast equation in (6.7) expressed in the fast time variable s (see (6.2)) is $\varepsilon^{-1/2}(u(x(s),y) + o(\varepsilon^{1/2}))$. Direct integration, after plugging in the value of $x(s)$, reveals that at $s=\pi$ the y variable is within a bounded distance, independent of ε, from 0 as required in order to establish the desired impulse.

Example 6.5. In some cases there is an alternative approach. Consider the system

$$\begin{aligned} \tfrac{dx}{dt} &= y + (2-e^{-|y|})x \\ \varepsilon\tfrac{dy}{dt} &= u \\ x(0) &= 0, \quad y(0)=0 \\ u &\in [-1,1]. \end{aligned} \tag{6.9}$$

Suppose again that an impulse has to be generated at $t=0$ reaching the boundary condition $(x(0+), y(0+)) = (1,0)$. One can follow the track set in Solution Scheme 6.1 and derive the feedback $u(x,y) = (1+e^{-y}x)^{-1}(-x+1/2)$ as a solution. Alternatively, we observe that the right hand side of the x-equation is a bounded perturbation of $\frac{dx}{dt} = y$. A bounded input operating on time intervals tending to zero has a negligible effect on the slow variable. A direct inspection then shows that the solution derived in Section 3 for the problem (6.9) without the perturbation, namely, $u = -x + 1/2$, generates the desired impulse for (6.9).

In conclusion I wish to point out that the present contribution provides only preliminary observations, concerning the possibility of generating an impulse with a singularly perturbed variable. The examples show that in some cases the impulses can be designed and examined. Many issues, however, have not been touched. These include primarily, optimality considerations, the description of a limit problem (possibly of a measure-differential equation or inclusion) and, as in the general problem (2.1), allowing the control to affect the slow state directly.

References

1. Bressan A, Rampazzo F (1991) Impulsive systems with commutative vector fields, J Optim Theory and Appl 71:67-83
2. Bressan A, Rampazzo F (1994) Impulsive control systems without commutativity assumptions, J Optim Theory Appl 81:435-457
3. Hartman P (1964) Ordinary Differential Equations. John Wiley New York
4. Kokotovic P, Khalil H, O'Reilly J (1986) Singular Perturbation Methods in Control Analysis and Design. Academic Press London
5. Liu W, Sussmann H (1999) Continuous dependence with respect to the input of trajectories of control-affine systems, SIAM J Control Optim 37:777-803
6. McShane E (1967) Relaxed controls and variational problems, SIAM J Control Optim 5:438-485
7. Motta M, Rampazzo F (1996) Dynamic programming for nonlinear systems driven by ordinary and impulse controls, SIAM J Control Optim 34:199-225
8. Murray J (1986) Existence theorems for optimal control and the calculus of variations when the state can jump, SIAM J Control Optim 24: 412-438
9. Warga J (1965) Variational problems with unbounded controls, SIAM J Control 3:424-438
10. Wolenski P (2002) Invariance properties for impulse systems. In: Proceedings of the 41st IEEE Conference on Decision and Control 1113-1116

Optimal Control of Differential-Algebraic Inclusions

Boris Mordukhovich[1] and Lianwen Wang[2]

[1] Department of Mathematics, Wayne State University. boris@math.wayne.edu
[2] Department of Mathematics and Computer Science, Central Missouri State University. lwang@cmsu1.cmsu.edu

1 Introduction

In this paper we consider the following dynamic optimization problem (P) governed by *differential-algebraic inclusions*:

$$\text{minimize} \quad J[x,z] := \varphi(x(a),x(b)) + \int_a^b f(x(t),x(t-\Delta),\dot{z}(t),t)\,dt \tag{1}$$

subject to the constraints

$$\dot{z}(t) \in F(x(t),x(t-\Delta),t), \quad \text{a.e. } t \in [a,b], \tag{2}$$

$$z(t) = x(t) + Ax(t-\Delta), \quad t \in [a,b], \tag{3}$$

$$x(t) = c(t), \quad t \in [a-\Delta,a), \tag{4}$$

$$(x(a),x(b)) \in \Omega \subset \mathbb{R}^{2n}, \tag{5}$$

where $x : [a-\Delta,b] \to \mathbb{R}^n$ is continuous on $[a-\Delta,a)$ and $[a,b]$ (with a possible jump at $t=a$), and where $z(t)$ is absolutely continuous on $[a,b]$. We always assume that $F : \mathbb{R}^n \times \mathbb{R}^n \times [a,b] \rightrightarrows \mathbb{R}^n$ is a set-valued mapping of closed graph, that Ω is a closed set, that $\Delta > 0$ is a constant delay, and that A is a constant $n \times n$ matrix.

Differential-algebraic control systems are attractive mathematically (since they are essentially different from standard control systems even in the case of smooth dynamics) and very important for applications, especially in process systems engineering. Necessary optimality conditions for controlled differential-algebraic *equations* with *no delays* are obtained in [11], where one can find detailed discussions and references on this topic. Let us also mention the research [1] on the so-called *implicit control systems* related to controlled differential-algebraic equations without delays. Necessary optimality conditions derived in these papers are based on reductions to standard (while nonsmooth) control systems by using uniform inverse mapping and implicit function theorems as well as on powerful techniques of nonsmooth analysis.

M. de Queiroz et al. (Eds.): Optimal Control, Stabilization and Nonsmooth Analysis, LNCIS 301, pp. 73–83, 2004.

Note, however, that such reductions require rather restrictive assumptions of the 'index one' type.

We are not familiar with any results in the literature on optimal control problems governed by differential-algebraic *inclusions* in either nondelayed or delayed settings. It seems that necessary optimality conditions for delayed differential-algebraic systems have not been specifically studied even in the case of controlled equations with smooth dynamics. On the other hand, differential-algebraic systems with delays are similar in many aspects to the so called *neutral functional-differential systems* that contain time-delays not only in state but also in velocity variables. Neutral systems have drawn much attention in the theory and applications of optimal control in the case of smooth dynamics. Necessary optimality conditions for nonsmooth neutral problems were first obtained in [9, 10] in the framework of neutral functional-differential inclusions. The techniques and constructions of [9, 10] are essentially used in what follows.

In this paper we derive necessary optimality conditions for the above problem (P) by the *method of discrete approximations* developed in [7]. The results obtained are given in the forms of both Euler-Lagrange and Hamiltonian inclusions in terms of basic generalized differential constructions of variational analysis. We skip most of the proofs, which are similar to those given in [10] for the case of neutral systems.

2 Discrete Approximations of Differential-Algebraic Inclusions

This section deals with discrete approximations of an arbitrary admissible trajectory to the differential-algebraic system (2)–(4) without taking into account endpoint constraints. Let $(\bar{x}, \bar{z})$ be an admissible pair to (2)–(4), i.e., $\bar{x}(\cdot)$ is continuous on $[a-\Delta, a)$ and $[a, b]$ (with a possible jump at $t = a$) and $\bar{z}(\cdot)$ is absolutely continuous on $[a, b]$. The following assumptions are imposed throughout the paper:

(H1) There are an open set $U \subset \mathbb{R}^n$ and numbers $\ell_F, m_F > 0$ such that $\bar{x}(t) \in U$ for all $t \in [a-\Delta, b]$, the sets $F(x, y, t)$ are closed, and one has

$$F(x, y, t) \subset m_F \mathbb{B},$$
$$F(x_1, y_1, t) \subset F(x_2, y_2, t) + \ell_F(|x_1 - x_2| + |y_1 - y_2|)\mathbb{B}$$

for all $(x, y, t) \in U \times U \times [a, b]$, $(x_1, y_1), (x_2, y_2) \in U \times U$, and $t \in [a, b]$, where $\mathbb{B}$ stands for the closed unit ball in $\mathbb{R}^n$.

(H2) $F(x, y, \cdot)$ is a.e. Hausdorff continuous on $t \in [a, b]$ uniformly in $U \times U$.

(H3) The function $c(\cdot)$ is continuous on $[a-\Delta, a]$.

Following [2], we consider the so-called *averaged modulus of continuity* for $F(x, y, t)$ with $(x, y) \in U \times U$ and $t \in [a, b]$ defined by

$$\tau(F;h) := \int_a^b \sigma(F;t,h)\,dt,$$

where $\sigma(F;t,h) := \sup\big\{\vartheta(F;x,y,t,h)\big|\ (x,y) \in U \times U\big\}$ with

$$\vartheta(F;x,y,t,h) :=$$
$$\sup\Big\{\text{haus}(F(x,y,t_1),F(x,y,t_2))\ \Big|\ (t_1,t_2) \in \big[t-\tfrac{h}{2},t+\tfrac{h}{2}\big] \cap [a,b]\Big\},$$

and where haus$(\cdot,\cdot)$ stands for the Hausdorff distance between two compact sets. It is proved in [2] that if $F(x,y,t)$ is Hausdorff continuous for a.e. $t \in [a,b]$ uniformly in $(x,y) \in U \times U$, then $\tau(F;h) \to 0$ as $h \to 0$.

Let us construct a sequence of discrete approximations of the given trajectory to the differential-algebraic inclusion replacing the derivative in (2) by the *Euler finite difference* $\dot{z}(t) \approx [z(t+h) - z(t)]/h$. For any $N \in \mathbb{N} := \{1,2,\ldots\}$, we set $h_N := \Delta/N$ and define the discrete mesh $t_j := a + jh_N$ for $j = -N,\ldots,k$, and $t_{k+1} := b$, where k is a natural number determined from $a + kh_N \le b < a + (k+1)h_N$. Then the corresponding *discrete systems* associated with (2)–(4) are given by

$$\begin{aligned} z_N(t_{j+1}) &\in z_N(t_j) + h_N F(x_N(t_j), x_N(t_j - \Delta), t_j),\ j = 0,\ldots,k,\\ z_N(t_j) &= x_N(t_j) + Ax_N(t_j - \Delta),\ j = 0,\ldots,k+1,\\ x_N(t_j) &= c(t_j),\ j = -N,\ldots,-1. \end{aligned} \tag{6}$$

Given discrete functions $x_N(t_j)$ and $z_N(t_j)$ satisfying (6), we consider their *extensions* to the continuous-time interval $[a-\Delta,b]$ such that $x_N(t)$ are defined piecewise-linearly on $[a,b]$ and piecewise-constantly, continuously from the right on $[a-\Delta,a)$, while the 'discrete velocities' $[z_N(t_{j+1}) - z_N(t_j)]/h_N$ are extended to $[a,b]$ by

$$v_N(t) := \frac{z_N(t_{j+1}) - z_N(t_j)}{h_N}, \quad t \in [t_j,t_{j+1}),\ j = 0,\ldots,k.$$

Let $W^{1,2}[a,b]$ be the standard Sobolev space with the norm

$$\|z(\cdot)\|_{W^{1,2}} := \max_{t\in[a,b]} |z(t)| + \Big(\int_a^b |\dot{z}(t)|^2 dt\Big)^{1/2}.$$

The following theorem establishes a *strong approximation* of any admissible trajectory for the differential-algebraic system by corresponding solutions to discrete approximations (6):

Theorem 1. *Let $(\bar{x},\bar{z})$ be an admissible pair for (2)–(4) under hypotheses (H1)–(H3). Then there is a sequence $\{\Theta_N(t_j)\mid j = -N,\ldots,k+1\}$, $N \in \mathbb{N}$, of solutions to discrete inclusions (refdis) such that $\Theta_N(t_0) = \bar{x}(a)$ for all $N \in \mathbb{N}$, the extensions $\Theta_N(t)$, $a - \Delta \le t \le b$, converge uniformly to $\bar{x}(\cdot)$ on $[a-\Delta,b]$ while $Z_N(t) := \Theta_N(t) + A\Theta_N(t-\Delta)$ converges to $\bar{z}(t) := \bar{x}(t) + A\bar{x}(t-\Delta)$ in the $W^{1,2}$-norm on $[a,b]$ as $N \to \infty$.*

3 Strong Convergence of Discrete Optimal Solutions

This section constructs a sequence of *well-posed* discrete approximations for problem (P) such that *optimal solutions* to discrete approximation problems *strongly converge* to the *reference* optimal solution $\bar{x}(\cdot)$ of (P).

Given $\bar{x}(t)$, $a-\Delta \le t \le b$, take its approximation $\Theta_N(t)$ from Theorem 1 and denote $\eta_N := |\Theta_N(t_{k+1}) - \bar{x}(b)|$. Consider the following *discrete-time dynamic optimization problem* (P_N):

$$\begin{aligned} &\text{minimize} \quad J_N[x_N, z_N] := \varphi(x_N(t_0), x_N(t_{k+1})) + |x_N(t_0) - \bar{x}(a)|^2 \\ &+h_N \sum_{j=0}^{k} f(x_N(t_j), x_N(t_j-\Delta), \frac{z_N(t_{j+1}) - z_N(t_j)}{h_N}, t_j) \\ &+\sum_{j=0}^{k} \int_{t_j}^{t_{j+1}} \left| \frac{z_N(t_{j+1}) - z_N(t_j)}{h_N} - \dot{\bar{z}}(t) \right|^2 dt \end{aligned} \tag{7}$$

subject to the dynamic constraints (6), the *perturbed* endpoint constraints

$$(x_N(t_0), x_N(t_{k+1})) \in \Omega_N := \Omega + \eta_N I\!B, \tag{8}$$

and the auxiliary constraints

$$|x_N(t_j) - \bar{x}(t_j)| \le \varepsilon, \quad j = 1, \ldots, k+1, \tag{9}$$

with some $\varepsilon > 0$. The latter auxiliary constraints are needed to guarantee the existence of optimal solutions in (P_N) and can be ignored in the derivation of necessary optimality conditions.

In what follows we select $\varepsilon > 0$ in (9) such that $\bar{x}(t) + \varepsilon I\!B \subset U$ for all $t \in [a-\Delta, b]$ and take sufficiently large N ensuring that $\eta_N < \varepsilon$. Note that problems (P_N) have feasible solutions, since the trajectories Θ_N from Theorem 1 satisfy all the constraints (6), (8), and (9). Therefore, by the classical Weierstrass theorem in finite dimensions, each (P_N) admits an optimal solution $\bar{x}_N(\cdot)$ under the following assumption imposed in addition to the above hypotheses (H1)–(H3):

(H4) φ is continuous on $U \times U$, $f(x, y, v, \cdot)$ is continuous for a.e. $t \in [a, b]$ uniformly in $(x, y, v) \in U \times U \times m_F I\!B$, $f(\cdot, \cdot, \cdot, t)$ is continuous on $U \times U \times m_F I\!B$ uniformly in $t \in [a, b]$, and Ω is locally closed around $(\bar{x}(a), \bar{x}(b))$.

To justify the strong convergence of $(\bar{x}_N, \bar{z}_N) \to (\bar{x}, \bar{z})$ in the sense of Theorem 1, we need to involve an important intrinsic property of problem (P) called *relaxation stability*. Let us consider, along with the original system (2), the *convexified* delayed differential-algebraic system

$$\begin{aligned} \dot{z}(t) &\in \mathrm{co}F(x(t), x(t-\Delta), t) \quad \text{a.e. } t \in [a, b], \\ z(t) &= x(t) + Ax(t-\Delta), \quad t \in [a, b], \end{aligned} \tag{10}$$

where 'co' stands for the convex hull of a set. Further, given the integrand f in (1), we take its restriction

$$f_F(x, y, v, t) := f(x, y, v, t) + \delta(v; F(x, y, t))$$

to F in (2), where $\delta(\cdot; F)$ stands for the indicator function of a set. Denote by $\hat{f}_F(x, y, v, t)$ the *convexification* of f_F in the v variable and define the *relaxed generalized Bolza problem* (R) for delayed differential-algebraic systems as follows:

$$\text{minimize } \hat{J}\,[x, z] := \varphi(x(a), x(b)) + \int_a^b \hat{f}_F(x(t), x(t - \Delta), \dot{z}(t), t)\, dt$$

over feasible pairs (x, z) with the same analytic properties as in (P) subject to the tail (4) and endpoint (5) constraints. Every feasible pair to (R) is called a *relaxed pair* to (P).

One clearly has $\inf(R) \leq \inf(P)$. The original problem (P) is said to be *stable with respect to relaxation* provided $\inf(P) = \inf(R)$. This property, which obviously holds under the convexity assumption on the sets $F(x, y, t)$, goes far beyond the convexity. General sufficient conditions for the relaxation stability of problem (P) governed by differential-algebraic inclusions can be obtained similarly to those presented in [3] for the case of neutral inclusions. The next theorem makes a bridge between optimal control problems governed by differential-algebraic and difference-algebraic control systems.

Theorem 2. *Let $(\bar{x}, \bar{z})$ be an optimal solution to problem (P), which is assumed to be stable with respect to relaxation. Suppose also that hypotheses (H1)–(H4) hold. Then any sequence $(\bar{x}_N(\cdot), \bar{z}_N(\cdot))$, $N \in \mathbb{N}$, of optimal solutions to (P_N) extended to the continuous interval $[a - \Delta, b]$ strongly converges to $(\bar{x}, \bar{z})$ as $N \to \infty$ in the sense that $\bar{x}_N(\cdot)$ converge to $\bar{x}(\cdot)$ uniformly on $[a - \Delta, b]$ and $\bar{z}_N(\cdot)$ converge to $\bar{z}(\cdot)$ in the $W^{1,2}$-norm on $[a, b]$.*

4 Variational Analysis in Finite Dimensions

To conduct a variational analysis of problems (P_N), which are intrinsically *nonsmooth*, we use appropriate tools of generalized differentiation introduced in [4] and then developed in many publications; see, e.g., the books [5, 13, 14, 15] for more details and references, where the notation $N(\cdot; \Omega)$, D^*F, and $\partial\varphi$ stand for the basic (limiting) normal cone to sets, the coderivatives of set-valued mappings, and the subdifferential of extended-real-valued functions.

The following two results are particularly important for our subsequent analysis. The first one obtained in [6] (see also [13, Theorem 9.40]), gives a complete coderivative characterization of the classical local Lipschitzian property of multifunctions.

Theorem 3. *Let $F : \mathbb{R}^n \to \mathbb{R}^m$ be a closed-graph multifunction that is locally bounded around $\bar{x}$. Then the following conditions are equivalent:*

(i) *F is locally Lipschitzian around $\bar{x}$.*
(ii) *There exist a neighborhood U of $\bar{x}$ and a number $\ell > 0$ such that*

$$\sup\Big\{|x^*|\;\Big|\; x^* \in D^*F(x,y)(y^*)\Big\} \le \ell|y^*| \quad x \in U,\; y \in F(x),\; y^* \in \mathbb{R}^m.$$

The next result (see, e.g., [5, Corollary 7.5] and [14, Theorem 3.17]) provides necessary optimality conditions for the following problem (MP) of *nonsmooth mathematical programming* with *many geometric constraints*, which is essential for the application to dynamic optimization:

$$\begin{aligned} &\text{minimize } \varphi_0(w) \quad \text{subject to} \\ &\varphi_j(w) \le 0, \quad j = 1, \ldots, r, \\ &g_j(w) = 0, \quad j = 0, \ldots, m, \\ &w \in \Lambda_j, \quad j = 0, \ldots, l, \end{aligned}$$

where $\varphi_j : \mathbb{R}^d \to \mathbb{R}$, $g_j : \mathbb{R}^d \to \mathbb{R}^n$, and $\Lambda_j \subset \mathbb{R}^d$.

Theorem 4. *Let $\bar{w}$ be an optimal solution to (MP). Assume that all φ_i are Lipschitz continuous, that g_j are continuously differentiable, and that Λ_j are locally closed near $\bar{w}$. Then there exist real numbers $\{\mu_j |\, j = 0, \ldots, r\}$ as well as vectors $\{\psi_j \in \mathbb{R}^n |\, j = 0, \ldots, m\}$ and $\{w_j^* \in \mathbb{R}^d |\, j = 0, \ldots, l\}$, not all zero, satisfying the relations:*

$$\mu_j \ge 0, \quad j = 0, \ldots, r, \tag{11}$$

$$\mu_j \varphi_j(\bar{w}) = 0, \quad j = 1, \ldots, r, \tag{12}$$

$$w_j^* \in N(\bar{w}; \Lambda_j), \quad j = 0, \ldots, l, \tag{13}$$

$$-\sum_{j=0}^{l} w_j^* \in \partial\Big(\sum_{j=0}^{r} \mu_j \varphi_j\Big)(\bar{w}) + \sum_{j=0}^{m} \nabla g_j(\bar{w})^* \psi_j. \tag{14}$$

To formulate some results of this paper in the case of *nonautonomous* continuous-time systems, we need certain extensions of the basic normal cone, subdifferential, and coderivative for the corresponding *moving* objects. These extensions, denoted by $\tilde{N}$, $\tilde{\partial}$, and $\tilde{D}^*$, reduce to the basic constructions under some natural assumptions; see [7, 8, 10] for more details and discussions.

5 Optimality Conditions for Difference-Algebraic Systems

In this section we derive necessary optimality conditions for the discrete approximation problems (P_N) by reducing them to those in Theorem 4 for nonsmooth mathematical programming. Given $N \in \mathbb{N}$, consider problem (MP) with the decision vector

$$w := (x_0^N, x_1^N, \ldots, x_{k+1}^N, v_0^N, v_1^N, \ldots, v_k^N) \in \mathbb{R}^{n(2k+3)}$$

and the following data:

$$\begin{aligned} \varphi_0(w) &:= \varphi(x_0^N, x_{k+1}^N) + |x_0^N - \bar{x}(a)|^2 + h_N \sum_{j=0}^{k} f(x_j^N, x_{j-N}^N, v_j^N, t_j) \\ &\quad + \sum_{j=0}^{k} \int_{t_j}^{t_{j+1}} \left| v_j^N - \dot{\bar{z}}(t) \right|^2 dt, \\ \Lambda_j &:= \left\{ (x_0^N, \ldots, v_k^N) \mid v_j^N \in F(x_j^N, x_{j-N}^N, t_j) \right\}, \quad j = 0, \ldots, k, \\ \varphi_j(w) &:= |x_j^N - \bar{x}(t_j)| - \varepsilon, \quad j = 1, \ldots, k+1, \\ \Lambda_{k+1} &:= \left\{ (x_0^N, \ldots, v_k^N) \mid (x_0^N, x_{k+1}^N) \in \Omega_N \right\}, \\ g_j(w) &:= z_{j+1}^N - z_j^N - h_N v_j^N, \quad j = 0, \ldots, k, \end{aligned}$$

where $z_j^N := x_j^N + A x_{j-N}^N$ for $j = 0, \ldots, k$, and where $x_j^N := c(t_j)$ for $j < 0$. Let

$$\bar{w}^N := (\bar{x}_0^N, \ldots, \bar{x}_{k+1}^N, \bar{v}_0^N, \ldots, \bar{v}_k^N)$$

be an optimal solution to (MP). Applying Theorem 4, we find real numbers μ_j^N and vectors

$$w_j^* \in \mathbb{R}^{n(2k+3)}, \quad j = 0, \ldots, k+1$$

as well as vectors $\psi_j^N \in \mathbb{R}^n$ for $j = 0, \ldots, k$, not all zero, such that conditions (11)–(14) are satisfied.

Taking $w_j^* = (x_{0,j}^*, \ldots, x_{k+1,j}^*, v_{0,j}^*, \ldots, v_{k,j}^*) \in N(\bar{w}^N; \Lambda_j)$ for $j = 0, \ldots, k$ and employing Theorem 2 on the convergence of discrete approximations, we have $\varphi_j(\bar{w}^N) < 0$ for $j = 1, \ldots, k+1$ whenever N is sufficiently large. Thus $\mu_j^N = 0$ for these indexes due to the complementary slackness conditions (12). Let

$$\lambda^N := \mu_0^N \geq 0.$$

By the subdifferential sum rule for φ_0 defined above, inclusion (14) in Theorem 4 implies these relationships:

$$\begin{aligned} -x_{0,0}^* - x_{0,N}^* - x_{0,k+1}^* &= \lambda^N u_0^N + \lambda^N h_N \vartheta_0^N + \lambda^N h_N \kappa_0^N \\ &\quad + 2\lambda^N (\bar{x}_0^N - \bar{x}(a)) - \psi_0^N + A^*(\psi_{N-1}^N - \psi_N^N), \\ -x_{j,j}^* - x_{j,j+N}^* &= \lambda^N h_N \vartheta_j^N + \lambda^N h_N \kappa_j^N + \psi_{j-1}^N - \psi_j^N \\ &\quad + A^*(\psi_{j+N-1}^N - \psi_{j+N}^N), \quad j = 1, \ldots, k-N, \\ -x_{k-N+1,k-N+1}^* &= \lambda^N h_N \vartheta_{k-N+1}^N + \psi_{k-N}^N - \psi_{k-N+1}^N - A^* \psi_k^N, \\ -x_{j,j}^* &= \lambda^N h_N \vartheta_j^N + \psi_{j-1}^N - \psi_j^N, \quad j = k-N+2, \ldots, k, \\ -x_{k+1,k+1}^* &= \lambda^N u_{k+1}^N + \psi_k^N, \\ -v_{j,j}^* &= \lambda^N h_N \iota_j^N + \lambda^N \theta_j^N - h_N \psi_j^N, \quad j = 0, \ldots, k \end{aligned}$$

with the notation

$$(u_0^N, u_{k+1}^N) \in \partial\varphi(\bar{x}_0^N, \bar{x}_{k+1}^N), \quad (\vartheta_j^N, \kappa_{j-N}^N, \iota_j^N) \in \partial f(\bar{x}_j^N, \bar{x}_{j-N}^N, \bar{v}_j^N, t_j),$$
$$\theta_j^N := 2\int_{t_j}^{t_{j+1}} \left(\bar{v}_j^N - \dot{\bar{z}}(t)\right) dt.$$

Based on the above relationships, we arrive at the following result.

Theorem 5. *Let* $(\bar{x}^N, \bar{z}^N)$ *be an optimal solution to problem* (P_N). *Assume that the sets* Ω *and* $\mathrm{gph}F_j$ *are closed and that the functions* φ *and* f_j *are Lipschitz continuous around the points* $(\bar{x}_0^N, \bar{x}_{k+1}^N)$ *and* $(\bar{x}_j^N, \bar{x}_{j-N}^N, \bar{v}_j^N)$, *respectively, for all* $j = 0, \ldots, k$. *Then there exist*

$$\lambda^N \geq 0,\ p_j^N\ (j = 0, \ldots, k+N+1) \text{ and } q_j^N\ (j = -N, \ldots, k+1),$$

not all zero, satisfying the conditions

$$p_j^N = 0, \quad j = k+2, \ldots, k+N+1,$$
$$q_j^N = 0, \quad j = k-N+1, \ldots, k+1,$$
$$(p_0^N + q_0^N, -p_{k+1}^N) \in \lambda^N \partial\varphi(\bar{x}_0^N, \bar{x}_{k+1}^N) + N((\bar{x}_0^N, \bar{x}_{k+1}^N); \Omega_N),$$
$$\Big(\frac{P_{j+1}^N - P_j^N}{h_N}, \frac{Q_{j-N+1}^N - Q_{j-N}^N}{h_N}, -\frac{\lambda^N \theta_j^N}{h_N} + p_{j+1}^N + q_{j+1}^N\Big)$$
$$\in \lambda^N \partial f_j(\bar{x}_j^N, \bar{x}_{j-N}^N, \bar{v}_j^N) + N((\bar{x}_j^N, \bar{x}_{j-N}^N, \bar{v}_j^N); \mathrm{gph}F_j), \quad j = 0, \ldots, k,$$

with the notation

$$P_j^N := p_j^N + A^* p_{j+N}^N, \quad Q_j^N := q_j^N + A^* q_{j+N}^N,$$
$$\bar{v}_j^N := \frac{(\bar{x}_{j+1}^N + A\bar{x}_{j-N+1}^N) - (\bar{x}_j^N + A\bar{x}_{j-N}^N)}{h_N}.$$

6 Optimality Conditions for Differential-Algebraic Inclusions

Our main result establishes the following necessary optimality conditions of the *Euler-Lagrange* type for the original problem (P) derived by the limiting procedure from discrete approximations with the use of Theorem 3:

Theorem 6. *Let* $(\bar{x}, \bar{z})$ *be an optimal solution to problem* (P) *under hypotheses (H1)–(H4), where* φ *and* $f(\cdot,\cdot,\cdot,t)$ *are assumed to be Lipschitz continuous. Suppose also that* (P) *is stable with respect to relaxation. Then there exist a number* $\lambda \geq 0$ *and piecewise continuous functions* $p : [a, b+\Delta] \to \mathbb{R}^n$ *and* $q : [a-\Delta, b] \to \mathbb{R}^n$ *such that* $p(t) + A^*p(t+\Delta)$ *and* $q(t-\Delta) + A^*q(t)$ *are absolutely continuous on* $[a, b]$ *and the following conditions hold:*

$$\lambda + |p(b)| = 1,$$
$$p(t) = 0, \quad t \in (b, b+\Delta],$$
$$q(t) = 0, \quad t \in (b-\Delta, b],$$
$$(p(a) + q(a), -p(b)) \in \lambda\partial\varphi(\bar{x}(a), \bar{x}(b)) + N((\bar{x}(a), \bar{x}(b)); \Omega),$$
$$\Big(\frac{d}{dt}[p(t) + A^*p(t+\Delta)], \frac{d}{dt}[q(t-\Delta) + A^*q(t)]\Big)$$
$$\in \text{co}\Big\{(u,w) \mid (u,w,p(t)+q(t)) \in \lambda\widetilde{\partial} f(\bar{x}(t), \bar{x}(t-\Delta), \dot{\bar{z}}(t), t)$$
$$+\widetilde{N}\big((\bar{x}(t), \bar{x}(t-\Delta), \dot{\bar{z}}(t)); \text{gph}F(\cdot,\cdot,t)\big)\Big\} \quad a.e. \ t \in [a,b].$$

Observe that for the Mayer problem (P_M), which is problem (P) with $f = 0$, the generalized Euler-Lagrange inclusion is equivalently expressed in terms of the extended coderivative with respect to the first two variables of $F = F(x,y,t)$, i.e., in the form

$$\Big(\frac{d}{dt}[p(t) + A^*p(t+\Delta)], \frac{d}{dt}[q(t-\Delta) + A^*q(t)]\Big)$$
$$\in \text{co}\tilde{D}^*_{x,y}F\big(\bar{x}(t), \bar{x}(t-\Delta), \dot{\bar{z}}(t)\big)\big(-p(t) - q(t)\big) \ \text{a.e.}, \quad t \in [a,b].$$

It turns out that the extended Euler-Lagrange inclusion obtained above implies, under the relaxation stability of the original problems, two other principal optimality conditions expressed in terms of the Hamiltonian function built upon the mapping F in (2). The first condition called the *extended Hamiltonian inclusion* is given below in terms of a partial convexification of the basic subdifferential for the Hamiltonian function. The second one is a counterpart of the classical *Weierstrass-Pontryagin maximum condition* for hereditary differential-algebraic inclusions. Recall that a counterpart of the maximum principle does *not* generally hold for nonconvex differential-algebraic control systems even in the case of smooth dynamics.

The following relationships between the extended Euler-Lagrange inclusion and Hamiltonian inclusion are based on Rockafellar's dualization theorem [12] relating subgradients of Lagrangians and Hamiltonians associated with general set-valued mappings (which are not necessarily associated with dynamical systems). For simplicity, we consider the case of the Mayer problem (P_M) for autonomous differential-algebraic inclusions. Then the Hamiltonian function for F is defined by

$$H(x,y,p) := \sup\big\{\langle p, v\rangle \big| \ v \in F(x,y)\big\}.$$

Corollary 1. *Let $(\bar{x}, \bar{z})$ be an optimal solution to the Mayer problem (P_M) for the autonomous hereditary differential-algebraic inclusion (2) under the assumptions of Theorem 6. Then there exist a number $\lambda \geq 0$ and piecewise continuous functions $p : [a, b+\Delta] \to \mathbb{R}^n$ and $q : [a-\Delta, b] \to \mathbb{R}^n$ such that $p(t) + A^*p(t+\Delta)$ and $q(t-\Delta) + A^*q(t)$ are absolutely continuous on $[a,b]$*

and, besides the necessary optimality conditions of Theorem 6, one has the extended Hamiltonian inclusion

$$\begin{aligned} &\Big(\frac{d}{dt}[p(t)+A^*p(t+\Delta)],\ \frac{d}{dt}[q(t-\Delta)+A^*q(t)]\Big) \\ &\in \operatorname{co}\Big\{(u,w)\ \Big|\ \Big(-u,-w,\dot{\bar{z}}(t)\Big)\in\partial H\big(\bar{x}(t),\bar{x}(t-\Delta),p(t)+q(t)\big)\Big\} \end{aligned} \tag{15}$$

and the maximum condition

$$\langle p(t)+q(t),\dot{\bar{z}}(t)\rangle = H\big(\bar{x}(t),\bar{x}(t-\Delta),p(t)+q(t)\big) \tag{16}$$

for almost all $t\in[a,b]$. *If moreover* F *is convex-valued around* $(\bar{x}(t),\bar{x}(t-\Delta))$, *then (15) is equivalent to the Euler-Lagrange inclusion*

$$\begin{aligned} &\Big(\frac{d}{dt}[p(t)+A^*p(t+\Delta)],\ \frac{d}{dt}[q(t-\Delta)+A^*q(t)]\Big) \\ &\in \operatorname{co}D^*F\big(\bar{x}(t),\bar{x}(t-\Delta),\dot{\bar{z}}(t)\big)\big(-p(t)-q(t)\big),\quad a.e.\ \ t\in[a,b], \end{aligned}$$

which automatically implies the maximum condition (16) in this case.

Acknowledgement. This research was partly supported by the National Science Foundation under grants DMS-0072179 and DMS-0304989.

References

1. Devdariani E, Ledyaev Y (1999), Maximum principle for implicit control systems, Appl Math Optim 40:79–103
2. Dontchev A, Farhi E (1989) Error estimates for discretized differential inclusions, Computing 41:349–358
3. Kisielewicz M (1991) Differential Inclusions and Optimal Control. Kluwer Dordrecht The Netherlands
4. Mordukhovich B (1976) Maximum principle in problems of time optimal control with nonsmooth constraints, J Appl Math Mech 40:960–969
5. Mordukhovich B (1988) Approximation Methods in Problems of Optimization and Control. Nauka Moscow 1988
6. Mordukhovich B (1993) Complete characterization of openness, metric regularity, and Lipschitzian properties of multifunctions, Trans Amer Math Soc 340:1–35
7. Mordukhovich B (1995) Discrete approximations and refined Euler–Lagrange conditions for nonconvex differential inclusions, SIAM J Control Optim 33:882–915
8. Mordukhovich B, Trieman J, Zhu Q (2004) An extended extremal principle with applications to multiobjective optimization, SIAM J Control Optim to appear
9. Mordukhovich B, Wang L (2002) Optimal control of hereditary differential inclusions. In: Proc 41st IEEE Conference on Decision and Control Las Vegas NV 1107–1112

10. Mordukhovich B, Wang L (2004) Optimal control of neutral functional-differential inclusions, SIAM J Control Optim to appear
11. Pinho M, Vinter R (1997) Necessary conditions for optimal control problems involving nonlinear differential algebraic equations, J Math Anal Appl 212:493–516
12. Rockafellar R (1996) Equivalent subgradient versions of Hamiltonian and Euler–Lagrange conditions in variational analysis, SIAM J Control Optim 34:1300–1314
13. Rockafellar R, Wets R (1998) Variational Analysis. Springer-Verlag Berlin
14. Smirnov G (2002) Introduction to the Theory of Differential Inclusions. American Mathematical Society Providence RI
15. Vinter R (2000) Optimal Control. Birkhäuser Boston

Necessary and Sufficient Conditions for Turnpike Optimality: The Singular Case

Alain Rapaport[1] and Pierre Cartigny[2]

[1] LASB-INRA, 2 pl. Viala, 34060 Montpellier, France. rapaport@ensam.inra.fr
[2] GREQAM, 2 rue de la Vieille Charité, 13002 Marseille, France. cartigny@ehess.cnrs-mrs.fr

1 Introduction

In this paper we consider a problem of calculus of variations in infinite horizon whose objective $J[.]$ is given by

$$J[x(.)] = \lim_{T \to +\infty} \int_0^T e^{-\delta t} l(x(t), \dot{x}(t)) dt, \tag{1}$$

where δ is a positive number and $l : \mathbb{R} \times \mathbb{R} \to \mathbb{R}$ satisfies

$$\forall u, \quad v \to l(u, v) \quad \text{is linear.} \tag{2}$$

Our interest is the maximization of J over some set of paths. Each of them is a subset of the space of absolutely continuous functions defined on $[0, +\infty)$ and is defined by constraints that will be made precise later. One encounters such models in environmental economics, in particular when dealing with fisheries management [7, 10]. Analogous models in finite horizon (i.e., defined on [0,T] with fixed T) have been also studied in aeronautic applications [3]. In [16], Miele proposed a method leading to a sufficient condition for optimality for a finite horizon (see also [7]) that is today well known and that we recall below. In [15] this method was extended to the infinite horizon case. Aside from this result, most of the other results deal with second order optimality conditions and are established in the case of a finite horizon [3]. The particularity of this problem comes from the fact that the Euler first order optimality condition is an algebraic equation instead a differential one, as is generally the case when assumption (2) is dropped. For this reason, we call the case studied in this paper *singular*. In the standard case, that is to say when the Euler condition is not reduced to an algebraic equation, many results are available in the literature, and most often established in a vectorial framework (i.e., l defined on $\mathbb{R}^n \times \mathbb{R}^n$).

During the 1970's, many results were obtained on the so called "Turnpike property" [9] that we recall now. For the sake of simplicity, we assume the

M. de Queiroz et al. (Eds.): Optimal Control, Stabilization and Nonsmooth Analysis, LNCIS 301, pp. 85–100, 2004.

existence of a steady state, i.e., an optimal stationary solution (and therefore an equilibrium of the Euler equation). Under regularity and concavity assumptions, and if some condition on Hessian sub-matrices of l is satisfied, then one can prove that this steady state possesses the saddle point stability. More precisely, the Euler equation

$$l_x(x(t),\dot{x}(t)) - \frac{d}{dt}l_{\dot{x}}(x(t),\dot{x}(t)) + \delta l_{\dot{x}}(x(t),\dot{x}(t)) = 0 \tag{3}$$

is equivalent to a system of first order differential equations, and an relevant equivalent system is the so called modified Hamiltonian system [9]

$$\begin{cases} \dot{x}(t) = H_p(x(t),p(t)) \\ \dot{p}(t) = \delta p(t) - H_x(x(t),p(t)) \end{cases} \tag{4}$$

where $H(x,p) = py - l(x,y)$ and $p = l_y(x,y)$. To each constant solution $\overline{x}$ of (3) corresponds an equilibrium $(\overline{x},\overline{p})$ of (4). Under assumptions mentioned above, this equilibrium is a saddle point with stable and unstable manifolds of the same dimension. Therefore for any initial condition x_0 in a neighborhood of $\overline{x}$, we can chose a unique p_0 such that (x_0,p_0) belongs to the stable manifold and then the corresponding solution $(x^*(.),p^*(.))$ of (4) converges to the equilibrium $(\overline{x},\overline{p})$. Then one can prove that the path $x^*(.)$ is an optimal solution [8]. This property is known, quite often in the economic literature, as a turnpike property. Quoting from the preface in [7], "the term (turnpike) was first coined by Samuelson in 1958 [22] where he showed that an efficient expanding economy would spend most of the time in the vicinity of a balanced equilibrium path (also called a von Neumann path)" (see also the introduction in [24]). Now, for our singular case the situation is different. Assuming (2), the Euler equation degenerates into an algebraic equation, as one can easily check in (3). Therefore the only (regular) candidates for optimality are given by the solutions of this equation, which are stationary paths. For the optimal solutions emanating from initial conditions that are not solutions of this algebraic equation, nothing can be obtained directly from this approach. Therefore specific developments have to be elaborated for this singular case.

In this setting, the method proposed by Miele [16] concerns the scalar case with inequality constraints such as $\alpha \leq \dot{x}(t) \leq \beta$ for any t, with $\alpha < 0$ and $\beta > 0$. If the algebraic equation possesses a unique solution $\overline{x}$, it is relevant to introduce the most rapid approach path for each initial condition x_0, denoted $MRAP(x_0,\overline{x})$, which joins x_0 to $\overline{x}$ as quickly as possible, and stays at $\overline{x}$ at any further time. This method, recalled in Section 2, consists of using the Green theorem to prove the optimality of $MRAP(x_0,\overline{x})$. More precisely, write $C(x) = 0$ for the algebraic equation obtained from the Euler equation; a sufficient condition for the MRAPs to be optimal for finite horizon problems, is the following :

$$(\overline{x} - x)C(x) \geq 0, \tag{5}$$

which has to be fulfilled for any x. Indeed Hartl and Feichtinger have shown that this condition does not suffice for the optimality of MRAPs in the infinite horizon [15]. They proposed an additional condition which, in addition to (5), gives a system of sufficient conditions of optimality. The optimality of MRAPs is also called a turnpike property and moreover $\overline{x}$ is called a turnpike. We would like to emphasize that this terminology agrees with the concept introduced in the standard case: the optimal solutions remain as long as possible near a particular solution.

In this paper our objective is twofold:

1. To establish necessary and sufficient optimality conditions of MRAPs,
2. To obtain optimality results in the presence of multiple solutions of the algebraic equation.

Our approach is based on a characterization of the value function in terms of generalized solutions (cf. [11]) of a Hamilton-Jacobi equation (cf. [17]).

Roughly speaking, we first prove that the value function of the problem is the unique viscosity solution of a particular Hamilton-Jacobi equation, among a relevant class of functions. Then we consider the cost function along the MRAPs as a function of the initial condition, and prove that it belongs to this relevant class of curves. Then, we give necessary and sufficient conditions for this cost function to be a generalized solution of the particular Hamilton-Jacobi equation. Therefore we deduce the optimality of the MRAPs. We recall that uniqueness of generalized solutions of first order partial differential equations without boundary condition, as it is the case for infinite horizon problems, is hardly guaranteed for unbounded functions without additional assumptions. This is the reason why we consider a transformation of the value function, which allows us to consider bounded functions. We recall also that no uniqueness result is known in the literature for the class of C^1 functions. This means that even when the value function turns out to be C^1, we still have to consider a larger class of functions. For this kind of problem, Lipschitz continuous and bounded functions appear to be a well suited class, for which the viscosity solutions machinery provides uniqueness results. Furthermore, when more than one solution $\overline{x}$ provides optimal MRAPs, the value function is no longer differentiable, thus the required use of generalized solutions of the Hamilton-Jacobi equation, such as viscosity solutions.

Before closing this introduction we again emphasize that the main difficulty we encounter comes from the infinite horizon framework. Despite some recent papers that concern general calculus of variations problems in infinite horizon (with or without the condition (2)), many questions remain open. In the preface of [7] or in [12], a list of interesting problems with infinite horizon can be found. Indeed if the calculus of variations is well established in finite horizon, the situation is not the same when one deals with infinite horizon. When the admissible paths are piecewise continuously differentiable (PC^1), it is proved in [5, 6] that standard finite horizon optimal conditions, such as the Euler one, can be used in the infinite horizon framework. Similar results are given in [4] when one deals with bounded differentiable functions with bounded derivatives. In these papers, it is also made precise that admissibility of a curve means that the improper integral (1) converges. On the other hand we have presented our problem in a calculus of variations framework. Obviously similar difficulties appear when one deals with optimal control theory, the first order Euler optimality conditions corresponding to the first order Pontryagin's optimality conditions. We finally recall that the Pontryagin's Maximum Principle given in a finite horizon framework has been extended by Halkin to the infinite horizon case [14]. (For the weakly overtaking optimality case, see [7].)

The paper is organized as follows. In Section 2 we recall the standard use of the Green theorem to derive sufficient condition of optimality for MRAPs, when the Euler algebraic equation possesses a unique solution. Our main new results are presented in Sections 3 and 4. Their proofs can be consulted in the quoted papers [18, 19]. The aim of the present paper is first to present these new results together with the 'well-known' sufficient condition obtained by the Green theorem method. Next it is to propose a rapid survey on the turnpike property, which allows us to distinguish the "asymptotic turnpike property" for the standard case and the "exact turnpike property" for the singular one. In the paper, we write functions $l(.)$ that fulfill the property (2) in the following way:

$$l(u,v) = A(u) + B(u)v \tag{6}$$

where $A, B : \Omega \subset \mathbb{R} \to \mathbb{R}$, and the maximization problem is :

$$\max_{x(.)\in Adm(x_0)} \quad J[x(.)], \tag{7}$$

where $Adm(x_0)$ stands for the admissible paths that will be made precise later. The problem is considered in two different cases :

1. The unbounded case : $\Omega = \mathbb{R}$ and the velocities satisfy $\dot{x} \in [-1, +1]$.
2. The bounded case : $\Omega = (a, b)$ and the velocities satisfy the differential inclusion $\dot{x} \in [f^-(x), f^+(x)]$, with some conditions on $f^-(.), f^+(.)$.

In Section 5, two examples of each case are presented. Different possible occurrences of turnpikes (one or several) are given. Our approach provides in particular a criterion for the choice of turnpikes that are in competition. We also obtain situations in which our approach gives the optimality of single MRAPs, whereas the Hartl and Feichtinger sufficient conditions result does not.

2 The Euler Equation and the Use of the Green Theorem

Under the linearity assumption (2) and notation (6), the Euler equation (3) degenerates into the following algebraic equation :

$$C(x) := A'(x) + \delta B(x) = 0. \tag{8}$$

We denote by E the set of solutions of (8):

$$E := \{x \in \mathbb{R} \text{ s.t. } C(x) = 0\}.$$

We recall the classical method, introduced by Miele [16], that works in the unbounded case when E is reduced to a singleton $\{\overline{x}\}$. We assume here that the functional $J[.]$ converges for any admissible path.

Take any initial condition x_0 and consider an optimal path $x^*(.)$ from x_0. We assume that $x^*(.)$ is bounded (see [15]). Let T be a finite positive number and let x_T be $x^*(T)$. Then the restriction of the path $x^*(.)$ to the time interval $[0, T]$ is necessarily optimal for the problem with the finite horizon criterion

$$J_T[x(.)] = \int_0^T e^{-\delta t} \left[A(x(t)) + B(x(t))\dot{x}(t)\right] dt$$

under the two-points end condition

$$x(0) = x_0 \quad \text{and} \quad x(T) = x_T. \tag{9}$$

Let $\widetilde{x}(.)$ be the (unique) most rapid approach path to $\bar{x}$ on $[0, T]$ that fulfills the constraint (9) (where we assume that T is large enough for the existence of such a path). Consider any admissible path $x(.)$ defined on $[0, T]$ satisfying the constraints (9). Let $\bigcup_i [t_i, t_{i+1})$ be a partition of $[0, T)$ such that, for any i, $x(t_{i+1}) = \bar{x}$ or $t_{i+1} = T$, and

$$\{\, x(t) \geq \bar{x},\ \forall t \in [t_i, t_{i+1})\,\} \text{ or } \{\, x(t) \leq \bar{x},\ \forall t \in [t_i, t_{i+1})\,\}.$$

Then, on each interval $[t_i, t_{i+1}]$, one has $x(t_i) = \widetilde{x}(t_i)$, $x(t_{i+1}) = \widetilde{x}(t_{i+1})$ and

$$\{\, x(t) \leq \widetilde{x}(t),\ \forall t \in [t_i, t_{i+1}]\,\} \quad \text{or} \quad \{\, x(t) \geq \widetilde{x}(t),\ \forall t \in [t_i, t_{i+1}]\,\}.$$

In the plane (t, x), consider the points $P_i = (t_i, x(t_i))$ and the closed curve $P_i P_{i+1}$ made by the two curves $\{(t, x(t) \,|\, t \in [t_i, t_{i+1}]\}$, $\{(t, \widetilde{x}(t) \,|\, t \in [t_i, t_{i+1}]\}$. Then, one has

$$\int_{t_i}^{t_{i+1}} e^{-\delta t} \left[A(\widetilde{x}(t)) + B(\widetilde{x}(t))\dot{\widetilde{x}}(t)\right] dt - \int_{t_i}^{t_{i+1}} e^{-\delta t} \left[A(x(t)) + B(x(t))\dot{x}(t)\right] dt$$
$$= \sigma \oint_{P_i P_{i+1}} e^{-\delta t} A(x) dt + e^{--\delta t} B(x) dx,$$

where $\sigma = 1$ when $x(.) \leq \widetilde{x}(.)$ and $\sigma = -1$ otherwise. The closed curve $P_i P_{i+1}$ delimits a domain $\mathcal{D}_i$ in the plane (t, x), for which the Green theorem allows us to write

$$\oint_{PQ} e^{-\delta t} A(x) dt + e^{-\delta t} B(x) dx = \int\int_{\mathcal{D}_i} e^{-\delta t} C(x) dt dx.$$

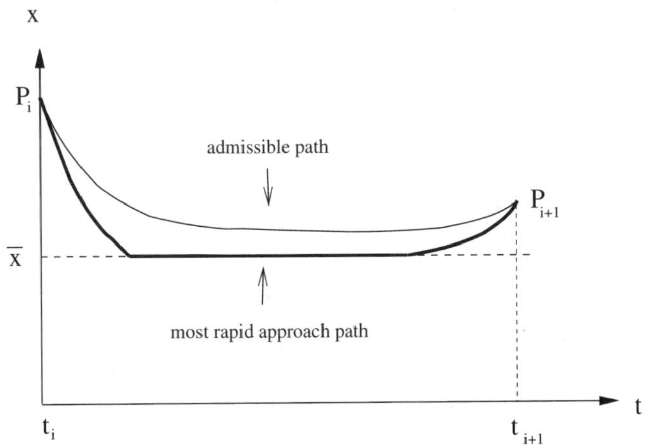

Fig. 1. Time interval $[t_i, t_{i+1}]$ for which P_i and P_{i+1} are above $\bar{x}$.

But the domain $\mathcal{D}_i$ is either entirely above the line $x = \bar{x}$, either entirely below (see Figure 1). Then, under condition (5), we have

$$\sigma \int\int_{\mathcal{D}_i} e^{-\delta t} C(x) dt dx \geq 0$$

in any case and we conclude that $J_T[\widetilde{x}(.)] \geq J_T[x(.)]$ for any $T > 0$, and any admissible curve $x(.)$. So, the condition (5) is sufficient for the optimality of the most rapid approach paths.

3 Main Results in the Unbounded Case

3.1 Statement of the Problem, Assumptions and Definitions

Let us consider the following set of admissible curves:

$$Adm_{x_0} = \{x(.) : [0, \infty) \to \mathbb{R},\ AC,\ x(0) = x_0,\ \dot{x}(t) \in [-1, +1] \ \ a.e.\}$$

whose elements are called the admissible paths. We assume that

($\mathbf{H_1}$) $A(.)$ is twice differentiable, $B(.)$ is differentiable, on $\Omega = \mathbb{R}$.
($\mathbf{H_2}$) There exist real numbers $k > 0$ and $\gamma < \delta$ such that for any x

$$\max\left(|A(x)|, |A'(x) + \delta B(x)|, |A''(x) + \delta B'(x)| \right) \leq k e^{\gamma |x|}$$

These growth assumptions are more general than the usual ones when one deals with necessary conditions (Euler condition, Maximum Principle). When the set E is nonempty, we consider stationary paths $\overline{x}(.)$ defined by

$$\overline{x}(t) = \overline{x}, \qquad \forall t \geq 0,$$

where $\overline{x} \in E$, which are candidate optimal solutions to the problem (7).

Given $\overline{x} \in E$ and any $x_0 \in \mathbb{R}$, we introduce the most rapid approach path from x_0 to $\overline{x}$, denoted by MRAP$(x_0, \bar{x})$ and defined as follows:

$$\text{if } x_0 \geq \bar{x}, \quad MRAP(x_0, \bar{x})(t) = \begin{cases} x_0 - t \text{ if } t \leq x_0 - \bar{x} \\ \bar{x} \qquad \text{if } t > x_0 - \bar{x} \end{cases}$$

$$\text{if } x_0 \leq \bar{x}, \quad MRAP(x_0, \bar{x})(t) = \begin{cases} x_0 + t \text{ if } t \leq \bar{x} - x_0 \\ \bar{x} \qquad \text{if } t > \bar{x} - x_0 \end{cases}$$

Such paths are clearly admissible (as we can easily establish the convergence of the improper integral $J[.]$).

Our goal is to obtain necessary and sufficient conditions for the optimality of $MRAP(x_0, \overline{x})$ with the help of a value function approach. Therefore, we begin by expressing the value of the objective $J[.]$ along the path $MRAP(x_0, \overline{x})(.)$:

$$J[MRAP(x_0, \overline{x})(.)] = \frac{A(x_0)}{\delta} + \frac{S(x_0, \bar{x})}{\delta} \tag{10}$$

where

$$S(x_0, \bar{x}) = \int_{x_0}^{\bar{x}} (A'(y) + \delta B(y)) e^{-\delta |x_0 - y|} dy = \int_{x_0}^{\bar{x}} e^{-\delta |x_0 - y|} C(y) dy.$$

We emphasize that the function $S(.,.)$, that will play a fundamental role in the sequel, depends only on the function $C(.)$, provided by the Euler equation.

Now, to make the turnpike property more precise, we introduce the following definition: The *optimality basin* of $\bar{x} \in E$ is the set

$$\mathcal{B}(\bar{x}) := \{x_0 \text{ s.t. } MRAP(x_0, \bar{x}) \text{ is optimal}\};$$

$\bar{x}$ is then called a turnpike exactly when $\mathcal{B}(\bar{x})$ is not empty.

Finally, we introduce the value function $V(.)$:

$$V(x_0) = \sup_{x(.) \in Adm(x_0)} J[x(.)]. \tag{11}$$

Proofs of Propositions 1, 2 and Corollary 1 given below can be found in [18].

3.2 A Particular Hamilton-Jacobi Equation

As we recalled in the introduction, the uniqueness of (generalized) solutions of first order partial differential equations defined on unbounded sets is known to be a delicate question. It can be obtained only for well chosen classes of functions (see for instance [2]). For this reason, we do not characterize the value function $V(.)$ itself as a solution of the Hamilton-Jacobi equation that we could associate to it, but a transformation of it, denoted by $Z(.)$ in the sequel. The function $Z(.)$ is a solution of another Hamilton-Jacobi equation, for which we are able to state a unique characterization result in the class of BUC (bounded and uniformly continuous) functions.

Proposition 1. *Under Assumptions* $(\mathbf{H_1})$ *and* $(\mathbf{H_2})$ *the function* $Z(.)$ *:*

$$Z(x) = e^{-\eta\sqrt{x^2+1}}\left(V(x) - \frac{A(x)}{\delta}\right),$$

where η *satisfies* $\gamma < \eta < \delta$, *is the unique bounded and uniformly continuous viscosity solution of the following Hamilton-Jacobi equation :*

$$\delta Z(x) - \left| Z'(x) + \eta\frac{x}{\sqrt{x^2+1}}Z(x) + \frac{e^{-\eta\sqrt{x^2+1}}}{\delta}(A'(x) + \delta B(x))\right| = 0, \quad x \in \mathbb{R} \tag{12}$$

3.3 Turnpike Optimality

Given the set E of solutions of the algebraic equation (8), assumed to be nonempty and finite, we define the following function $\overline{S}$:

$$\overline{S}(x) := \max_{\bar{x}\in E} S(x, \bar{x}).$$

The characterization (i.e., the necessary and sufficient condition) of the optimality of the MRAPs is given with the help of this function as is made precise in the next proposition.

Proposition 2. *Let us assume* $(\mathbf{H_1})$,$(\mathbf{H_2})$, *and suppose that* E *is nonempty and finite. Then the following statements are equivalent:*

i) For any x_0, *there exists a turnpike* $\bar{x} \in E$ *(i.e., there exists* $\bar{x}$ *such that* $x_0 \in \mathcal{B}(\bar{x})$*)*
ii) $\overline{S}(x) \geq 0, \quad \forall x \in \mathbb{R}$.

Moreover the value function of the problem is given by $V(x) = (A(x)+\overline{S}(x))/\delta$.

Roughly speaking, the proof consists of showing that the transformation of the function $\overline{S}(.)$ is a viscosity solution of the Hamilton-Jacobi equation (12), if and only if it is nonnegative. Therefore as this transformation belongs to the framework for which a uniqueness result is available, the proof follows.

We can now easily deduce the sufficient condition (5) described in Section 2 from this Proposition (see point 1. of the corollary below) when E is reduced to a singleton. Furthermore, we can also prove that the condition (5) is also a necessary condition but only locally (see point 2. of the corollary).

Corollary 1. *Assume* $(\mathbf{H_1})$-$(\mathbf{H_2})$,

1. *If the function* $C(.)$ *has exactly one zero* $\bar{x}$ *on* $\mathbb{R}$ *and fulfills the property (5) at any* $x \in \mathbb{R}$, *then the most rapid approach path to* $\bar{x}$ *is optimal from any initial condition* $x_0 \in \mathbb{R}$.
2. *For any* $\bar{x} \in E$ *such that* $\mathcal{B}(\bar{x})$ *is nonempty, there exists a neighborhood* $\mathcal{V}$ *of* $\bar{x}$ *such that the property (5) is fulfilled on* $\mathcal{V}$.

4 Main Results in the Bounded Case

4.1 Statement of the Problem and Assumptions

Let us first introduce the set of the admissible paths:

$$\begin{aligned} Adm(x_0) &= \\ &\{x(.) : [0, \infty) \to (a,b) A.C.,\ x(0) = x_0,\ \dot{x}(t) \in [f^-(x(t)), f^+(x(t))] \ a.e.\ \} \end{aligned}$$

where $a < b$ are fixed and $f^-(.), f^+(.)$ are real functions defined on $[a, b]$.

In the sequel we will use the following assumptions:

$(\mathbf{K_1})$ $A(.)$ is continuously differentiable and bounded, $B(.)$ is continuous on $\Omega = (a, b)$.
$(\mathbf{K_2})$ $\forall x \in [a, b], \quad f^-(x) \leq f^+(x)$.
$(\mathbf{K_3})$ $f^-(.), f^+(.)$ are Lipschitz on $[a, b]$.
$(\mathbf{K_4})$ $f^-(a) \geq 0, \quad f^+(b) \leq 0$.
$(\mathbf{K_5})$ $C(.)f^-(.)$ and $C(.)f^+(.)$ are bounded and Lipschitz on (a, b).

We remark that under the last assumption, $C(x(t))\dot{x}(t)$ remains bounded for all $t \geq 0$. Therefore we deduce the convergence of the improper integrals in (1). Note also that we do not require $B(.)$ to be bounded. We begin by giving a property of the admissible curves. We can establish that under $(\mathbf{K_2})$-$(\mathbf{K_4})$ the interval (a, b) is invariant under the solutions of the differential inclusion $\dot{x}(t) \in [f^-(x(t)), f^+(x(t))]$. Moreover these solutions are defined on $[0, +\infty)$.

Proofs of Propositions 3, 4 and Corollary 2 given below can be found in [19].

4.2 The Hamilton-Jacobi Equation

As in the preceding section we denote by V the value function associated to the problem (7). Also we do not characterize the value function V itself, but a transformation of it which is solution of a particular Hamilton-Jacobi equation, for which we are able to state our results.

Proposition 3. *We assume* $(\mathbf{K_1})$-$(\mathbf{K_5})$. *The function* $Z(.) = \delta V(.) - A(.)$ *is the unique Hölder continuous viscosity solution on* (a, b) *of the following Hamilton-Jacobi equation*

$$\delta Z(x) - \max[(C(x) + Z'(x))f^-(x), (C(x) + Z'(x))f^+(x)] = 0 \qquad (13)$$

The result on the uniqueness of the solution of the Hamilton-Jacobi equation for the bounded case is not the same as in the preceding unbounded case. We now have to consider a class of functions that are continuous on the closure of (a,b). (The existence of a continuous extension of the value function on $[a,b]$ is guaranteed by its Hölder continuity on (a,b).)

4.3 Turnpike Optimality

For any admissible trajectory $x(.) \in Adm(x_0)$ with $x_0 \in (a,b)$ and any $\bar{x}$ in E, we define the time to reach $\bar{x}$, as follows :

$$\tau_{\bar{x}}[x(.)] := \inf\{t \geq 0 \mid x(\tau) = \bar{x},\ \forall \tau \geq t\} \text{ or } +\infty.$$

Then, for any initial condition $x_0 \in (a,b)$ and any $\bar{x} \in E$, we say that an admissible trajectory $\widetilde{x}(.)$ is a *most rapid approach path* from x_0 to $\bar{x}$, $MRAP(x_0, \bar{x})$ in short, provided

$$\tau_{\bar{x}}[\widetilde{x}(.)] = \min_{x(.) \in Adm(x_0)} \tau_{\bar{x}}[x(.)] < +\infty.$$

Finally, for any $\bar{x} \in E$, we define its optimality basin (which can be empty) :

$$\mathcal{B}(\bar{x}) := \{x_0 \in (a,b) \mid \text{there exists } x^*(.) \text{ optimal from } x_0 \text{ and } MRAP(x_0, \bar{x})\}.$$

As in the preceding case, $\overline{x}$ is called a turnpike when $\mathcal{B}(\bar{x})$ is not empty.

For ensuring the existence of $MRAP(x_0, \bar{x})$ for any $(x_0, \bar{x}) \in (a,b) \times E$, we consider the following assumption :

($\mathbf{K_6}$) $f^-(x) < 0$ and $f^+(x) > 0$, $\forall x \in (a,b)$,

and introduce the following function $S : (a,b)^2 \mapsto \mathbb{R}$:

$$S(x,\bar{x}) := \begin{cases} \displaystyle\int_x^{\bar{x}} C(y) e^{-\delta \int_x^y \frac{dz}{f^+(z)}} dy & \text{if } x < \bar{x} \\ 0 & \text{if } x = \bar{x} \\ \displaystyle\int_x^{\bar{x}} C(y) e^{-\delta \int_x^y \frac{dz}{f^-(z)}} dy & \text{if } x > \bar{x} \end{cases} \tag{14}$$

Finally, we consider the following hypothesis on the set E of candidate turnpikes :

($\mathbf{K_7}$) E is non-empty and finite.

Under hypotheses $(\mathbf{K_6})$ and $(\mathbf{K_7})$, we introduce the function

$$\overline{S}(x) := \max_{\bar{x}\in E} S(x,\bar{x}), \quad x \in (a,b),$$

which plays the role of a candidate solution of the Hamilton-Jacobi equation (13). Indeed, under hypotheses $(\mathbf{K_1})$-$(\mathbf{K_4})$ and $(\mathbf{K_6})$-$(\mathbf{K_7})$, one obtains that the function $\overline{S}(.)$ is a continuous viscosity solution of (13) if and only if it is non negative.

Proposition 4. *Under hypotheses $(\mathbf{K_1})$-$(\mathbf{K_7})$, the following statements are equivalent:*

(i) For any $x_0 \in (a,b)$, there exists a turnpike $\bar{x} \in E$ (i.e., $\bar{x}$ such that $x_0 \in \mathcal{B}(\bar{x})$).
(ii) $\overline{S}(x) \geq 0, \quad \forall x \in (a,b)$.

Moreover the value function of the problem is given by $V(x) = (A(x)+\overline{S}(x))/\delta$.

As in the preceding section we can now easily deduce the well known result in the turnpike literature, when the function C has only one change of sign. Furthermore, we can also prove that the usual sufficient condition, which is required to be fulfilled globally, is also a necessary condition but only locally.

Corollary 2. *Assume (K_1)-(K_7). Then:*

1. *If the function $C(.)$ has exactly one zero $\bar{x}$ on (a,b) and fulfills the property (5) at any $x \in (a,b)$, then the most rapid approach path to $\bar{x}$ is optimal from any initial condition $x_0 \in (a,b)$.*
2. *For any $\bar{x} \in E$ such that $\mathcal{B}(\bar{x})$ is nonempty, there exists a neighborhood $\mathcal{V}$ of $\bar{x}$ such that the property (5) is fulfilled on $\mathcal{V}$.*

5 Examples

5.1 Example 1

Let $0 < a < b$ and consider the problem

$$\max_{x(.)} \int_0^{+\infty} e^{-t} x^2(t) \left[2\dot{x}(t)(a+b-x(t)) - ab\right] dt \;\text{ with }\; \dot{x}(t) \in [-1,+1] \text{ a.e.}$$

Then the Euler equation gives

$$C(x) = -2x(ab + (a+b)x - x^2) = -2x(a-x)(b-x),$$

which admits three solutions : $\bar{x} \in \{0, a, b\}$ (see Figure 2). We observe that $\bar{x} = 0$ and $\bar{x} = b$ satisfy (only locally) the classical condition (5). From Proposition 2, we prove now that depending on the values of a and b,

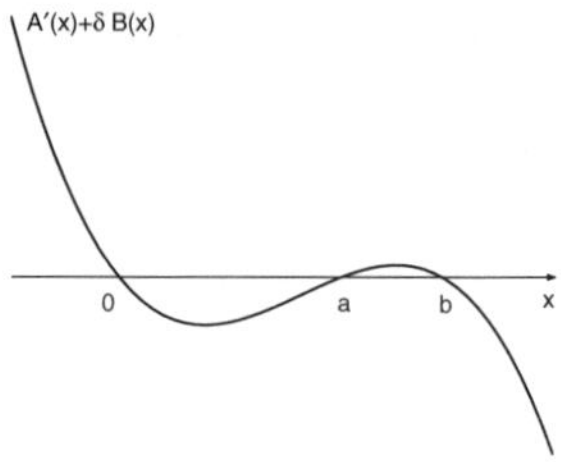

Fig. 2. The function $x \mapsto C(x)$.

- $MRAP(x_0, 0)$ or $MRAP(x_0, b)$ is optimal for any initial condition x_0,
- there exists $x^* \in (0, b)$ such that $MRAP(x_0, 0)$ (resp. $MRAP(x_0, b)$) is optimal for $x_0 \leq x^*$ (resp. $x_0 \geq x^*$).

It is easy to check the following properties :

$$\begin{cases} x \in [0, b] \Rightarrow S(x, a) \leq 0 \\ x \leq a \Rightarrow \{ \; S(x, 0) \geq 0 \; \text{ and } \; S(x, b) \geq S(0, b) \; \} \\ x \geq b \Rightarrow S(x, 0) \geq S(b, 0) \\ x \geq a \Rightarrow S(x, b) \geq 0 \end{cases}$$

Therefore we can conclude by Proposition 2 and Corollary 1 that :

1 $MRAP(x_0, a)$ is never optimal for any initial condition x_0.
2 $MRAP(x_0, 0)$ is optimal for all x_0 as soon as $S(b, 0) \geq 0$.
3 $MRAP(x_0, b)$ is optimal for all x_0 as soon as $S(0, b) \geq 0$.
4 If $x \rightarrow \max\{S(x, 0), S(x, b)\}$ is nonnegative for all $x \in [0, b]$, then $MRAP(x_0, 0)$ or $MRAP(x_0, b)$ is optimal.

For $x \in [0, b]$, we compute the functions

$$S(x, 0) = e^{-x} \int_0^x 2y(a - y)(b - y)e^y dy$$

$$S(x, b) = -e^x \int_x^b 2y(a - y)(b - y)e^{-y} dy$$

for different values of a and b, and obtain the three following possible cases :

1. For $a = 2$ and $b = 3$, we have $S(b, 0) = 2(22e^{-3} - 1) > 0$ and we can conclude that the paths that are going with a most rapid velocity to $\bar{x} = 0$ are optimal.
2. For $a = 1$ and $b = 4$, we obtain $S(0, b) = 64e^{-4} > 0$ and we can conclude that the paths that are going with a most rapid velocity to $\bar{x} = b$ are optimal.

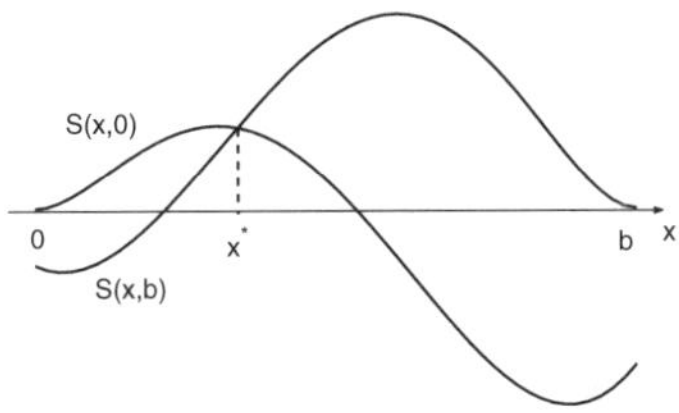

Fig. 3. Functions $x \mapsto S(x,0)$ and $x \mapsto S(x,b)$ for $a=2$ and $b=5$.

3. For $a=2$ and $b=5$, we obtain $S(b,0)=2(30e^{-5}-5)<0$ and $S(0,b)=2(37e^{-5}-2)<0$ but we observe that $\max\{S(x,0),S(x,b)\}$ is nonnegative on $[0,b]$. Let $x^* \in (0,b)$ be such that $S(x^*,0)=S(x^*,b)$ (see Figure 3). Then there is a competition between the two turnpikes $\bar{x}=0$ and $\bar{x}=b$: for $x_0 \leq x^*$ (resp. $x_0 \geq x^*$), it is optimal to go as quickly as possible to $\bar{x}=0$ (resp. $\bar{x}=b$). Let us emphasize that in this last case there is no longer uniqueness of the turnpike for the initial condition $x_0=x^*$. It can be also shown that it corresponds to a non-differentiability point of the value function.

5.2 Example 2

We revisit the well known model of optimal harvesting introduced by Clark [10]. Let $x(t)$ be the biomass stock, whose time evolution obeys the differential equation

$$\dot{x}(t)=f(x(t))-qx(t)e(t), \quad x(0)=x_0,$$

where $f(.)$ is the growth function, q the capturability parameter and $e(.)$ the harvesting effort. The usual bio-economic criterion for choosing an optimal harvesting policy $e(.) \in [0,E_{\max}]$ is to maximize the benefit

$$J[x_0,e(.)]=\int_0^{+\infty} e^{-\delta t}\left[pqx(t)-c\right]e(t)dt,$$

where $\delta>0$ is the discount factor, p is the unit price and c is the unitary effort cost. When f is the logistic law $f(x)=rx(1-x/K)$ and the price p is constant, there exists a unique solution of the function $C(.)$ and the condition (5) is fulfilled everywhere (see [10]). Now, we consider an extension of this model with the following properties :

1. There exists a depensation in the growth function. For instance, $f(.)$ has the form :

$$f(x)=rx^{\gamma}\left(1-\frac{x}{K}\right), \quad (\gamma>1)$$

2. The unit price is not constant, and is decreasing with the available stock. For instance, $p(.)$ could be the following expression :

$$p(x) = \frac{\bar{p}}{1+\alpha x^{\beta}}, \quad (\alpha > 0,\ \beta > 1)$$

It is easy to check that this problem can be expressed in the form (1), with $A(x) = (p(x)qx - c)f(x)/(qx)$ and $B(x) = (p(x)qx - c)/(qx)$ defined on $\Omega = (0, K)$. The extremal velocities are : $f^{-}(x) = f(x) - qE_{\max}x$ and $f^{+}(x) = f(x)$. Here $B(.)$ is not bounded on Ω but it is straightforward to check that assumptions $(\mathbf{K_1})$-$(\mathbf{K_7})$ are fulfilled. The function $C(.)$ can have more than one zero, depending on the values of the parameters. The computation of the zeros of the functions $C(.)$ and the numerical comparison of the function $S(.,\bar{x})$ have been achieved for the values given in Table 1, while δ has been taken varying between 0.3 and 1.5. Applying the results of Section 4, the conclusions are summarized in Table 2.

Table 1. Values of the parameters

r	K	γ	$\bar{p}$	α	β	c/q
5	1	4	5	0.2	2	1

Table 2. Conclusions

δ	$\overline{X}$	Conclusion
0.3	$\{\bar{x} \simeq 0.779\}$	$MRAP(x,\bar{x})$ is optimal from any x
0.6	$\{\bar{x}_1 \simeq 0.221, \bar{x}_2 \simeq 0.749\}$	$MRAP(x,\bar{x}_2)$ is optimal from any x
0.9	$\{\bar{x}_1 \simeq 0.211, \bar{x}_2 \simeq 0.710\}$	$MRAP(x,\bar{x}_1)$ is optimal from any $x \leq x^*$
		$MRAP(x,\bar{x}_2)$ is optimal from any $x \geq x^*$
1.2	$\{\bar{x}_1 \simeq 0.208, \bar{x}_2 \simeq 0.630\}$	$MRAP(x,\bar{x}_1)$ is optimal from any x
1.5	$\{\bar{x} \simeq 0.207\}$	$MRAP(x,\bar{x})$ is optimal from any x

So, we notice the existence of a bifurcation of the optimal steady-state stock level, depending on the discount factor δ.

6 Conclusion

For singular scalar problems of calculus of variations in infinite horizon, we have proposed a new necessary and sufficient condition for the optimality of the MRAPs. This condition generalizes the standard one which is only sufficient and valid only in the case of a unique solution of the singular Euler equation. Our result is established with the characterization of a transformation of the value function, in terms of viscosity solutions of a particular Hamilton-Jacobi equation, which provides global optimality conditions. Our condition also applies when one is dealing with singular Euler equations that possess more than one solution. The lack of differentiability of the value function is connected to the existence of competition between most rapid approaches towards several singular arcs.

References

1. Bardi M, Capuzzo-Dolcetta I (1997) Optimal Control and Viscosity Solutions of Hamilton-Jacobi-Bellman Equations. Birkhaüser Boston
2. Barles G (1994) Solutions de viscosité des équations de Hamilton-Jacobi. Springer-Verlag New York
3. Bell D, Jacobson D (1975) Singular Optimal Control Problems. Academic Press London
4. Blot J, Cartigny P (1995) Bounded solutions and oscillations of concave Lagrangian systems in presence of a discount rate, J Analysis Appl 14:731–750
5. Blot J, Cartigny P (2000) Optimality in infinite horizon variational problems under signs conditions, J Optimization Theory Appl 106:411–419
6. Blot J, Michel P (1996) First-order necessary conditions for the infinite-horizon variational problems, J Optimization Theory Appl 88:339–364
7. Carlson D, Haurie A, Leizarowitz A (1991) Infinite Horizon Optimal Control Second edition. Springer Verlag Berlin
8. Cartigny P (1994) Stabilité des solutions stationnaires et oscillations dans un problème variationnel approche Lagrangienne, Compte-Rendus de l'Académie des Sciences I 318:869–872
9. Cass D, Shell K (1976) The Hamiltonian Approach to Dynamic Economics. Academic Press New York
10. Clark C (1976) Mathematical Bioeconomics The Optimal Management of Renewable Resources. Wiley New York
11. Crandall M, Lions P-L (1983) Viscosity solutions of Hamilton-Jacobi equations, Transactions of the American Mathematical Society 277:1–42
12. Ekeland I (1986) Some variational problems arising from mathematical economics. In: Lectures Notes in Mathematics 1330. Springer Verlag Berlin
13. Feinstein C, Oren S (1985) A funnel turnpike theorem for optimal growth problems with discounting, J Economic Dynamics Control 9:25–39
14. Halkin H (1974) Necessary conditions for optimal control problems with infinite horizon, Econometrica 42:267–273
15. Hartl R, Feichtinger G (1987) A new sufficient condition for most rapid approach paths, Journal of Optimization Theory and Applications 54:403–411
16. Miele A (1962) Extremization of linear integrals by Green's Theorem. In: Leitmann G (ed) Optimization Techniques. Academic Press New York 69–98
17. Rapaport A, Cartigny P (2002) Théorème de l'autoroute et l'equation d'Hamilton-Jacobi, CR Mathématique Académie des Sciences Paris 335:1091–1094
18. Rapaport A, Cartigny P (2003) Turnpike theorems by a value function approach, ESAIM Control Optimisation and Caclulus of Variations to appear

19. Rapaport A, Cartigny P (2003) Competition between most rapid approach paths necessary and sufficient conditions, Journal of Optimization Theory and Applications submitted
20. Rockafellar R (1973) Saddle points of hamiltonian systems in convex problem of Lagrange, J Optim Theory Appl 12:367–389
21. Rockafellar R (1976) Saddle points of Hamiltonian systems in convex Lagrange problems having nonzero discount rate, Journal of Economic Theory 12:71–113
22. Samuelson P (1965) A Catenary turnpike theorem involving consumption and the golden rule, American Economic Review 55:486–496
23. Zaslavski A (1997) Turnpike property of optimal solutions of infinite-horizon variational problems, SIAM J Control Optim 35:1169–1203
24. Zaslavski A (1998) Turnpike theorem for convex infinite dimensional discrete-time control system, Journal of Convex Analysis 5:237–248

Min–Plus Eigenvector Methods for Nonlinear H_∞ Problems with Active Control*

Gaemus Collins[1] and William McEneaney[2]

[1] University of California, San Diego, Department of Mathematics, 0112, 9500 Gilman Dr., La Jolla, CA. gcollins@ucsd.edu
[2] University of California, San Diego, Depts. of Mathematics and of Mechanical and Aerospace Engineering. Mail: Dept. of Mathematics, UCSD, 9500 Gilman Dr., La Jolla, CA. wmceneaney@ucsd.edu

1 Introduction

We consider the H_∞ problem for a nonlinear system. The corresponding dynamic programming equation (DPE) is a fully nonlinear, first-order, steady-state partial differential equation (PDE), possessing a term which is quadratic in the gradient (for background, see [2], [4], [13], [25] among many notable others). The solutions are typically nonsmooth, and further, there are multiple viscosity solutions – that is, one does not even have uniqueness among the class of viscosity solutions (cf. [17]). Note that the nonuniqueness is not only due to an additive constant, but is much more difficult. If one removes the additive constant issue by requiring the solution to be zero at the origin, then, for example, the linear-quadratic case typically has two classical solutions and an infinite number of viscosity solutions [17]. Since the H_∞ problem is a differential game, the PDE is, in general, a Hamilton-Jacobi-Isaacs (HJI) PDE. The computation of the solution of a nonlinear, steady-state, first-order PDE is typically quite difficult, and possibly even more so in the presence of the non-uniqueness mentioned above. Some previous works in the general area of numerical methods for these problems are [3], [6], [7], [14], and the references therein.

In recent years, a new set of methods based on max-plus linearity have been developed for the H_∞ problem in the design case. In that problem class, a feedback control and disturbance attenuation parameter are chosen, and the associated PDE is solved to determine whether this is indeed an H_∞ controller with that disturbance attenuation level [18], [19], [21], [23]. Note that in that problem class, the PDE is a Hamilton-Jacobi-Bellman (HJB) PDE. Here, we address a class of active control H_∞ problems where the controller is able

*Sponsors: NSF Grant DMS-0307229 (McEneaney), DARPA Contract F30602-01-C-0201 (Collins and McEneaney)

M. de Queiroz et al. (Eds.): Optimal Control, Stabilization and Nonsmooth Analysis, LNCIS 301, pp. 101–120, 2004.

to dominate the disturbance in a certain way. In this class, the HJI PDE will have an equivalent representation as an HJB PDE. A similar method will be applied, but the required mathematical proofs are necessarily quite different. In particular, in the former class, one made heavy use of the nominal stability of the dynamics, while in this latter class, we use the equivalent control formulation.

2 Background

In [22]–[23], McEneaney et al develop a new method for solving steady-state Hamilton–Jacobi–Bellman (HJB) partial differential equations (PDE) using max-plus based methods. (This work was predated by work on the time-dependent case [11]. The first author to note the max-plus linearity that is now being exploited in these approaches was Maslov [15].)

In [21] the dynamics

$$\begin{aligned} \dot{X} &= f(X) + \sigma(X)w \\ X(0) &= x \end{aligned} \tag{1}$$

were considered, where X is the state taking values in $\mathbb{R}^n$, f represents the nominal dynamics, the disturbance w lies in $\mathcal{W} := \{w : [0,\infty) \to \mathbb{R}^m : w \in L_2[0,T]\ \forall T < \infty\}$, and σ is an $n \times m$ matrix–valued multiplier on the disturbance. In this introductory paper, the following assumptions were also made. First, one assumes that all the functions f, σ and l (given below) are smooth. One also assumes that there exist $K, c, M \in (0,\infty)$ such that

$$\begin{aligned} &(x-y)^T(f(x)-f(y)) \le -c|x-y|^2 \qquad \forall x,y \in \mathbb{R}^n \\ &f(0) = 0 \\ &|f_x(x)| \le K \qquad \forall x \in \mathbb{R}^n. \end{aligned}$$

Note that this automatically implies the closed–loop stability criterion of H_∞ control. Further one assumes that

$$\begin{aligned} &|\sigma(x)| \le M \qquad \forall x \in \mathbb{R}^n \\ &|\sigma^{-1}(x)| \le M \qquad \forall x \in \mathbb{R}^n \\ &|\sigma_x(x)| \le K \qquad \forall x \in \mathbb{R}^n. \end{aligned}$$

Here, of course, the use of σ^{-1} indicates the Moore-Penrose pseudo-inverse, and it is implicit in the bound on $\sigma^{-1}(x)$ that σ is uniformly nondegenerate. Also, $l(x)$ is the running cost (to which the L_2–norm disturbance penalty was added). It is also assumed that there exist $\beta, \alpha < \infty$ such that

$$\begin{aligned} &l_{xx}(x) \le \beta \qquad \forall x \in \mathbb{R}^n \\ &0 \le l(x) \le \alpha|x|^2 \qquad \forall x \in \mathbb{R}^n. \end{aligned}$$

Under these assumptions, it was shown [17] that the HJB equation

$$0 = -\left[\frac{1}{2\gamma^2}\nabla W^T(x)\sigma(x)\sigma^T(x)\nabla W(x) + f^T(x)\nabla W(x) + l(x)\right]$$

has a unique viscosity solution in the class of continuous viscosity solutions satisfying

$$0 \le W(x) \le c\frac{(\gamma-\delta)^2}{2M^2}|x|^2$$

for sufficiently small $\delta > 0$, and that this solution is the value function (available storage) given by

$$W(x) = \sup_{T<\infty}\sup_{w\in L_2^{loc}}\int_0^T l(X(t)) - \frac{\gamma^2}{2}|w(t)|^2\,dt, \tag{2}$$

provided that γ satisfies

$$\frac{\gamma^2}{2M^2} > \frac{\alpha}{c^2}.$$

Further, W is the unique solution in the class of continuous functions satisfying the same quadratic growth bound, of fixed point problem $W = S_\tau[W]$ ([23], [21]), where

$$\mathcal{S}_\tau[\phi](x) := \sup_{w\in L_2}\left\{\int_0^\tau \left[l(X(t)) - \frac{\gamma^2}{2}|w(t)|^2\right]dt + \phi(X(\tau))\right\}$$

and $X(\cdot)$ solves (1). Here S_τ is a linear operator on functions over the max–plus semifield, so $W = S_\tau[W]$ is equivalent to the linear eigenvalue problem $0 \otimes W = S_\tau[W]$.

Note that [22]–[23] focus mainly on the above problem where there is only a maximizing controller (the disturbance w). It is shown that this W is semiconvex, and that W can be represented in an expansion of semiconvex basis functions as

$$W(x) = \bigoplus_{i=1}^{\infty}\left[a_i \otimes \psi_i(x)\right].$$

Truncating this basis expansion, the problem $0 \otimes W = S_\tau[W]$ yields the matrix–vector eigenvalue problem $0 \otimes a = B \otimes a$ where B corresponds to a discretized version of the dual of S_τ. This can then be solved by the power method, and convergence is obtained as the number of basis functions increases. Interestingly, the max-plus power method converges to the truncation of the particular viscosity solution of interest, the available storage.

3 The Nonlinear H-Infinity Problem with Active Control

The analysis in [21] concentrated on the case where one is testing a control which is already included in the dynamics in order to determine the value function. Here we consider the case where one has not a priori chosen a particular control, but is instead attempting to compute the value function of the active control problem for some given value of disturbance attenuation parameter γ. In this case, the problem takes the form of a game. We specifically consider the case with dynamics

$$\dot{X} = A(X) + D(X)u + \sigma(X)w$$
$$X(0) = x$$

where $w(t) : \mathbb{R} \to \mathbb{R}^k$ is the disturbance process, taking values from $\mathcal{W} = \{w : [0,\infty) \to \mathbb{R}^k \,|\, w \in L_2[0,T]\ \forall T < \infty\}$, $u(t) : \mathbb{R} \to \mathbb{R}^j$ is the control coming from $\mathcal{U} := \{u : [0,\infty) \to \mathbb{R}^j \,|\, u \in L_2[0,T]\ \forall T < \infty\}$, $\sigma : \mathbb{R}^n \to \mathbb{R}^{n\times k}$ is a matrix–valued multiplier on the disturbance, $D : \mathbb{R}^n \to \mathbb{R}^{n\times j}$ is a matrix–valued multiplier on the control, and $A : \mathbb{R}^n \to \mathbb{R}^n$ dictates the nominal dynamics.

Consider a payoff and value given by

$$J_1(T,x,u,w) = \int_0^T l(X(t)) + \frac{\eta^2}{2}|u(t)|^2 - \frac{\gamma^2}{2}|w(t)|^2 dt,$$
$$W_1(x) = \inf_{\lambda\in\Lambda} \sup_{T<\infty} \sup_{w\in\mathcal{W}} J_1(T,x,\lambda[w],w),$$

where $l(x)$ is the running cost (e.g. $l(x) = |x|^2$), γ is the attenuation level, and Λ is the set of non-anticipative strategies for the controller. A non-anticipative strategy in this case is defined according to the Elliott–Kalton definition [8] as a mapping $\lambda : \mathcal{W} \to \mathcal{U}$ such that if $w(r) = \hat{w}(r)$ for almost every $r \in [0,t]$, then $\lambda[w](r) = \lambda[\hat{w}](r)$ for almost every $r \in [0,t]$, for all $t > 0$.

The corresponding Dynamic Programming Equation (DPE) is

$$\begin{aligned}
0 &= -A(x)\cdot \Delta W_1 - l(x) \\
&\quad - \inf_{u\in\mathbb{R}^j}\left\{(D(x)u)\cdot\nabla W_1(x) + \frac{\eta^2}{2}|u(t)|^2\right\} \\
&\quad - \sup_{w\in\mathbb{R}^k}\left\{(\sigma(x)w)\cdot\nabla W_1 - \frac{\gamma^2}{2}|w|^2\right\} \\
&= -\Bigg[A(x)\cdot\nabla W_1 + l(x) \\
&\quad - (\nabla W_1)^T(x)\left(\frac{D(x)D^T(x)}{2\eta^2} - \frac{\sigma(x)\sigma^T(x)}{2\gamma^2}\right)\nabla W_1(x)\Bigg]
\end{aligned}$$

with $W_1(0) = 0$. Assume the controller dominates the disturbance so that

$$\frac{QQ^T}{2} := \frac{DD^T}{2\eta^2} - \frac{\sigma\sigma^T}{2\gamma^2}$$

is uniformly bounded and nondegenerate in the sense that

$$\alpha_1|y|^2 \le y^T\frac{Q(x)Q^T(x)}{2}y \le \alpha_2|y|^2 \qquad \forall x,y\in\mathbb{R}^n, \tag{A1}$$

and assume the long–term average cost per unit time is zero (which follows from the assumptions below). Then the DPE, which can be rewritten in the form

$$0 = -\left[A(x)\cdot\nabla W_1 + l(x) - (\nabla W_1)^T(x)\left(\frac{QQ^T}{2}\right)\nabla W_1(x)\right] \tag{3}$$

$$= -\inf_{v\in\mathcal{V}}\left\{(A(x)+Q(x)v)\cdot\nabla W_1 + l(x) + \frac{1}{2}|v|^2\right\} \tag{4}$$

$$=: -H(x,\nabla W_1), \tag{5}$$

also corresponds to the pure control problem

$$\dot{X} = A(X) + Q(X)v \qquad X(0) = x \tag{6}$$

with payoff and value

$$J_2(T,x,v) = \int_0^T l(X(t)) + \frac{1}{2}|v(t)|^2 dt, \tag{7}$$

$$W_2(x) = \inf_{v\in\mathcal{V}}\sup_{T<\infty} J_2(T,x,v), \tag{8}$$

where $v(t) : \mathbb{R} \to \mathbb{R}^n$ is the control taking values from $\mathcal{V} = \{v : [0,\infty) \to \mathbb{R}^n \,|\, v \in L_2[0,T]\ \forall T < \infty\}$, and $Q : \mathbb{R}^n \to \mathbb{R}^{n\times m}$ is a matrix–valued multiplier on the control. We will work from this formulation from now on.

We assume that $A(0) = 0$, that the functions A, Q and l are $C^2(\mathbb{R}^n)$, and that there exist $c_1 \in (0,\infty)$ such that

$$\begin{array}{l} |A_x(x)| \le c_1 \quad \forall x\in\mathbb{R}^n, \\ |A_{xx}(x)| \le c_1 \quad \forall x\in\mathbb{R}^n, \\ |Q_x(x)| \le c_1 \quad \forall x\in\mathbb{R}^n, \\ |(Q^{-1})_x(x)| \le c_1 \quad \forall x\in\mathbb{R}^n, \\ |(Q^{-1})_{xx}(x)| \le c_1 \quad \forall x\in\mathbb{R}^n. \end{array} \tag{A2}$$

Assumption (A1) implies

$$|Q(x)|, |Q^{-1}(x)| \le c_2 \quad \forall x\in\mathbb{R}^n, \tag{9}$$

for some $c_2 \in (0,\infty)$. Here $l(x)$ is the running cost, for which we require that there exist $c_3, c_4, c_5 \in (0,\infty)$ such that

$$|l_{xx}(x)| \le c_3 \quad \forall x\in\mathbb{R}^n, \tag{A3}$$

and

$$c_4|x|^2 \le l(x) \le c_5|x|^2 \quad \forall x\in\mathbb{R}^n. \tag{A4}$$

The quadratic bounds in $(A4)$ are chosen for convenience; it is expected that most of the results will hold for more general polynomial bounds (but which still enforce $l(0) = 0$ and $l(x) > 0$ for $x \neq 0$). By the nonnegativity of l and controllability implied by (9), one can rewrite (8) as

$$W_2(x) = \inf_{v \in \mathcal{V}} \left\{ \int_0^\infty l(X(t)) + \frac{1}{2}|v(t)|^2 dt \right\}. \tag{10}$$

The control problem (6)–(10) is different from the control problem considered in (1)–(2) in a nontrivial way. Not only is the controller minimizing instead of maximizing, but the nature of the dynamics is quite different. In the problem (1)–(2), the nominal dynamics are stable. However, in this problem, the uncontrolled dynamics may be unstable. If the nominal dynamics are not stable, then the system will be stabilized by the minimizing controller by virtue of the fact that the running cost, l, is strictly positive away from the origin. (Again, controllability is guaranteed by the above assumptions.)

First, consider W_2. We begin with a result which serves to produce some a priori bounds on the behavior of near-optimal trajectories of the system.

Lemma 1. *Given $R < \infty$, there exists $M_R < \infty$ such that $W_2(x) \leq M_R$ for all $x \in B_R(0)$. Further, there exists $\hat{M}_R < \infty$ such that for any $\epsilon \in (0,1)$ the following statements hold: If v_ϵ is ϵ-optimal for infinite horizon problem (10), then $|v_\epsilon|_{L_2[0,\infty)} \leq \hat{M}_R$, and $|X_\epsilon|_{L_2[0,\infty)} \leq \hat{M}_R$ where*

$$\dot{X}_\epsilon = A(X_\epsilon) + Q(X_\epsilon)v_\epsilon.$$

If v_ϵ is ϵ-optimal for finite horizon problem

$$V(T,x) = \inf_{v \in \mathcal{V}} \left\{ \int_0^T l(X(t)) + \frac{1}{2}|v(t)|^2 dt \right\},$$

then $|v_\epsilon|_{L_2[0,T]} \leq \hat{M}_R$ and $|X_\epsilon|_{L_2[0,T]} \leq \hat{M}_R$, where $\dot{X}_\epsilon = A(X_\epsilon) + Q(X_\epsilon)v_\epsilon$.

Proof. If $x = 0$, then by virtue of the fact that $l(0) = 0$ and $l(x) > 0$ for $x \neq 0$, the optimal v, v^*, satisfies $v^* \equiv 0$, and hence $W_2(0) = V(T,0) = 0$. Otherwise, let

$$\tilde{v}(t) := \begin{cases} Q^{-1}(X(t))\left(-\frac{x}{|x|} - A(X(t))\right), & 0 \leq t \leq |x| \\ 0, \quad t > |x|. \end{cases}$$

Then

$$\dot{\tilde{X}} = A(\tilde{X}) + Q(\tilde{X})\tilde{v} = \begin{cases} -\frac{x}{|x|}, & \text{if } 0 \leq t \leq |x| \\ A(\tilde{X}), & \text{if } t > |x| \end{cases}$$

which is solved by

$$\tilde{X}(t) = \begin{cases} x\left(1 - \frac{t}{|x|}\right), & \text{if } t \leq |x| \\ 0, & \text{if } t > 0, \end{cases}$$

and this implies $\tilde{v} \in C[0,\infty)$ with $\tilde{v}(t) = 0$ for $t \geq |x|$. Therefore

$$J_2(T,x,\tilde{v}) = \int_0^{|x|} l\left(x - \frac{tx}{|x|}\right) + \frac{1}{2}|\tilde{v}(t)|^2\, dt$$

for all $T > |x|$. Since $|Q|, |Q^{-1}|$ are bounded by (9), $\tilde{v} \in \mathcal{V}$. Thus, there exists some M_R (independent of $T > |x|$) such that

$$\begin{aligned} V(T,x) &= \inf_{v\in\mathcal{V}} J_2(T,x,v) \leq M_R < \infty, \\ W_2(x) &= \inf_{v\in\mathcal{V}} \sup_{T<\infty} J_2(T,x,v) \leq M_R < \infty \end{aligned} \tag{11}$$

for all $T \in [0,\infty)$ and $x \in B_R(0)$. Using this same $\tilde{v}$ for $T \in (0,|x|]$, one obtains (11) for all $T \in (0,\infty)$ and $x \in B_R(0)$.

If v_ϵ is ϵ-optimal for V, then

$$\int_0^T \frac{1}{2}|v_\epsilon(t)|^2 dt \leq \int_0^T l(X_\epsilon) + \frac{1}{2}|v_\epsilon(t)|^2 dt < V(T,x) + \epsilon \leq M_R + 1$$

and

$$\int_0^T c_4|X_\epsilon(t)|^2 dt \leq \int_0^T l(X_\epsilon) + \frac{1}{2}|v_\epsilon(t)|^2 dt < V(T,x) + \epsilon \leq M_R + 1.$$

Consequently,

$$|v_\epsilon|_{L_2[0,T]} \leq (2(M_R+1))^{1/2}$$

and

$$|X_\epsilon|_{L_2[0,T]} \leq [(1/c_4)(M_R+1)]^{1/2}.$$

Similarly, if v_ϵ is ϵ-optimal for W_2, then

$$|v_\epsilon|_{L_2[0,\infty)} \leq (2(M_R+1))^{1/2}$$

and

$$|X_\epsilon|_{L_2[0,\infty)} \leq [(1/c_4)(M_R+1)]^{1/2}.$$

Simply let

$$\hat{M}_R := \max\left\{[2(M_R+1)]^{1/2}, [(1/c_4)(M_R+1)]^{1/2}\right\}$$

to complete the proof. □

The following result is entirely standard, and we do not include a proof:

Lemma 2. *W_2 satisfies the dynamic programming principle (DPP):*

$$W_2(x) = \inf_{v\in\mathcal{U}} \left\{\int_0^\tau l(X(t)) + \frac{1}{2}|v(t)|^2 dt + W_2(X(\tau))\right\} \tag{12}$$

for any $\tau \in (0,\infty)$.

Theorem 1. *W_2 is a continuous viscosity solution to the Hamilton–Jacobi–Bellman equation (5).*

Before the proof, we define what it means for W_2 to be a continuous viscosity solution of a Hamilton–Jacobi equation (cf. [9]). If $H : \mathbb{R}^n \times \mathbb{R}^n \to \mathbb{R}$ is continuous, a continuous function $w : \mathbb{R}^n \to \mathbb{R}$ is called a *continuous viscosity solution* of

$$-H(x, \nabla w) = 0, \qquad w(0) = 0 \tag{13}$$

provided that $w(0) = 0$ and for each $\phi \in C^1(\mathbb{R}^n)$: (a) if $w - \phi$ attains a local maximum at $(t, x) = (0, x_0)$, then

$$-H(x_0, \nabla\phi(x_0)) \leq 0,$$

and (b) if $w - \phi$ attains a local minimum at $(t, x) = (0, x_0)$, then

$$-H(x_0, \nabla\phi(x_0)) \geq 0.$$

Proof. First, by Corollary 2, Section 4.1, W_2 is locally Lipschitz continuous.

Now, suppose $W_2 - \phi$ has a local maximum at x, but

$$-H(x, \nabla\phi(x)) \geq \epsilon > 0. \tag{14}$$

Then there must exist a $v_* \in \mathbb{R}^n$ such that

$$(A(x) + Q(x)v_*) \cdot \nabla\phi(x) + l(x) + \frac{1}{2}|v_*|^2 \leq -\frac{\epsilon}{2} < 0. \, .$$

Let $v_0(t) \equiv v_*$ and $\dot{X}_0 = A(X_0) + Q(X_0)v_0$. Since H, v_0, X_0 are continuous, there exists a $\tau > 0$ such that

$$(A(X_0(t)) + Q(X_0(t))v_0(t)) \cdot \nabla\phi(X_0(t)) + l(X_0(t)) + \frac{1}{2}|v_0(t)|^2 \leq -\frac{\epsilon}{4} < 0$$

for all $0 < t < \tau$. Therefore

$$\int_0^\tau [A(X_0(t)) + Q(X_0(t))v_0(t)] \cdot \nabla\phi(X_0(t)) + l(X_0(t)) + \frac{1}{2}|v_0(t)|^2 \, dt < 0. \tag{15}$$

Rearranging (12) yields

$$0 = \inf_{v \in \mathcal{V}} \left\{ \int_0^\tau l(X(t)) + \frac{1}{2}|v(t)|^2 \, dt + W_2(X(\tau)) - W_2(x) \right\}. \tag{16}$$

Since $W_2 - \phi$ has a local maximum at x,

$$W_2(X(t)) - W_2(x) \leq \phi(X(t)) - \phi(x). \tag{17}$$

Substituting (17) into (16), one finds

$$0 \leq \inf_{v \in \mathcal{V}} \left\{ \int_0^\tau l(X(t)) + \frac{1}{2}|v(t)|^2 \, dt + \phi(X(\tau)) - \phi(x) \right\}$$
$$\leq \int_0^\tau l(X_0(t)) + \frac{1}{2}|v_0(t)|^2 dt + \phi(X_0(\tau)) - \phi(x). \tag{18}$$

However,

$$\phi(X_0(\tau)) - \phi(x) = \int_0^\tau \nabla\phi(X_0(t)) \cdot [A(X_0(t)) + Q(X_0(t))v_0(t)] \, dt, \tag{19}$$

and substituting (19) into (18) yields

$$0 < \int_0^\tau l(X_0(t)) + \frac{1}{2}|v_0(t)|^2 + [A(X_0(t)) + Q(X_0(t))v_0(t)] \cdot \nabla\phi(X_0(t)) \, dt,$$

which contradicts (15). We are forced to reject (14) in favor of the desired result.

The proof that $W_2 - \phi$ having a local minimum at x implies that

$$-H(x, \nabla\phi(x)) \geq 0$$

is similar. □

Remark 1. Note that the proof of Theorem 1 also demonstrates that any continuous solution of DPP (12) is a continuous viscosity solution of (5).

To introduce the next section of the paper, we define the semigroup S_τ by

$$S_\tau[\phi](x) := \inf_{v \in \mathcal{U}} \left\{ \int_0^\tau l(X(t)) + \frac{1}{2}|v(t)|^2 dt + \phi(X(\tau)) \right\}$$

where $X(\cdot)$ satisfies (6), and note, from the DPP (Lemma 2), that

$$W_2(x) = S_\tau[W_2](x). \tag{20}$$

Remark 1 implies that a continuous solution of fixed point problem (20) is a continuous viscosity solution of (5).

Remark 2. As noted in the above proof (and proved below), W_2 is locally Lipschitz continuous. We anticipate that one can show that W_1 is also locally Lipschitz continuous under the above assumptions, but do not include a proof.

Remark 3. By Soravia [24], one also has that the game value function, W_1, and the control value function, W_2, are both minimal, nonnegative, continuous viscosity supersolutions of (5). Consequently, $W_1 \equiv W_2$. Therefore the development to follow for computation of W_2 via (20) will also serve to compute W_1.

4 Solving $W_2 = S\ [W_2]$ in the Min–Plus Algebra

We proceed to illustrate a method for obtaining the correct solution of (20) (i.e., the one corresponding to $W_1 = W_2$) via a technique using the min–plus algebra. The first major point is the linearity of the S_τ operator over this semifield. The triplet $\mathcal{A} = (\mathbb{R} \cup \{+\infty\}, \oplus, \otimes)$ with the operations defined by

$$a \oplus b := \min\{a, b\} \quad \text{(additive identity is } +\infty)$$
$$a \otimes b := a + b \ (\quad \text{multiplicative identity is } 0)$$

would form a field, except that most elements do not have additive inverses. The name *semifield* is often given to such structures, but this particular one is known in the literature as the *min–plus algebra.*

Theorem 2. *S_τ is a linear operator on functions over the min–plus algebra; that is,*

$$S_\tau[\phi \oplus \psi](x) = S_\tau[\phi](x) \oplus S_\tau[\psi](x)$$
$$S_\tau[c \otimes \phi](x) = c \otimes S_\tau[\psi](x)$$

for all functions in $\mathcal{X} = \{\psi : \mathbb{R}^n \to \mathcal{A} \,|\, S_\tau[\psi](x) < \infty \quad \forall x \in \mathbb{R}^n\}$.

(Thus, $\mathcal{X}$ is a semi-vector space (moduloid) over $\mathcal{A}$). The proof of this linearity is quite easy, and we do not include it. However, see [11], [15], [21], [23] for similar proofs. The fixed point problem $S_\tau[W_2] = W_2$ can now be reposed as an eigenvalue problem

$$S_\tau[W_2] = 0 \otimes W_2. \tag{21}$$

We aim to solve the eigenvalue problem with the *power method*, an iterative technique for solving *matrix* eigenvalue problems (not operator eigenvalue problems). First it will be shown that W_2 (and therefore $S_\tau[W_2]$) is semiconcave in x. The space of semiconcave functions over the min-plus algebra has a basis, and $\mathcal{S}_\tau$ has an “infinite matrix” representation using this. A discretization (matrix representation) of the $\mathcal{S}_\tau$ operator can then be obtained by truncating its basis expansion.

4.1 The Semiconcavity of W_2

Lemma 3. *Given $\epsilon \in (0,1)$, $R < \infty$, and $|x| < R$, let $v_{\epsilon,T}$ be ϵ-optimal for (22) below. Then there exists $D_R < \infty$ (independent of ϵ, x, and T) such that*

$$|X(t)| \le D_R \qquad \forall t \in [0, T].$$

The proof (not included) uses the bounds obtained in Lemma 1.

Definition 1. *We say that a function, $\phi : \mathbb{R}^n \to \mathbb{R}$ is* semiconcave *provided given any $R < \infty$, there exists $C_R < \infty$ such that $\phi(x) - \frac{C_R}{2}|x|^2$ is concave over $B_R = B_R(0)$. The minimal such C_R for a given R will be referred to as the* semiconcavity constant *for B_R.*

Theorem 3. *The value function for the finite time horizon problem,*

$$V(T,x) := \inf_{v\in\mathcal{V}} \left\{ \int_0^T l(X(t)) + \frac{1}{2}|v(t)|^2 dt \right\}, \tag{22}$$

is semiconcave in x, and the semiconcavity constant over any B_R is bounded uniformly in T.

Proof. We will show that second differences of V in x are bounded from above. Let $R < \infty$ and $x \in B_R$. Recall

$$J_2(T,x,v) = \int_0^T l(X(t)) + \frac{1}{2}|v(t)|^2 dt, \tag{23}$$

so that $V(T,x) := \inf_{v\in\mathcal{V}} J_2(T,x,v)$. Let $\eta \in \mathbb{R}^n$ with $|\eta| = 1$. Let $\delta,\epsilon \in (0,1)$. Let v_ϵ be ϵ-optimal for (22). Let

$$\begin{aligned}
\dot{X}^\epsilon &= A(X^\epsilon) + Q(X^\epsilon)v^\epsilon,\\
X^\epsilon(0) &= x,\\
\dot{X}^\epsilon_+ &= A(X^\epsilon_+) + Q(X^\epsilon_+)v^\epsilon_+,\\
X^\epsilon_+(0) &= x + \delta\eta,\\
\dot{X}^\epsilon_- &= A(X^\epsilon_-) + Q(X^\epsilon_-)v^\epsilon_-,\\
X^\epsilon_-(0) &= x - \delta\eta,
\end{aligned}$$

where the controls $v^\epsilon_+, v^\epsilon_-$ are given as follows: Let

$$v^\epsilon_+(t) = \begin{cases} Q^{-1}(X^\epsilon_+(t))\left[A(X^\epsilon(t)) - A(X^\epsilon_+(t)) + Q(X^\epsilon(t))v^\epsilon(t) - \delta\eta\right] & t \le 1\\ v^\epsilon(t) & t > 1\end{cases}$$

and

$$v^\epsilon_-(t) = \begin{cases} Q^{-1}(X^\epsilon_-(t))\left[A(X^\epsilon(t)) - A(X^\epsilon_-(t)) + Q(X^\epsilon(t))v^\epsilon(t) + \delta\eta\right] & t \le 1\\ v^\epsilon(t) & t > 1.\end{cases}$$

With these choices of $v^\epsilon_+, v^\epsilon_-$, one has

$$X^\epsilon_+(t) - X^\epsilon(t) = \delta(1-t)\eta, \qquad X^\epsilon_-(t) - X^\epsilon(t) = -\delta(1-t)\eta \tag{24}$$

for $t < 1$ and

$$X^\epsilon_+(t) = X^\epsilon(t) = X^\epsilon_-(t)$$

for $t \ge 1$. Since we are interested in the long-term semiconcavity, we will assume $T \ge 1$.

Now $V(T,x) \ge J_2(T,x,v_\epsilon) - \epsilon$ by the choice of v^ϵ, so that

$$\begin{aligned}
&V(T,x+\delta\eta) - 2V(t,x) + V(T,x-\delta\eta)\\
&\quad\le\ J_2(T,x+\delta\eta,v^\epsilon_+) - 2J_2(T,x,v^\epsilon) + J_2(T,x-\delta\eta,v^\epsilon_-) + 2\epsilon
\end{aligned}$$

$$\leq \int_0^1 l(X_+^\epsilon(t)) - 2l(X^\epsilon(t)) + l(X_-^\epsilon(t))\,dt$$

$$+ \frac{1}{2}\int_0^1 |v_+^\epsilon(t)|^2 - 2|v^\epsilon(t)|^2 + |v_-^\epsilon(t)|^2\,dt + 2\epsilon$$

which by (24) and $(A3)$

$$\leq c_3\delta^2 + \frac{1}{2}\int_0^1 \left(|v_+^\epsilon(t)|^2 - |v^\epsilon(t)|^2\right) + \left(|v_-^\epsilon(t)|^2 - |v^\epsilon(t)|^2\right)dt + 2\epsilon. \quad (25)$$

Considering the integral of the second difference in v^ϵ on the right-hand side of (25), (where, for simplicity of exposition, we drop the notation indicating the t-dependence), one has

$$\int_0^1 |v_+^\epsilon|^2 - 2|v^\epsilon|^2 + |v_-^\epsilon|^2\,dt$$

$$= \int_0^1 \left|Q^{-1}(X_+^\epsilon)[A(X^\epsilon) - A(X_+^\epsilon) - \delta\eta + Q(X^\epsilon)v^\epsilon]\right|^2 - 2|v^\epsilon|^2$$

$$+\left|Q^{-1}(X_-^\epsilon)[A(X^\epsilon) - A(X_-^\epsilon) + \delta\eta + Q(X^\epsilon)v^\epsilon]\right|^2 dt$$

$$= \int_0^1 \left|Q^{-1}(X_+^\epsilon)[A(X^\epsilon) - A(X_+^\epsilon) - \delta\eta]\right|^2$$

$$+\left|Q^{-1}(X_-^\epsilon)[A(X^\epsilon) - A(X_-^\epsilon) + \delta\eta]\right|^2 dt$$

$$+ \int_0^1 2[A(X^\epsilon) - A(X_+^\epsilon) - \delta\eta]^T Q^{-T}(X_+^\epsilon)Q^{-1}(X_+^\epsilon)Q(X^\epsilon)v^\epsilon$$

$$+2[A(X^\epsilon) - A(X_-^\epsilon) + \delta\eta]^T Q^{-T}(X_-^\epsilon)Q^{-1}(X_-^\epsilon)Q(X^\epsilon)v^\epsilon\,dt$$

$$+ \int_0^1 \left|Q^{-1}(X_+^\epsilon)Q(X^\epsilon)v^\epsilon\right|^2 - 2|v^\epsilon|^2 + \left|Q^{-1}(X_-^\epsilon)Q(X^\epsilon)v^\epsilon\right|^2 dt. \quad (26)$$

Considering the three integrals on the right-hand side of (26) separately, we first have

$$\int_0^1 \left|Q^{-1}(X_+^\epsilon)[A(X^\epsilon) - A(X_+^\epsilon) - \delta\eta]\right|^2$$

$$+\left|Q^{-1}(X_-^\epsilon)[A(X^\epsilon) - A(X_-^\epsilon) + \delta\eta]\right|^2 dt$$

$$\leq 2\int_0^1 \left|Q^{-1}(X_+^\epsilon)[A(X^\epsilon) - A(X_+^\epsilon)]\right|^2 + \left|Q^{-1}(X_+^\epsilon)\delta\eta\right|^2$$

$$+\left|Q^{-1}(X_-^\epsilon)[A(X^\epsilon) - A(X_-^\epsilon)]\right|^2 + \left|Q^{-1}(X_-^\epsilon)\delta\eta\right|^2 dt$$

which using (9) and Assumption $(A2)$,

$$\leq 2c_2^2 \int_0^1 \left|A(X^\epsilon) - A(X_+^\epsilon)\right|^2 + \left|A(X^\epsilon) - A(X_-^\epsilon)\right|^2 + 2\delta^2|\eta|^2\,dt$$

$$\leq 2c_1^2c_2^2 \int_0^1 \left|X^\epsilon - X_+^\epsilon\right|^2 + \left|X^\epsilon - X_-^\epsilon\right|^2 dt + 4c_2^2\delta^2$$

and by (24)

$$= 4c_1^2 c_2^2 \int_0^1 |\delta(1-t)\eta|^2\, dt + 4c_2^2\delta^2 \le 4(c_1^2+1)c_2^2\delta^2. \tag{27}$$

The second integral on the right-hand side of (26) can be bounded as follows:

$$\begin{aligned}
&\int_0^1 2[A(X^\epsilon) - A(X_+^\epsilon) - \delta\eta]^T Q^{-T}(X_+^\epsilon) Q^{-1}(X_+^\epsilon) Q(X^\epsilon) v^\epsilon \\
&\qquad +2[A(X^\epsilon) - A(X_-^\epsilon) + \delta\eta]^T Q^{-T}(X_-^\epsilon) Q^{-1}(X_-^\epsilon) Q(X^\epsilon) v^\epsilon\, dt \\
&= 2\int_0^1 [[A(X^\epsilon) - A(X_+^\epsilon)]^T Q^{-T}(X_+^\epsilon) Q^{-1}(X_+^\epsilon) \\
&\qquad +[A(X^\epsilon) - A(X_-^\epsilon)]^T Q^{-T}(X_-^\epsilon) Q^{-1}(X_-^\epsilon)] Q(X^\epsilon) v^\epsilon \\
&\qquad -\delta\eta^T [Q^{-T}(X_+^\epsilon) Q^{-1}(X_+^\epsilon) - Q^{-T}(X_-^\epsilon) Q^{-1}(X_-^\epsilon)] Q(X^\epsilon) v^\epsilon\, dt \\
&= 2\int_0^1 [[A(X^\epsilon) - A(X_+^\epsilon)]^T [Q^{-T}(X_+^\epsilon) Q^{-1}(X_+^\epsilon) - Q^{-T}(X_-^\epsilon) Q^{-1}(X_-^\epsilon)] \\
&\qquad +[A(X^\epsilon) - A(X_+^\epsilon) + A(X^\epsilon) - A(X_-^\epsilon)]^T \\
&\qquad \cdot Q^{-T}(X_-^\epsilon) Q^{-1}(X_-^\epsilon)] Q(X^\epsilon) v^\epsilon\, dt \\
&-2\int_0^1 \delta\eta^T [Q^{-T}(X_+^\epsilon) Q^{-1}(X_+^\epsilon) - Q^{-T}(X_-^\epsilon) Q^{-1}(X_-^\epsilon)] Q(X^\epsilon) v^\epsilon\, dt.
\end{aligned} \tag{28}$$

Combining Assumption $(A2)$ and (9) yields

$$\left|(Q^{-T}(x) Q^{-1}(x))_x\right| \le 2c_1c_2 \qquad \forall x \in \mathbb{R}^n. \tag{29}$$

Therefore, the second integral on the right-hand side of (28) is

$$\begin{aligned}
&\le 4c_1c_2\delta \int_0^1 |\eta||X_+^\epsilon - X_-^\epsilon||Q(X^\epsilon)||v^\epsilon|\, dt \\
&\le 4c_1c_2^2\delta \int_0^1 |\eta||2\delta(1-t)\eta||v^\epsilon|\, dt \le 8c_1c_2^2 \hat{M}_R \delta^2.
\end{aligned} \tag{30}$$

The first integral on the right-hand side of (28) becomes

$$\begin{aligned}
&2\int_0^1 [[A(X^\epsilon) - A(X_+^\epsilon)]^T [Q^{-T}(X_+^\epsilon) Q^{-1}(X_+^\epsilon) - Q^{-T}(X_-^\epsilon) Q^{-1}(X_-^\epsilon)] \\
&\qquad -[A(X_+^\epsilon) - 2A(X^\epsilon) + A(X_-^\epsilon)]^T Q^{-T}(X_-^\epsilon) Q^{-1}(X_-^\epsilon)] Q(X^\epsilon) v^\epsilon\, dt \\
&\le 2\int_0^1 |A(X^\epsilon) - A(X_+^\epsilon)| \\
&\qquad \cdot |Q^{-T}(X_+^\epsilon) Q^{-1}(X_+^\epsilon) - Q^{-T}(X_-^\epsilon) Q^{-1}(X_-^\epsilon)||Q(X^\epsilon)||v^\epsilon|\, dt \\
&+2\int_0^1 |A(X_+^\epsilon) - 2A(X^\epsilon) + A(X_-^\epsilon)||Q^{-T}(X_-^\epsilon)||Q^{-1}(X_-^\epsilon)||Q(X^\epsilon)||v^\epsilon|\, dt.
\end{aligned}$$

Employing $(A2)$, (9), and (29), one finds

$$\leq 4c_1^2c_2^2 \int_0^1 |X^\epsilon - X_+^\epsilon||X_+^\epsilon - X_-^\epsilon||v^\epsilon| + 2c_1c_2^3 \int_0^1 |X_+^\epsilon - X_-^\epsilon|^2|v^\epsilon|\,dt$$
$$\leq (4c_1^2c_2^2 + 8c_1c_2^3)\delta^2|v^\epsilon|_{L^1[0,1]} \leq (4c_1^2c_2^2 + 8c_1c_2^3)\hat{M}_R\delta^2. \tag{31}$$

Combining (28), (30) and (31), one finds that the second integral on the right-hand side of (26) satisfies

$$\int_0^1 2[A(X^\epsilon) - A(X_+^\epsilon) - \delta\eta]^T Q^{-T}(X_+^\epsilon)Q^{-1}(X_+^\epsilon)Q(X^\epsilon)v^\epsilon$$
$$+2[A(X^\epsilon) - A(X_-^\epsilon) + \delta\eta]^T Q^{-T}(X_-^\epsilon)Q^{-1}(X_-^\epsilon)Q(X^\epsilon)v^\epsilon\,dt,$$

which, upon application of Hölder's Inequality, is shown to be

$$\leq [8c_1c_2^2 + 4c_1^2c_2^2 + 8c_1c_2^3]\hat{M}_R\delta^2. \tag{32}$$

The third integral in (26) involves the second differences of v^ϵ.

$$\int_0^1 \left|Q^{-1}(X_+^\epsilon)Q(X^\epsilon)v^\epsilon\right|^2 - 2|v^\epsilon|^2 + \left|Q^{-1}(X_-^\epsilon)Q(X^\epsilon)v^\epsilon\right|^2 dt$$
$$= \int_0^1 v^{\epsilon T}Q^T(X^\epsilon)Q^{-T}(X_+^\epsilon)Q^{-1}(X_+^\epsilon)Q(X^\epsilon)v^\epsilon$$
$$-2v^{\epsilon T}Q^T(X^\epsilon)Q^{-T}(X^\epsilon)Q^{-1}(X^\epsilon)Q(X^\epsilon)v^\epsilon$$
$$+v^{\epsilon T}Q^T(X^\epsilon)Q^{-T}(X_-^\epsilon)Q^{-1}(X_-^\epsilon)Q(X^\epsilon)v^\epsilon\,dt. \tag{33}$$

Consider any C^2 function $F : \mathbb{R}^n \to \mathbb{R}$ and $x \in \mathbb{R}^n$, and let $x_+ = x + \delta\eta$, $x_- = x - \delta\eta$. Denote the first and second directional derivatives in direction η by F_η and $F_{\eta\eta}$. Then

$$F(x_+) - 2F(x) + F(x_-) = \int_0^\delta F_\eta(x + r\eta) - F_\eta(x - \delta\eta + r\eta)\,dr$$
$$= \int_0^\delta \int_0^\delta F_{\eta\eta}(x - (\delta + r)\eta + \rho\eta)\,d\rho\,dr$$
$$\leq \max_{|r|\leq\delta} |F_{\eta\eta}(x + r\eta)|\delta^2. \tag{34}$$

Taking $F(x) = \nu^T[Q^{-T}(x)Q^{-1}(x)]\nu$ with $\nu \in \mathbb{R}^n$, one has by $(A2)$ and (9)

$$|F_{\eta\eta}(x)| \leq |[Q^{-T}(x)Q^{-1}(x)]_{xx}|\,|\nu|^2 \leq 2(c_1c_2 + c_1^2)|\nu|^2. \tag{35}$$

Combining (33) with (34) and (35) (for $x = X^\epsilon$, $x_+ = X_+^\epsilon$, $x_- = X_-^\epsilon$ and $\nu = Q(X^\epsilon)v^\epsilon$) yields

$$\int_0^1 \left|Q^{-1}(X_+^\epsilon)Q(X^\epsilon)v^\epsilon\right|^2 - 2|v^\epsilon|^2 + \left|Q^{-1}(X_-^\epsilon)Q(X^\epsilon)v^\epsilon\right|^2 dt$$
$$\leq \int_0^1 2(c_1c_2 + c_1^2)|Q(X^\epsilon)v^\epsilon|^2\delta^2\,dt$$
$$\leq 2c_2^2c_1(c_2 + c_1)\delta^2 \int_0^1 |v^\epsilon|^2\,dt \leq 2\widehat{M}_R^2c_2^2c_1(c_2 + c_1)\delta^2. \tag{36}$$

Combining (27), (32) and (36) bounds the right hand side of (26); therefore there exists $C < \infty$ such that

$$\int_0^1 |v_+^\epsilon|^2 - 2|v^\epsilon|^2 + |v_-^\epsilon|^2 \, dt < C\delta^2.$$

Substituting this into (25) yields

$$V(T, x - \delta\eta) - 2V(t, x) + V(T, x + \delta\eta) \leq (c_3 + C)\delta^2 + 2\epsilon. \tag{37}$$

Since $\delta, \epsilon \in (0, 1)$ were arbitrary, this implies semi-concavity. □

Lemma 4. *$V(T, x) \to W_2(x)$ uniformly on compact sets.*

Proof. Note that $V(T, x)$ is monotonically increasing in T and bounded above by $W_2(x)$. The proof that, in fact, $V(T, x) \to W_2(x)$ uniformly on compact sets is similar to that appearing in the proof of Theorem 2.6 in [17]. □

Combining Theorem 3 and Lemma 4 implies the following:

Corollary 1. *W_2 and $S_\tau[W_2]$ are semiconcave.*

Concavity (or convexity) and local boundedness imply locally Lipschitz continuity [10]. By Lemma 1, W_2 is locally bounded. Noting that semiconcave functions differ from concave functions by only a quadratic term on any ball B_R, one has the following:

Corollary 2. *W_2 and $S_\tau[W_2]$ are locally Lipschitz continuous.*

4.2 A Basis Representation of W_2, $\mathcal{S}_\tau$

Semiconcave locally bounded functions are not only locally Lipschitz (cf. [10]), but also have a duality relationship (see [11] for proof of a similar result in the semiconvex case):

Theorem 4. (Semiconcave Duality) *Let $\mathcal{S} := \{\phi : \mathbb{R}^n \to \mathbb{R} \mid \forall R < \infty \;\; \exists C_R < \infty$ such that $\phi(x) - \frac{1}{2}C_R|x|^2$ is concave on $B_R(0)\}$, the space of semiconcave functions on $B_R(0)$, and let $\phi \in \mathcal{S}$. Let C be a symmetric matrix such that $C - C_R I > 0$, where C_R is as in the requirement for $\phi \in \mathcal{S}$. Let L_R be the Lipschitz constant for $\phi + \frac{1}{2}C_R|x|^2$ over $B_R(0)$. Then*

$$\begin{aligned} \phi(x) &= \min_{\tilde{x} \in B_{L_R/C_R}} \left[\frac{1}{2}(x - \tilde{x})^T C (x - \tilde{x}) + a(\tilde{x}) \right] \\ &= \min_{\tilde{x} \in \mathbb{R}^n} \left[\frac{1}{2}(x - \tilde{x})^T C (x - \tilde{x}) + a(\tilde{x}) \right] \end{aligned}$$

where

$$a(\tilde{x}) = - \min_{x \in B_R} \left[\frac{1}{2}(x - \tilde{x})^T C (x - \tilde{x}) - \phi(x) \right].$$

Semiconcave duality points to a natural basis for the space of semiconcave functions. Let $\{x_i\}_{i=1}^{\infty}$ be a countable dense set in $B_{L_R/C_R}(0)$ and $C - C_R I > 0$ be a symmetric matrix (where $C_R > 0$ is the semiconcavity constant for ϕ over $B_R(0)$). Define

$$\psi_i(x) := \frac{1}{2}(x - x_i)^T C(x - x_i).$$

Now any semiconcave function $\phi \in \mathcal{S}$ (with semiconcavity and Lipschitz constants over $B_R(0)$ being C_R and L_R, respectively) can be represented on $B_R(0)$ as

$$\begin{aligned}\phi(x) &= \inf_{i \in \mathbb{N}} \left[a(x_i) + \frac{1}{2}(x - x_i)^T C(x - x_i) \right] \\ &= \bigoplus_{i=1}^{\infty} \left[a(x_i) \otimes \psi_i(x) \right].\end{aligned}$$

In particular, W_2 has the following basis expansion:

Theorem 5. *Given $R < \infty$, there exists a semiconcavity constant $C_R < \infty$ and a Lipschitz constant $L_R < \infty$ for W_2 over $B_R(0)$. Let $C - C_R I > 0$, and let $\{x_i\}_{i=1}^{\infty}$ be dense in $B_{(L_R + |C|)/C_R}(0)$. Let $\psi_i(x) = \frac{1}{2}(x - x_i)^T C(x - x_i)$ for all i. Then*

$$W_2(x) = \bigoplus_{i=1}^{\infty} \left[b_i \otimes \psi_i(x) \right]$$

where $b_i = b(x_i) = -\min_{\hat{x} \in B_R} \left[\psi_i(\hat{x}) - W_2(\hat{x}) \right]$.

Also, $\mathcal{S}_\tau$ also has an infinite matrix representation as

$$\mathcal{S}_\tau \psi_j(x) = \bigoplus_{i=1}^{\infty} \left[B_{ij}^\tau \otimes \psi_i(x) \right]$$

where $B_{ij}^\tau := -\min_{\hat{x} \in B_R} [\psi_i(\hat{x}) - \mathcal{S}_\tau \psi_j(\hat{x})]$.

The representations of semiconcave functions in the $\{\psi_i\}$ basis can be simplified in the semiconcave dual space. A semiconcave function $\phi(x)$ (with semiconcavity and Lipschitz constants over $B_R(0)$ being C_R and L_R, respectively) with basis representation

$$\phi(x) = \bigoplus_{i=1}^{\infty} \left[a_i \otimes \psi_i(x) \right]$$

is represented in the dual space simply by the "infinite vector" or sequence of coefficients

$$a = [a_1 \quad a_2 \quad a_3 \quad \ldots]^T.$$

The dual representations of W_2 and $\mathcal{S}_\tau$ are therefore

$$b = \begin{bmatrix} b_1 \\ b_2 \\ b_3 \\ \cdot \\ \cdot \\ \cdot \end{bmatrix}, \quad B^\tau = \begin{bmatrix} B_{11}^\tau & B_{12}^\tau & B_{13}^\tau & \ldots \\ B_{21}^\tau & B_{22}^\tau & B_{23}^\tau & \ldots \\ B_{31}^\tau & B_{32}^\tau & B_{33}^\tau & \ldots \\ \cdot & \cdot & \cdot & \\ \cdot & \cdot & \cdot & \\ \cdot & \cdot & \cdot & \end{bmatrix}.$$

For this preliminary paper, as in [21], some strong assumptions have been made; one of which is that W_2 has a *finite* expansion in the $\{\psi_i\}$ basis,

$$W_2(x) = W_2^k(x) := \bigoplus_{i=1}^{k} [b_i \otimes \psi_i(x)]\,.$$

Although it is unrealistic to assume a finite basis expansion, we are confident that if W_2^k is a truncated basis expansion of W_2, then $W_2^k \to W_2$ uniformly on a compact set as $k \to \infty$. This has been proven in the problem class considered previously [18], [19]. Similarly, assume $\mathcal{S}_\tau$ also has a finite basis expansion

$$S_\tau \psi_j(x) = S_\tau^k \psi_j(x) = \bigoplus_{i=1}^{k} \left[B_{ij}^\tau \otimes \psi_i(x)\right]$$

where $B_{ij}^\tau = -\min_{\hat{x}\in B_R}[\psi_i(\hat{x}) - S_\tau\psi_j(\hat{x})]$, as before. In the dual space, let

$$b^k = \begin{bmatrix} b_1 \\ b_2 \\ \cdot \\ \cdot \\ b_k \end{bmatrix}, \quad B^{\tau,k} = \begin{bmatrix} B_{11}^\tau & B_{12}^\tau & \ldots & B_{1k}^\tau \\ B_{21}^\tau & B_{22}^\tau & \ldots & B_{2k}^\tau \\ \cdot & & & \cdot \\ \cdot & & & \cdot \\ B_{k1}^\tau & B_{k2}^\tau & \ldots & B_{kk}^\tau \end{bmatrix}.$$

The current picture is

(semiconcave function space)		(semiconcave dual space)	
$W_2(x), \mathcal{S}_\tau$	$\longleftrightarrow$	b, B^τ	(continuous)
$\downarrow$		$\downarrow$	
$W_2^k, \mathcal{S}_\tau^k$	$\longleftrightarrow$	$b^k, B^{\tau,k}$	(discrete)

Using the linearity of $\mathcal{S}_\tau$ in the min–plus semifield, the eigenvalue problem (21) can be approximated by

$$\bigoplus_{j=1}^{k} \left[b_j \otimes \left[\bigoplus_{i=1}^{k} B_{ij}^{\tau,k} \otimes \psi_i(x) \right] \right] = 0 \otimes \left(\bigoplus_{i=1}^{k} [b_i \otimes \psi_i(x)] \right),$$

which implies that

$$\bigoplus_{j=1}^{k} B_{ij}^{\tau,k} \otimes b_j = 0 \otimes b_i \quad \forall i \in \{1, 2, \ldots, k\}.$$

We consolidate these equalities into

$$B^{\tau,k} \otimes b^k = 0 \otimes b^k,$$

where the scalar-vector multiplication on the right is understood componentwise, and the matrix-vector multiplication on the left is defined by

$$\left(B^{\tau,k} \otimes b^k\right)_i := \bigoplus_{j=1}^{k} B_{ij}^{\tau,k} \otimes b_j.$$

Theorem 6. *(a) Assume W_2 is a solution to $S_\tau[W_2] = W_2$, and any expansion*

$$W_2 = \bigoplus_{i=1}^{k} c_i \otimes \psi_i(x)$$

on $B_R(0)$ requires that $c_i < \infty \quad \forall i$. Then for $c = [c_1, c_2, ..., c_k]^T$,

$$c = B^{\tau,k} \otimes c.$$

(b) Assume $b^k = B^{\tau,k} \otimes b^k$ and

$$W_2(x) = \bigoplus_{i=1}^{k} \left[b_i^k \otimes \psi_i(x)\right]$$

on $B_R(0)$; then on $B_R(0)$

$$S_\tau[W_2] = W_2.$$

The theorem allows interchangeability of the eigenvalue problem (21) between the space of semiconcave functions and the semiconcave dual space. In the dual space, we can solve $b^k = B^{\tau,k} \otimes b^k$ numerically with the power method. We note that in H_∞ design problems ([22]–[23]) and mixed L_2/L_∞ problems [5], the power method has been shown to yield the correct solution (i.e. the value function). We anticipate that this property will hold for the class of problems under consideration here.

For the max-plus problem (1)–(2), [21] illustrates the feasibility of this approach, while convergence and error analysis can be found in [18], [20], and [22]. Gaining similar insight into the choice of basis functions and time stepping for the min-plus problem [19], [21] considered here is a direction for future work.

References

1. Baccelli F, Cohen G, Olsder G, Quadrat J-P (1992) Synchronization and Linearity. John Wiley New York

2. Ball J, Helton J (1989) H_∞ control for nonlinear plants: connections with differential games. In: Proc 28th IEEE Conf Dec Control 956–962
3. Bardi M, Capuzzo-Dolcetta I (1997) Optimal Control and Viscosity Solutions of Hamilton-Jacobi-Bellman Equations. Birkhauser Boston
4. Basar T, Bernhard P (1991) H_∞–Optimal Control and Related Minimax Design Problems. Birkhäuser Boston
5. Dower P, McEneaney W (2003) A max-plus affine power method for approximation of a class of mixed L_2/L_∞ value functions. In: Proc 42nd IEEE CDC to appear
6. Dupuis P, Boué M (1999) Markov chain approximations for deterministic control problems with affine dynamics and quadratic cost in the control, SIAM J Numerical Analysis, 36:667–695
7. Dupuis P, Szpiro A (2001) Convergence of the optimal feedback policies in a numerical method for a class of deterministic optimal control problems, SIAM J Control Optim 40:393-420
8. Elliott R, Kalton N (1972) The existence of value in differential games (Memoirs Amer Math Soc 126). Amer Math Soc Providence RI
9. Evans L, Sougandis P (1984) Differential games and representation formulas for solutions of Hamilton-Jacobi-Isaacs equations, Indiana Univ Math J 33:773-796
10. Fleming W (1977) Functions of Several Variables. Springer–Verlag
11. Fleming W, McEneaney W (2000) A max-plus-based algorithm for a Hamilton-Jacobi-Bellman equation of nonlinear filtering, SIAM J Control Optim 38:683-710
12. Fleming W, Rishel R (1975) Deterministic and Stochastic Optimal Control. Springer-Verlag New York
13. James M (1993) A partial differential inequality for dissipative nonlinear systems, Systems Control Lett 21:315–320
14. Kushner H, Dupuis P (1992) Numerical methods for Stochastic Control Problems in Continuous Time. Springer Verlag New York
15. Maslov V (1987) On a new principle of superposition in optimization problems, Russian Math Surveys 42:43–54.
16. McEneaney W (1995) Robust control and differential games on a finite time horizon, Math Control Signals Systems 8:138-166
17. McEneaney W (1998) A uniqueness result for the Isaacs equation corresponding to nonlinear H_∞ control, Math Control Signals Systems 11:303-334
18. McEneaney W (2000) Convergence and error analysis for a max-plus algorithm. In: Proc 39th IEEE Conf on Decision and Control 1194-1199
19. McEneaney W (2001) Error Analysis of a Max-Plus Algorithm for a First-Order HJB equation Stochastic Theory and Control. In: Pasik-Duncan B (ed) Proceedings of a Workshop held in Lawrence Kansas (Lecture Notes in Control and Information Sciences 280). Springer-Verlag New York
20. McEneaney W (2001) Error Analysis of a Max–plus Algorithm for a First-Order HJB Equation. In: Proceedings Workshop on Max-Plus Algebras First Symposium on System Structure and Control Prague
21. McEneaney W (2003) Max-plus eigenvector representations for solution of nonlinear H_∞ problems basic concepts, IEEE Trans Auto Control 48:1150-1163
22. McEneaney W (2004) Max-plus eigenvector representations for solution of nonlinear H_∞ problems convergence and error analysis, SIAM J Control Optim to appear

23. McEneaney W, Horton M (1998) Max-plus eigenvector representations for nonlinear H_∞ value functions. In: Proc 37th IEEE Conf Decision Control 3506-3511
24. Soravia P (1996) H_∞ control of nonlinear systems: differential games and viscosity solutions, SIAM J Control Optim 34:1071–1097
25. van der Schaft A (1992) L_2–Gain analysis of nonlinear systems and nonlinear state feedback H_∞ control, IEEE Trans Auto Control 37:770–784

Optimization and Feedback Control of Constrained Parabolic Systems under Uncertain Perturbations

Boris S. Mordukhovich[1] and Ilya Shvartsman[2]

[1] Wayne State University. boris@math.wayne.edu
[2] Imperial College London. i.shvartsman@imperial.ac.uk

1 Introduction

This paper concerns a *minimax control design* problem for a class of *parabolic systems* with nonregular boundary conditions and *uncertain distributed perturbations* under *pointwise control and state constraints*. We deal with boundary controllers acting through *Dirichlet boundary conditions* that are the most challenging for parabolic dynamics.

The original motivation for this problem comes from an environmental problem of groundwater control [4, 5], which has important applications to agriculture, ecology, and other practical areas. The goal of the control is to neutralize the adverse effect of uncertain disturbances (in particular, weather conditions) on the dynamics of the groundwater level. In practice we usually do *not* have information about the magnitude of the disturbance and neither do we know its probability distribution. The only thing we know about the disturbance is the *range* of its possible values. Thus the above problem belongs to the class of *feedback control* problems with the groundwater level as the feedback parameter. Here we study a more general class of *multidimensional* parabolic control systems that covers a fairly broad range of practical applications.

A natural approach to control design of such uncertain systems is *minimax synthesis*, which guarantees the *best* system performance under the *worst* perturbations and ensures an acceptable (at least *stable*) behavior for any admissible perturbations. This approach is related to H_∞-control and differential games; see, e.g., [1, 2]. However, we are not familiar with any results in these theories that may be applied to the parabolic systems with hard control and state constraints under consideration. In [3] the reader can find a number of feedback boundary control results for unconstrained parabolic systems based on Riccati equations.

In this paper we develop an efficient design procedure to solve minimax control problems for hard-constrained parabolic systems. This procedure takes

M. de Queiroz et al. (Eds.): Optimal Control, Stabilization and Nonsmooth Analysis, LNCIS 301, pp. 121–132, 2004.

into account *monotonicity properties* of the parabolic dynamics and *asymptotic characteristics* of transients on the *infinite horizon*. It was initiated in [5] for the case of one-dimensional heat-diffusion equations and then developed in [6, 8], where some *first-order* approximation results have been obtained for controls acting in Dirichlet as well as mixed boundary conditions.

This paper involves both first-order and *second-order* approximations to justify an appropriate structure and compute optimal parameters of *suboptimal* controls to the original state-constrained parabolic problem. In this way we *minimize* an energy-type cost functional in the case of *maximal perturbations* and ensure the desired state performance within the required constraints for all admissible disturbances. Based on a *variational approach*, we obtain verifiable conditions for *stability in the large* of the *highly nonlinear closed-loop* control system that excludes unacceptable self-vibrating regimes.

Our design and justification procedures involve *multistep approximations* and results from the optimal control theory for ordinary differential equations. As a by-product of this approach, we obtain a complete *measure-free* solution for a class of *state-constrained* optimal control problems related to approximations of the parabolic dynamics. The paper also presents some results of *numerical simulation* allowing us to compare suboptimal solutions obtained via first-order and second-order approximation procedures.

2 Problem Formulation and Basic Properties

It is well known (see, e.g., [4] and the references therein) that in many practical situations the groundwater level $y(t)$ can be modelled by the one-dimensional heat-diffusion equation

$$\frac{\partial y}{\partial t} = a\frac{\partial^2 y}{\partial x^2} + w(t), \quad 0 < x < L, \; t > 0$$

with the initial and boundary conditions given by

$$y(x,0) \equiv 0, \quad 0 \le x \le L,$$

$$y(0,t) = y(L,t) = u(t), \quad t \ge 0,$$

where $w(t)$ reflects uncertain disturbances and $u(t)$ stands for boundary controls. The constant a is proportional to the square of the interval length L^2 and is determined experimentally. The goal is to find a *feedback control* of the type $u(y(t, L/2))$ to *minimize* the energy-type cost functional

$$J(u) = \max_{w(\cdot)} \int_0^T |u(y(t, L/2))| dt$$

in the case of the *most unfavorable* boundary disturbances without violating the *state constraints*

$$|y(t, L/2)| \le \eta \;\; \forall \quad t \ge 0.$$

In this paper we study a more general *multidimensional* version of this problem.

Consider a self-adjoint and uniformly strongly elliptic operator defined by

$$A := -\sum_{i,j=1}^{n} a_{ij}(x)\frac{\partial^2}{\partial x_i \partial x_j} - c, \tag{1}$$

where $c \in \mathbb{R}$, $a_{ij} \in C^\infty(\bar{\Omega})$,

$$\begin{array}{l} \sum_{i,j=1}^{n} a_{ij}(x)\xi_i\xi_j \ge \nu \sum_{i=1}^{n} \xi_i^2, \;\; \nu > 0 \\ \text{for all } \; x \in \Omega, \; \xi = (\xi_1, \; \xi_2, \; \ldots, \; \xi_n) \in \mathbb{R}^n, \end{array}$$

and where Ω is a bounded open region in $\mathbb{R}^n$ with a sufficiently smooth boundary Γ. Given positive numbers $\underline{\alpha}$, $\bar{\alpha}$, $\underline{\beta}$, and $\bar{\beta}$, we define the sets of *admissible controls* $u(t)$ and admissible *uncertain disturbances* $w(t)$ by

$$\begin{array}{l} U_{ad} := \big\{u \in L^2(0,T) \;\big|\; u(t) \in [-\bar{\alpha}, \underline{\alpha}] \;\text{ a.e. }\; t \in [0,T]\big\}, \\ W_{ad} := \big\{w \in L^2(0,T) \;\big|\; w(t) \in [-\underline{\beta}, \bar{\beta}] \;\text{ a.e. }\; t \in [0,T]\big\}. \end{array}$$

Suppose that x_0 is a given point in Ω at which we are able to collect information about the system performances, and let $\eta > 0$. Consider the following *minimax feedback control problem* (P):

$$\text{minimize } \; J(u) = \max_{w(\cdot)\in W_{ad}} \int_0^T |u(y(t,x_0))|dt$$

over $u(\cdot) \in U_{ad}$ subject to the system

$$\begin{cases} \dfrac{\partial y}{\partial t} + Ay = w(t) \;\text{ a.e. in }\; Q := (0,T) \times \Omega, \\ y(0,x) = 0, \; x \in \Omega, \\ y(t,x) = u(t), \; (t,x) \in \Sigma := (0,T] \times \Gamma, \end{cases} \tag{2}$$

the pointwise state constraints

$$|y(t,x_0)| \le \eta \quad \text{for all} \quad t \in [0,T], \tag{3}$$

and the feedback control law

$$u(t) = u(y(t,x_0)) \tag{4}$$

acting through the Dirichlet boundary conditions in (2).

Problem (P) formulated above is one of the *most difficult* control problems unsolved in the general theory. Our purpose is to develop an approach that takes into account *specific features of parabolic systems* and allows us to find a feasible *suboptimal* feedback control. To furnish this, we employ the spectral representation of solutions to the parabolic system (2) with Dirichlet boundary conditions.

Let $\lambda \in \mathbb{R}$ be an eigenvalue of the operator A in (1) and let $\phi \in L^2(\Omega)$ be the corresponding eigenfunction satisfying the condition $\phi|_\Gamma = 0$. It is well known that, under the assumptions made, the following properties hold:

(a) All the eigenvalues λ_i, $i = 1, 2, \ldots$ of A form a nondecreasing sequence that accumulates only at $+\infty$, and the first eigenvalue λ_1 is simple.
(b) The corresponding orthonormal system of eigenfunctions is complete in the space $L^2(\Omega)$.

Note that the special one-dimensional case of $\Omega = [0, L]$ and

$$A = -a\frac{\partial^2}{\partial^2 x}$$

one has

$$\lambda_i = \frac{\pi^2 a}{L^2} i^2 \quad \text{and} \quad \phi_i(x) = \sqrt{2/L}\sin\frac{\pi i}{L}x, \; i \in \mathbb{N}.$$

Let $y \in L^2(Q)$ be a *generalized solution* to (2), which uniquely exists for each $(u, w) \in U_{ad} \times W_{ad}$; see, e.g., [3]. Based on the properties (a) and (b) and taking into account that both admissible controls and perturbations in (2) depend only on t, we conclude that the generalized solution $y(\cdot)$ admits the representation

$$\begin{aligned} y(t,x) = \textstyle\sum_{i=1}^\infty \mu_i \phi_i(x) \Big(\int_0^t w(\theta) e^{-\lambda_i(t-\theta)} d\theta \\ + (c+\lambda_i) \int_0^t u(\theta) e^{-\lambda_i(t-\theta)} d\theta \Big) \end{aligned} \tag{5}$$

where

$$\mu_i = \int_\Omega \phi_i(x) dx$$

and the series is strongly convergent in $L^2(Q)$. This allows us to deduce, using the maximum principle for parabolic equations, the following *monotonicity property* of solutions to the parabolic system (2) with respect to both controls and perturbations:

Theorem 1. *Let $(u_i, w_i) \in L^2(0,T) \times L^2(0,T)$ and let $y_i(\cdot)$, $i = 1, 2$, be the corresponding generalized solutions to (2).* Then

$$y_1(t,x) \geq y_2(t,x) \quad \text{a.e. in } Q$$

if $u_1(t) \geq u_2(t)$ and $w_1(t) \geq w_2(t)$ a.e. in $[0,T]$.

One can see from Theorem 1 that *the bigger magnitude of a perturbation, the more the control of the opposite sign that should be applied* to neutralize this perturbation and to keep the corresponding transient within the state constraint (3). This leads us to consider feedback control laws (4) satisfying the *compensation property*

$$u(y) \leq u(\tilde{y}) \text{ if } y \geq \tilde{y} \text{ and } y \cdot u(y) \leq 0 \quad \text{for all} \quad y, \tilde{y} \in \mathbb{R}. \tag{6}$$

The latter property implies that

$$\int_0^T |u(y(t))|dt \geq \int_0^T |u(\tilde{y}(t))|dt \text{ if } y(t) \geq \tilde{y}(t) \geq 0 \text{ or } y(t) \leq \tilde{y}(t) \leq 0$$

for all $t \in [0, T]$, i.e., the compensation of bigger (by magnitude) perturbations requires more cost with respect to the maximized cost functional in (P). This allows us to seek a *suboptimal control structure* in (P) by examining the control response to feasible perturbations of the *maximal magnitudes* $w(t) = \bar{\beta}$ and $w(t) = -\underline{\beta}$ on the whole time interval $t \in [0, T]$.

3 Optimal Control under Maximal Perturbations in Approximation Problems

Taking into account the symmetry of (P) relative to $y = 0$, we consider the case of *upper level maximal perturbations* $w(\cdot) = \bar{\beta}$ and the corresponding set of admissible time-dependent controls

$$\overline{U}_{ad} := \{u(\cdot) \in U_{ad} \mid -\bar{\alpha} \leq u(t) \leq 0 \quad \text{a.e.} \quad t \in [0, T]\}.$$

To find an optimal control $\bar{u}(t)$ in response to the maximal perturbations, we have the following *open-loop* control problem $(\overline{\mathrm{P}})$:

$$\text{minimize } \bar{J}(u) = -\int_0^T u(t)dt \tag{7}$$

over $u(\cdot) \in \overline{U}_{ad}$ subject to system (2) with $w(\cdot) = \bar{\beta}$ and the constraint

$$y(t) \leq \eta \quad \text{for all} \quad t \in [0, T]. \tag{8}$$

This is a *state-constrained Dirichlet boundary control* problem, which was considered in [7] in more generality. In [7] we obtained necessary optimality conditions for $(\overline{\mathrm{P}})$ that involve the adjoint operator to the so-called Dirichlet map and Borel measures. Those conditions are rather complicated and do not allow us to compute or even properly estimate an optimal control.

Following [5, 6], let us explore another approach to solve problem $(\overline{\mathrm{P}})$. It leads to *suboptimal feasible solutions* of a simple structure that can be used to design and justify a required feedback law in the original minimax control problem (P). To furnish this, we approximate $(\overline{\mathrm{P}})$ by optimal control problems for ODE systems obtained from the spectral representation (5) with $x = x_0$ and $w(\cdot) = \bar{\beta}$. In what follows we suppose, additionally to the basic assumptions in Section 2, that the *first eigenvalue* λ_1 in (a) is *positive.* Define $y_i(t)$ by

$$y_i(t) := \mu_i\phi_i(x_0)\left(\int_0^t \bar{\beta}e^{-\lambda_i(t-\theta)}d\theta + (c+\lambda_i)\int_0^t u(\theta)e^{-\lambda_i(t-\theta)}d\theta\right)$$

and directly observe that each $y_i(t)$, $i \in \mathbb{N} := 1, 2, \ldots$, satisfies the ordinary differential equation

$$\begin{aligned} &\dot{y_i} = -\lambda_i y_i + \mu_i\phi_i(x_0)(\bar{\beta} + (c+\lambda_i)u(t)) \;\text{ a.e. } t \in [0,T], \\ &y_i(0) = 0,\; i \in \mathbb{N} \end{aligned} \tag{9}$$

Thus the original PDE system (2) is *equivalent* to an infinite-dimensional system of ODEs (9) with

$$y(t, x_0) = \sum_{i=1}^{\infty} y_i(t).$$

To determine a suboptimal control to $(\overline{\mathrm{P}})$, we take finitely many terms in (9) and find an optimal control for such approximations. Below we present some results for the first-order (one-term) and second-order (two-term) approximations. Considering the *one-term approximation*, we arrive at the following problem:

$(\overline{\mathrm{P}}_1)$ minimize the cost functional (7) along the controlled differential equation

$$\dot{y} = -\lambda_1 y + \mu_1\phi_1(x_0)(\bar{\beta} + (c+\lambda_1)u(t)), \;\; y(0) = 0, \;\; \text{a.e. } t \in [0,T] \tag{10}$$

subject to $u(\cdot) \in \overline{U}_{ad}$ and the state constraint (8).

The next theorem we present provides a complete *exact solution* of the state-constrained problem $(\overline{\mathrm{P}}_1)$ with no measure involved.

Theorem 2. *Let* $\mu_1\phi_1(x_0)\bar{\beta} > \lambda_1\eta$. *Assume in addition that either*

$$\mu_1\phi_1(x_0)(\bar{\beta} - \bar{\alpha}(c+\lambda_1)) \le \lambda_1\eta \;\;\textit{or}\;\; \tau_1 := \frac{1}{\lambda_1}\ln\frac{\mu_1\phi_1(x_0)\bar{\beta}}{\mu_1\phi_1(x_0)\bar{\beta} - \lambda_1\eta} \ge T. \tag{11}$$

Then the system (8), (10) is controllable, i.e., there is $u(\cdot) \in \bar{U}_{ad}$ *such that the corresponding trajectory of (10) satisfies the state constraint (8). Moreover, problem* $(\overline{P}_1)$ *admits an optimal control of the form*

$$\bar{u}_1(t) = \begin{cases} 0 & \text{if } t \in [0, \bar{\tau}_1), \\ \dfrac{\lambda_1\eta - \mu_1\phi_1(x_0)\bar{\beta}}{\mu_1\phi_1(x_0)(c+\lambda_1)} & \text{if } t \in [\bar{\tau}_1, T], \end{cases} \tag{12}$$

where $\bar{\tau}_1 = \min\{\tau_1, T\}$ *with* τ_1 *computed in (11).*

To *prove* the theorem, we first approximate ($\overline{\mathrm{P}}_1$) by a parametric family of optimal control problems with *no state constraints.* The latter problems can be completely solved by using the *Pontryagin maximum principle*, which provides *necessary and sufficient* conditions for optimality in this case. In this way we derive that optimal controls to approximating problems are piecewise constant and contain both *bang-bang and singular modes.* Passing to the limit, we justify all the results of Theorem 2 and come to a rather *surprising conclusion* that the optimal control (12) for the state-constrained problem happens to be *simpler* than the ones for the unconstrained approximations.

Next let us consider the *two-term approximation* of the system (9), which leads to the following problem:

($\overline{\mathrm{P}}_2$)minimize the cost functional (7) along the controlled differential system

$$\begin{aligned}\dot{y_1} &= -\lambda_1 y_1 + \mu_1\phi_1(x_0)(\bar{\beta} + (c+\lambda_1)u(t)), \;\; y_1(0) = 0,\\ \dot{y_2} &= -\lambda_2 y_2 + \mu_2\phi_2(x_0)(\bar{\beta} + (c+\lambda_2)u(t)), \;\; y_2(0) = 0 \;\text{ a.e. } t \in [0,T]\end{aligned}$$

subject to $u(\cdot) \in \overline{U}_{ad}$ and the state constraint (8).

Employing the technique similar to the one used in the proof of Theorem 2, we get the following solution of problem ($\overline{\mathrm{P}}_2$):

Theorem 3. *Let*

$$\frac{\mu_1\phi_1(x_0)}{\lambda_1} + \frac{\mu_2\phi_2(x_0)}{\lambda_2} > \frac{\eta}{\bar{\beta}}.$$

Assume in addition that either

$$\frac{\mu_1\phi_1(x_0)(\bar{\beta} - \bar{\alpha}(c+\lambda_1))}{\lambda_1} + \frac{\mu_2\phi_2(x_0)(\bar{\beta} - \bar{\alpha}(c+\lambda_2))}{\lambda_2} \le \eta$$

or $\tau_1 \ge T$, where τ_1 is the solution of the transcendental equation

$$\frac{\mu_1\phi_1(x_0)}{\lambda_1}(1 - e^{-\lambda_1\tau_1}) + \frac{\mu_2\phi_2(x_0)}{\lambda_2}(1 - e^{-\lambda_2\tau_1}) = \frac{\eta}{\bar{\beta}}.$$

Also suppose that x_0 satisfies the inequality

$$(\lambda_1 + c)\mu_1\phi_1(x_0) + (\lambda_2 + c)\mu_2\phi_2(x_0) > 0,$$

which automatically holds in the special case of $\Omega = [0,L]$ and $c = 0$ for x_0 near the boundary. Then if $\bar{\alpha}$ in $\bar{U}_{ad} = [-\bar{\alpha}, 0]$ is sufficiently large in absolute value, the problem ($\overline{P}_2$) admits an optimal control of the form

$$\bar{u}_1(t) = \begin{cases} 0 & \text{if } t \in [0, \bar{\tau}_1), \\ -u_0 e^{-\alpha(t-\tau_1)} - u_1 & \text{if } t \in [\bar{\tau}_1, T] \end{cases} \tag{13}$$

with the parameters computed by

$$\alpha = \frac{\lambda_1 a_2 + \lambda_2 a_1}{a_1 + a_2}; \quad a_i = (\lambda_i + c)\mu_i\phi_i(x_0), \;\; i = 1, 2,$$

$$u_0 = \frac{\lambda_2 - \lambda_1}{a_1 + a_2}\left(y_1(\tau_1) - \frac{a_1(\lambda_2 \eta - \mu_2 \varphi_2(x_0)\bar{\beta}) + a_2 \mu_1 \varphi_1(x_0)\bar{\beta}}{\lambda_1 a_2 + \lambda_2 a_1} \right),$$

$$u_1 = \frac{\lambda_1 \lambda_2}{\lambda_1 a_2 + \lambda_2 a_1}\left(\frac{\mu_1 \phi_1(x_0)\bar{\beta}}{\lambda_1} + \frac{\mu_2 \phi_2(x_0)\bar{\beta}}{\lambda_2} - \eta \right),$$

and $\bar{\tau}_1 = \min\{\tau_1, T\}$.

Observe that the control $\bar{u}_1(t)$ given by (13) agrees with (12) when $\mu_2\phi_2(x_0) = 0$.

4 Suboptimal Control for Parabolic System under Maximal Perturbations

Let us optimize the control structure (12) subject to (1), (2), and $u(\cdot) \in \overline{U}_{ad}$ assuming that the point x_0 is chosen so that the solution $t = t_1$ of the equation

$$\sum_{i=1}^{\infty} \frac{\mu_i \phi_i(x_0)}{\lambda_i} e^{-\lambda_i t} = \sum_{i=1}^{\infty} \frac{\mu_i \phi_i(x_0)}{\lambda_i} - \frac{\eta}{\beta}$$

satisfies the strict inequality

$$t_1 > \frac{1}{\lambda_1} \ln \frac{\bar{\beta}(1 + c\gamma)}{(c + \lambda_1)(\gamma\bar{\beta} - \eta)},$$

which particularly holds in the special case of $\Omega = [0, L]$, $A = -a\frac{\partial^2}{\partial x^2}$ (i.e. $c = 0$) with any $a > 0$ for x_0 around $L/2$. This gives the control function

$$u(t) = \begin{cases} 0 & \text{for } 0 \le t < \tau, \\ -\bar{u} & \text{for } \tau \le t \le T \end{cases} \tag{14}$$

with the optimal parameters

$$\bar{u} := \frac{\gamma\bar{\beta} - \eta}{1 + c\gamma}, \quad \gamma := \sum_{i=1}^{\infty} \frac{\mu_i \phi_i(x_0)}{\lambda_i}, \tag{15}$$

and τ satisfying the equation

$$\sum_{i=1}^{\infty} \frac{\mu_i \phi_i(x_0)}{\lambda_i} e^{-\lambda_i T}\left[(c + \lambda_i)(\gamma\bar{\beta} - \eta)e^{\lambda_i \tau} - \bar{\beta}(1 + c\gamma)\right] = 0. \tag{16}$$

We can prove that (16) has a *unique solution* $\tau = \bar{\tau}(T) \in (0, T)$ for all T sufficiently large and that any control (14), (15) is *feasible* to $(\overline{\text{P}})$ for all positive $\tau \le \bar{\tau}(T)$. Moreover, the switching time $\tau = \bar{\tau}(T)$ is *optimal* in $(\overline{\text{P}})$ and $\bar{\tau}(T) \downarrow \bar{\tau}$ as $T \to \infty$, where the *asymptotically optimal* switching time $\bar{\tau}$ is computed by

$$\bar{\tau}_1 = \frac{1}{\lambda_1} \ln \frac{\bar{\beta}(1+c\gamma)}{(c+\lambda_1)(\gamma\bar{\beta}-\eta)}. \tag{17}$$

Optimizing the control structure (13) with respect to the parameters u_0, u_1, α, τ_1 is essentially more difficult than the above optimization with respect to the two parameters $\bar{u}, \tau$. Putting

$$u_1 := \frac{\gamma\bar{\beta}-\eta}{1+c\gamma}, \quad \gamma := \sum_{i=1}^{\infty} \frac{\mu_i \phi_i(x_0)}{\lambda_i}$$

in (13) and substituting it into (5), we obtain the equation

$$\sum_{i=1}^{\infty} \frac{\mu_i \phi_i(x_0)}{\lambda_i} \Big[(c+\lambda_i)(\gamma\bar{\beta}-\eta)e^{-\lambda_i(T-\tau)} - \bar{\beta}(1+c\gamma)e^{-\lambda_i T} + \frac{u_0(\lambda_i+c)}{\alpha-\lambda_i}(e^{-\alpha(T-\tau)} - e^{-\lambda_i(T-\tau)})\Big] = 0.$$

It is *no longer true* that the solution τ_2 of the latter equation can be asymptotically found from the condition of vanishing of its first term. To estimate the situation, *numerical simulation* is performed; see the next section. It has been found that if $\bar{\alpha}$ is sufficiently large in $\bar{U}_{ad} = [-\bar{\alpha}, 0]$, then there are parameters u_0, u_1, α, τ_1 such that the corresponding control (13) gives a *smaller value* of the cost functional in the case of maximal disturbance in comparison with the one provided by (14)–(16).

5 Numerical Simulation

Numerical simulation is performed for the one-dimensional equation

$$\frac{\partial y}{\partial t} = a\frac{\partial^2 y}{\partial x^2} + \bar{\beta}, \quad 0 < x < \pi, \ t > 0$$

with the initial and boundary conditions

$$y(x,0) \equiv 0, \quad 0 \le x \le \pi$$

$$y(0,t) = y(\pi,t) = u(t), \quad t \ge 0,$$

and the constraint $|y(t, \frac{\pi}{2})| \le \eta$.

The suboptimal control (14),(15) derived from the one-term approximation has the asymptotically optimal switching time $\bar{\tau}_1$ given by formula (17).

Let us consider the structure of suboptimal control (13) derived from the two-term approximation. If $\alpha > 0$ and $u_0 > 0$, then it is clear from physical considerations that the corresponding asymptotically optimal switching time $\bar{\tau}_2$ is greater than $\bar{\tau}_1$ (the larger the magnitude of the feasible control, the later it may be applied). It can be shown that the difference in costs provided by the control structure (13) and the first-order one with the switching time (17) tends to

$$-\bar{u}(\bar{\tau}_2 - \bar{\tau}_1) + \frac{u_0}{\alpha}$$

as $T \to \infty$. If the gain $\bar{u}(\bar{\tau}_2 - \bar{\tau}_1)$ exceeds the loss

$$\frac{u_0}{\alpha},$$

then the structure (13) is better than the one with (17).

In our numerical simulation we considered the extreme case when u_0 and α are very large positive numbers. Then

$$u(t) = \begin{cases} 0, & t < \bar{\tau}_2, \\ +\infty, & t = \bar{\tau}_2, \\ \gamma\bar{\beta} - \eta, & t > \bar{\tau}_2, \end{cases} \tag{18}$$

and

$$\lim_{\varepsilon \downarrow 0} \int_{\bar{\tau}_2 - \varepsilon}^{\bar{\tau}_2 + \varepsilon} u(t)dt = J,$$

where J is a parameter corresponding to the intensity of the impulse.

It happens (see the table below for selected data) that for any admissible values of $a, \bar{\beta}, \eta$ there always exists a value of J that provides a *smaller cost* for control (18), which can be thought of as a *second-order suboptimal control.* The last column in the table corresponds to the cost under the first order suboptimal control (14)–(16):

$$\int_{\bar{\tau}_1}^{T} u(t)dt = \bar{u}(T - \bar{\tau}_1) = (\bar{\beta}\gamma - \eta)(T - \bar{\tau}_1).$$

Observe that the relative gain for large T is *very small and does not have much of a practical meaning*; this can serve as a *justification of the first-order approximation* and the three-positional control law considered in the next section. The data in this table correspond to $\bar{\beta} = 2$.

a	η	optimal value of J	gain in the cost	$(\bar{\beta}\gamma - \eta)\ (T - \bar{\tau}_1)$
1	1	0.165	0.010	$1.47(T - \bar{\tau}_1)$
1.2	1	0.076	0.012	$0.88(T - \bar{\tau}_1)$
1.2	0.1	0.140	0.022	$1.78(T - \bar{\tau}_1)$
2	0.1	0.032	0.015	$0.77(T - \bar{\tau}_1)$

6 Feedback Control Design

The obtained results allow us to justify the *three-positional control law*

$$u(y) = \begin{cases} -\bar{u} & \text{if } y \geq \bar{\sigma}, \\ 0 & \text{if } -\underline{\sigma} < y < \bar{\sigma}, \\ \underline{u} & \text{if } y \leq -\underline{\sigma} \end{cases} \tag{19}$$

as a *suboptimal feedback structure* in (P) with the compensation property (6). Now using the *monotonicity* of transients with respect to both controls and perturbations as well as their asymptotic properties as $t \to \infty$, we arrive at the following theorem:

Theorem 4. *Let the feedback control parameters* $(\bar{u}, \underline{u}, \bar{\sigma}, \underline{\sigma})$ *in (19) be computed by the formulas (15),*

$$\bar{\sigma}(T) := \bar{\beta}(\gamma - \sum_{i=1}^{\infty} \frac{\mu_i \phi_i(x_0)}{\lambda_i} e^{-\lambda_i \bar{\tau}(T)}),$$

and their counterparts for $\underline{\beta}$. *Then the control law (19) is feasible for any perturbations* $w(\cdot) \in W_{ad}$ *being optimal in the case of maximal perturbations when* T *is sufficiently large. Moreover,* $\bar{\sigma}(T) \downarrow \bar{\sigma}$ *and* $\underline{\sigma}(T) \downarrow \underline{\sigma}$ *as* $T \to \infty$, *where the positive numbers*

$$\bar{\sigma} := \bar{\beta} \left(\gamma - \sum_{i=1}^{\infty} \frac{\mu_i \phi_i(x_0)}{\lambda_i} \left[\frac{(c + \lambda_1)(\gamma \bar{\beta} - \eta)}{\bar{\beta}(1 + c\gamma)} \right]^{\frac{\lambda_i}{\lambda_1}} \right), \tag{20}$$

$$\underline{\sigma} := \underline{\beta} \left(\gamma - \sum_{i=1}^{\infty} \frac{\mu_i \phi_i(x_0)}{\lambda_i} \left[\frac{(c + \lambda_1)(\gamma \underline{\beta} - \eta)}{\underline{\beta}(1 + c\gamma)} \right]^{\frac{\lambda_i}{\lambda_1}} \right) \tag{21}$$

form the maximal dead region $[-\underline{\sigma}, \bar{\sigma}]$ *under which feedback (19)keeps the state constraints (3) on the infinite horizon* $[0, \infty)$ *for any admissible perturbations.*

We finally observe that the feedback control (19) with the parameters calculated in Theorem 4 does not guarantee the *robust stability* of the highly nonlinear (discontinuous) closed-loop system (2), (4), (19) under any admissible perturbations. Indeed, this system may have a *self-vibrating regime* (i.e., its zero-equilibrium is not stable in the large) if the dead region $[-\underline{\sigma}, \bar{\sigma}]$ is not sufficiently wide. The next theorem excludes such a possibility and ensures the required *robust stability of the closed-loop control system.* Its proof is based on a *variational approach* and *turnpike asymptotic properties of the parabolic dynamics.*

Theorem 5. *The closed-loop control system (2), (4), (19) with arbitrary parameters* $(\bar{u}, \underline{u}, \bar{\sigma}, \underline{\sigma})$ *is stable in the large if*

$$\bar{\sigma} + \underline{\sigma} \geq \min\{\bar{u}, \underline{u}\} \Big[\frac{\mu_1 \phi_1(x_0)(c + \lambda_1)}{\lambda_1} - (1 + c\gamma) \Big] > 0. \tag{22}$$

When $\bar{\beta} \leq \underline{\beta}$, (22) can be written in the *simplified form*

$$2\bar{\sigma}_1 + \underline{\sigma}_1 \geq \eta$$

directly through the *suboptimal dead region bounds*

$$\bar{\sigma}_1 := \bar{\beta}\Big(\gamma - \frac{\mu_1\phi_1(x_0)(c+\lambda_1)(\gamma\bar{\beta}-\eta)}{\lambda_1\bar{\beta}(1+c\gamma)}\Big),$$

$$\underline{\sigma}_1 := \underline{\beta}\Big(\gamma - \frac{\mu_1\phi_1(x_0)(c+\lambda_1)(\gamma\underline{\beta}-\eta)}{\lambda_1\underline{\beta}(1+c\gamma)}\Big),$$

which correspond to the first terms in (20) and (21).

Acknowledgement. This research was partly supported by the National Science Foundation under grants DMS-0072179 and DMS-0304989.

References

1. Basar T, Bernard P (1995) H_∞-Optimal Control and Related Minimax Design Problems. Birkhäuser Boston
2. Krasovskii, A Subbotin (1998) Game-Theoretical Control Problems. Springer New York
3. Lasiecka I, Triggiani R (2000) Control Theory for Partial Differential Equations. Cambridge University Press Cambridge UK
4. Mordukhovich B (1986) Optimal control of the groundwater regime on engineering reclamation systems, Water Resources 12:244–253
5. Mordukhovich B (1999) Minimax design for a class of distributed parameter systems, Autom Remote Control 50:262–283
6. Mordukhovich B (1999) Minimax design of constrained parabolic systems. In: S. Chen et al. (eds) Control of Distributed Parameter and Stochastic Systems. Kluwer Boston
7. Mordukhovich B, Zhang K (1997) Minimax control of parabolic systems with Dirichlet boundary conditions and state constraints, Appl Math Optim 36:323–360
8. Mordukhovich B, Zhang K (1999) Robust suboptimal control of constrained parabolic systems under uncertainty conditions. In: Leitmann L, Udwadia E, Kryazhimskii A (eds) Dynamics and Control. Gordon and Breach Amsterdam

Part II

Stabilization and Lyapunov Functions

Interconnections of Monotone Systems with Steady-State Characteristics*

David Angeli[1] and Eduardo Sontag[2]

[1] Dip. Sistemi e Informatica, University of Florence, 50139 Firenze, Italy.
angeli@dsi.unifi.it

[2] Department of Mathematics, Rutgers University, New Brunswick, NJ 08901.
sontag@control.rutgers.edu

Summary. One of the key ideas in control theory is that of viewing a complex dynamical system as an interconnection of simpler subsystems, thus deriving conclusions regarding the complete system from properties of its building blocks. Following this paradigm, and motivated by questions in molecular biology modeling, the authors have recently developed an approach based on components which are monotone systems with respect to partial orders in state and signal spaces. This paper presents a brief exposition of recent results, with an emphasis on small gain theorems for negative feedback, and the emergence of multi-stability and associated hysteresis effects under positive feedback.

1 Introduction

Tools from control theory have long played an important role in the analysis of dynamical systems properties such as stability, hysteresis, and oscillations. Key to the application of control tools, is the idea of viewing a complex dynamical system as feedforward and feedback interconnections of simpler subsystems, thus deriving conclusions regarding the complete system from properties of its building blocks. One may then analyze complicated structures on the basis of the behavior of elementary subsystems, each of which is "nice" in a suitable input/output sense (stable, passive, etc), in conjunction with the use of tools such as the small gain theorem to characterize interconnections.

This paper focuses on a special case of this idea, when the components are *monotone systems with a well-defined steady state response.* Although this work originated in our study of certain particular models in the area of cell signaling, a field of interest in contemporary molecular biology, and we will use such models in order to illustrate our results, the mathematical results

*Research supported in part by USAF Grant F49620-01-1-0063 and NSF Grant CCR-0206789

M. de Queiroz et al. (Eds.): Optimal Control, Stabilization and Nonsmooth Analysis, LNCIS 301, pp. 135–154, 2004.

and techniques that we have developed are of independent interest, and one can easily foresee many other areas of application.

Our framework is that of *monotone systems with inputs and outputs*, a class of systems introduced recently in [2] and further studied in [3]. They provide a natural generalization of classical (no inputs and outputs) monotone dynamical systems, defined by the requirement that trajectories must preserve a partial ordering on states; classical monotone systems include the subclass of *cooperative* systems, in which different state variables reinforce each other (positive feedback) as well as certain more general systems in which each pair of variables may affect each other in either positive or negative, or even mixed, forms. Among the classical references for monotone dynamical systems are the textbook by Smith [29] and the papers [16, 17] by Hirsch, which provide a rich and elegant theory dealing with the precise characterization of omega limit sets and other asymptotic behavior.

The extension to systems with inputs and outputs is by no means a purely academic exercise, but it is a necessary first step in order to analyze interconnections, especially those including feedback loops, built up out of monotone components. It is perhaps remarkable that this 'system-theoretic' view of dynamical systems fits perfectly with one of the main themes and challenges in current molecular biology, namely, the understanding of cell behavior in terms of common 'modules' (see e.g. [15]).

In this paper, we provide a brief exposition of several of our results for monotone i/o systems. Proofs are not included, as they can be found in [2, 3]. In addition, in order to make the presentation easier to follow, we often make simplifying assumptions, such as convexity of state spaces and input spaces, which can be substantially relaxed, as discussed in [2, 3]. We first review the basic setup, based upon monotone i/o systems with well-defined steady-state responses. Then we present a very simple yet powerful result for the analysis of global convergence of feedback interconnections in the special case where feedback does not introduce multiple steady states. This result plays the role of a small-gain theorem when applied to negative feedback loops. In general, however, positive feedback may lead to bifurcations of equilibria, and our second main result deals with the analysis of this type of situation, providing a characterization of locations and stability of steady states. A unifying theme to our work is the reduction of the analysis of high-dimensional dynamics to a simple planar graphical test.

In closing this introduction, we wish to comment upon the biological motivations for our work. The study of oscillations in biochemical systems is a major theme in molecular biology modeling; see, for instance, the book [13], which is one of the major references in the field. Our work in monotone systems originated in the analysis of the emergence of oscillations in certain biological inhibitory feedback loops, and specifically in a mathematical model of *mitogen-activated protein kinase (MAPK) cascades*, which represent a 'biological module' or subcircuit which is ubiquitous in eukaryotic cell signal transduction processes. In the paper [32], one of the authors presented a small-gain theorem which provided very tight estimates of parameters at which a Hopf bifurcation occurs, in a model of inhibitory MAPK cascades given in [20], and which had been previously studied numerically in [26]. Analyzing a more realistic model than the one in those references (multiple as opposed to single phosphorylation) led us the results presented in [2] and reviewed here. We will discuss that example, as well as a classical model of circadian oscillators. Multi-stability and associated hysteresis effects form the basis of many models in molecular biology, in areas such as cell differentiation, development, and periodic behavior described by relaxation oscillations. See for instance the classic work by Delbrück [8], who suggested in 1948 that multi-stability could explain cell differentiation, as well as references in the current literature (e.g., [5], [9], [10], [11], [19], [22], [24], [33]). We will discuss a system of differential equations which describes a biochemical network of the type which appears in cell signaling, as an illustration of our theorems on multi-stability.

2 Basic Definitions

We will only consider systems defined by ordinary differential equations. Thus, all state-spaces, as well as input and output signal spaces, are assumed to be subsets of Euclidean space. By an *ordered Euclidean space* we mean here an Euclidean space $\mathbb{R}^n$, for some positive integer n, together with an order $\succeq$ induced by a positivity cone K. That is, $K \subseteq \mathbb{R}^n$ is a nonempty, closed, convex, pointed ($K \bigcap -K = \{0\}$), and solid (K has a nonempty interior) cone, and $x_1 \succeq x_2$ (or "$x_2 \preceq x_1$") means that $x_1 - x_2 \in K$. Given a cone K, two notions of strict ordering are possible. By $x_1 \succ x_2$, one means that $x_1 \succeq x_2$ and $x_1 \neq x_2$, and by $x_1 \gg x_2$ that $x_1 - x_2 \in \text{int}(K)$. For example, with respect to the 'NorthEast' ordering given by the first orthant $K = \mathbb{R}^n_{\geq 0}$, $x_1 \succeq x_2$ means that each coordinate of x_1 is bigger or equal than the corresponding coordinate of x_2, $x_1 \gg x_2$ means that every coordinate of x_1 is strictly larger than the corresponding coordinate of x_2, and $x_1 \succ x_2$ which means that *some* coordinate is strictly larger.

(More general notions of ordered spaces, not necessarily induced by cones, may be considered as well; and indeed some of the results in [2] are proved for more abstract ordered spaces. In addition, one may wish to study monotone I/O systems on infinite-dimensional spaces, particularly when dealing with delay-differential systems or systems defined by partial differential equations.)

By a *state-space* we will mean a subset X of an ordered Euclidean space $(\mathbb{R}^n, K^X)$ such that X is the closure of an open subset of $\mathbb{R}^n$ and is convex. Similarly, by an *input set* $\mathcal{U}$ we mean a subset of an ordered space $(\mathbb{R}^m, K^{\mathcal{U}})$, and by an *output set* $\mathcal{Y}$ a subset of an $(\mathbb{R}^p, K^{\mathcal{Y}})$. When there is no risk of ambiguity and the meaning is clear from the context, we drop the superscripts and write just 'K' instead of K^X, $K^{\mathcal{U}}$, or $K^{\mathcal{Y}}$. For simplicity, we state many of our results for *single-input single-output* ("SISO") systems, those for which input and output signals are scalar: $m = p = 1$. By an *input* (function) we mean a locally essentially bounded Lebesgue measurable function $u(\cdot) : \mathbb{R}_{\geq 0} \to \mathcal{U}$, and we write $u_1 \succeq u_2$ provided that $u_1(t) \succeq u_2(t)$ for almost all $t \geq 0$.

By a (finite-dimensional continuous-time) *system*, we mean the usual concept from control theory (see e.g. [31]), namely a system with inputs and outputs

$$\dot{x} = f(x,u), \quad y = h(x) \tag{1}$$

specified by a state space X, an input set $\mathcal{U}$, and an output set $\mathcal{Y}$, such that solutions of (1) that start in X do not leave X (forward invariance of X). We assume that $f : X \times \mathcal{U} \to \mathbb{R}^n$ (which may be thought of as a vector field with parameters) is continuous in (x,u), and is locally Lipschitz continuous in x locally uniformly in u, and that the function $h : X \to \mathcal{Y}$ is continuous. When mentioning derivatives of f, we will implicitly assume that we have in addition required f to be differentiable, meaning that map f extends differentiably to an open subset of $\mathbb{R}^n \times \mathbb{R}^m$ that contains $X \times \mathcal{U}$. We will make the assumption that all solutions with initial states in X must be defined for all $t \geq 0$ (forward completeness), and denote by $x(t) = \phi(t,\xi,u)$ (or just "$x(t,\xi,u)$") the (unique) solution of $\dot{x}(t) = f(x(t),u(t))$ with initial condition $x(0) = \xi$ and input $u(\cdot)$ at time $t \geq 0$. We also sometimes use the notation $y(t,\xi,u)$ as a shorthand for $h(\phi(t,\xi,u))$. For classical dynamical systems $\dot{x} = f(x)$, f does not depend on inputs, and there is no mapping h; they can be thought of as the particular case in which $\mathcal{U}$ and $\mathcal{Y}$ reduce to a single point, so that all definitions to be given will also apply to them.

The system (1) is said to be *monotone* if the following property holds, with respect to the orders on states and inputs:

$$\xi_1 \succeq \xi_2 \quad \& \quad u_1 \succeq u_2 \quad \Rightarrow \quad x(t,\xi_1,u_1) \succeq x(t,\xi_2,u_2) \quad \forall t \geq 0$$

and also the mapping h is monotone (with respect to the orders on states and output values). It is an elementary fact (see e.g. [2]) that the above inequality is equivalent to:

$$\xi_1 \gg \xi_2 \quad \& \quad u_1 \succeq u_2 \quad \Rightarrow \quad x(t,\xi_1,u_1) \gg x(t,\xi_2,u_2) \quad \forall t \geq 0$$

(which follows from the fact that a set which is the closure of its interior is invariant under a controlled dynamics iff its interior is invariant). A monotone system is *strongly monotone* if the following stronger property holds:

$$\xi_1 \succ \xi_2 \quad \& \quad u_1 \succeq u_2 \quad \Rightarrow \quad x(t,\xi_1,u_1) \gg x(t,\xi_2,u_2) \quad \forall t > 0.$$

This is a fairly strong statement, saying that a very strict inequality among states holds immediately after the initial time, provided only that the initial states be weakly comparable to each other.

The basic components to be interconnected will not be merely monotone, but they will satisfy the additional requirement that they be monostable in the sense of having a well-defined steady-state response for each possible *constant* input. This property, which we define next, *when combined with monotonicity,* will strongly constrain the limiting behavior or arbitrary time-varying inputs.

Definition 1. *The system (1)* admits an input to state (I/S) static characteristic $k^X(\cdot) : \mathcal{U} \to X$ *if, for each constant input* $u(t) \equiv u \in \mathcal{U}$, *there exists a unique globally asymptotically stable equilibrium* $k^X(u)$. *The characteristic is said to be* non-degenerate *if, in addition, the Jacobian* $D_x f(k^X(u), u)$ *is nonsingular, for all such* u. *If (1) admits an I/S characteristic, its* input/output (I/O) characteristic *is by definition the composition* $k^{\mathcal{Y}} := h \circ k^X$.

2.1 Infinitesimal Characterizations

It is obviously important to be able to check monotonicity without having to actually solve the differential equations. We quote two results from [2]; in both cases we assume we are given appropriate positivity cones. The first one uses (contingent) tangent cones, in the sense of nonsmooth analysis, and admits generalizations beyond orders defined by positivity cones:

Theorem 1. *The system (1) is monotone if and only if* h *is monotone and the following property holds:*

$$\xi_1 \succeq \xi_2 \text{ and } u_1 \succeq u_2 \Rightarrow f(\xi_1, u_1) - f(\xi_2, u_2) \in \mathcal{T}_{\xi_1 - \xi_2} K\,, \tag{2}$$

where $\mathcal{T}_\xi K$ *denotes the (contingent) tangent cone to* K *at the point* ξ.

The second one relies upon convex analysis, and is based upon viewing monotone systems with inputs (and outputs) as those for which the vector field f is *quasi-monotone* in the sense of differential equations (see e.g. [25]):

Theorem 2. *The system (1) is monotone if and only if* h *is monotone and the following property holds:*

$$\begin{aligned} \xi_1 \succeq \xi_2,\ u_1 \succeq u_2,\ \zeta \in K^*,\ \text{and } \langle \zeta, \xi_1 \rangle = \langle \zeta, \xi_2 \rangle \\ \Rightarrow \langle \zeta, f(\xi_1, u_1) \rangle \ \geq\ \langle \zeta, f(\xi_2, u_2) \rangle \end{aligned} \tag{3}$$

where K^* *is the set of all* $\zeta \in \mathbb{R}^n$ *so that* $\langle \zeta, k \rangle \geq 0$ *for all* $k \in K$.

Condition (3) may be separated into the conjunction of: for all ξ and all $u_1 \succeq u_2$, $f(\xi, u_1) - f(\xi, u_2) \in K$, and for all u, $\xi_1 \succeq \xi_2$, and $\langle \zeta, \xi_1 \rangle = \langle \zeta, \xi_2 \rangle$, $\langle \zeta, f(\xi_1, u) \rangle \geq \langle \zeta, f(\xi_2, u) \rangle$. A similar separation is possible in Theorem 1. In both cases, it suffices to check the properties for $\xi_1 - \xi_2 \in \partial K$ instead of arbitrary points.

2.2 Orthant Orders

We assume in this section, to simplify the exposition, that both the interior of the state space X and of the input set $\mathcal{U}$ are convex; also, f is supposed to be continuously differentiable.

A special class of positivity cones are orthants. Any orthant K in $\mathbb{R}^n$ has the form

$$K^{(\varepsilon)} = \{x \in \mathbb{R}^n \mid (-1)^{\varepsilon_i} x_i \geq 0\,,\; i = 1, \ldots, n\}$$

for some binary vector $\varepsilon = (\varepsilon_1, \ldots, \varepsilon_n) \in \{0,1\}^n$. A special case is when states, inputs, and outputs are ordered with $K =$ the main orthant (all $\varepsilon_i = 0$); systems that are monotone with respect to these orthants are called *cooperative systems*.

Proposition 1. *The system (1) is cooperative if and only if the following properties hold:*

$$\frac{\partial f_i}{\partial x_j}(x,u) \geq 0 \quad \forall x \in X,\ \forall u \in \mathcal{U},\ \forall i \neq j$$
$$\frac{\partial f_i}{\partial u_j}(x,u) \geq 0 \quad \forall x \in X,\ \forall u \in \mathcal{U},\ \forall i,j$$
$$\frac{\partial h_i}{\partial x_j}(x) \geq 0 \quad \forall x \in X,\ \forall i,j\,.$$

The term "cooperative" arises from the fact that the various variables "help" each other (rates of change depend increasingly on the remaining variables).

Under appropriate changes of variables, one may reduce the study of monotone systems with respect to arbitrary orthants to the study of cooperative systems; see [2] for details. The following corollary results from this reduction:

Corollary 1. *The system (1) is monotone with respect to the orders induced from orthants $K^X = K^{(\varepsilon)}$, $K^{\mathcal{U}} = K^{(\delta)}$, and $K^{\mathcal{Y}} = K^{(\mu)}$, if and only if the following properties hold:*

$$(-1)^{\varepsilon_i+\varepsilon_j}\frac{\partial f_i}{\partial x_j}(x,u) \geq 0 \quad \forall x \in X,\ \forall u \in \mathcal{U},\ \forall i \neq j$$
$$(-1)^{\varepsilon_i+\delta_j}\frac{\partial f_i}{\partial u_j}(x,u) \geq 0 \quad \forall x \in X,\ \forall u \in \mathcal{U},\ \forall i,j$$
$$(-1)^{\varepsilon_i+\mu_j}\frac{\partial h_i}{\partial x_j}(x) \geq 0 \quad \forall x \in X,\ \forall i,j\,.$$

It is important to be able to decide, just from the system equations, if there is *some* orthant K such that a given system is monotone with respect to K. Graphical conditions are very useful for that. It is not difficult to see that the conditions given in [21] for classical monotone systems can be generalized as follows. We associate to a given system (1) a signed digraph, with vertices $x_1, x_2 \dots x_n$, $u_1, u_2, \dots u_m$, $y_1, y_2 \dots y_p$ and edges constructed according to the following set of rules and constraints (if the rules do not all apply, the graph is undefined).

For edges between nodes x_k, the graph is defined only for systems so that, for any pair of integers $1 \leq i, j \leq n$ with $i \neq j$, either (a) $\partial f_i / \partial x_j(x, u) \equiv 0$, or (b) $\partial f_i / \partial x_j(x, u) \geq 0$ for all x, u and $\partial f_i / \partial x_j(x, u) > 0$ for some x, u, or (c) $\partial f_i / \partial x_j(x, u) \leq 0$ for all x, u and $\partial f_i / \partial x_j(x, u) < 0$ for some x, u; in that case, we draw no edges, a positive edge, or a negative edge, respectively, from x_j to x_i. For edges from nodes u_j to nodes x_i, similar rules apply, where we consider the Jacobians $\partial f_i / \partial u_j$. Finally, for edges from nodes x_j to nodes y_i, we proceed analogously using the Jacobians $\partial h_i / \partial x_j$.

A *cycle*, not necessarily directed, is a sequence of vertices $v_{n_0}, v_{n_1} \dots v_{n_L}$ such that $v_{n_0} = v_{n_L}$ and the constraint that, for each $0 \leq k \leq L - 1$, either there is an edge from v_k to v_{k+1} or there is an edge from v_{k+1} to v_k. The sign of a cycle is defined as the product of the signs of the edges comprising it, and the sign of a path is defined to be the product of the signs of its edges.

Proposition 2. *The system (1) which admits an incidence graph according to the above set of rules, is monotone with respect to some orthants K^X, $K^{\mathcal{U}}$ and $K^{\mathcal{Y}}$ if and only if its graph does not contain any negative cycles.*

Graphical conditions for strong monotonicity can also be given. Suppose that we strengthen the conditions to require either $\partial f_i / \partial x_j(x, u) \equiv 0$, $\partial f_i / \partial x_j(x, u) > 0$ for all x, u, or $\partial f_i / \partial x_j(x, u) < 0$ for all x, u, and similarly for edges from nodes u_j to nodes x_i and for edges from nodes x_j to nodes y_i. Assuming that a graph in this sense is well-defined, results are provided in [3] which allow one to check if a closure under unity feedback is strongly monotone. The results are stated in terms of "excitability" and "transparency" notions associated to the input/output system. For the purposes of treating the examples in this paper, we only require the more classical version of these results, for systems $\dot{x} = f(x)$ with no inputs:

Proposition 3. *Suppose that the system $\dot{x} = f(x)$ is monotone with respect to an orthant K. If either of these conditions hold:*

1. *the Jacobian $\partial f_i / \partial x_j(x)$ is an irreducible matrix, for each $x \in X$, or*
2. *the Jacobian $\partial f_i / \partial x_j(x)$ is an irreducible matrix, for each $x \in \operatorname{inter} X$ and every trajectory lies in the interior of X for all $t > 0$,*

then the system is strongly monotone with respect to K.

See Theorem 1.1 in [16] or Chapter 4 in [29] for a proof. (These references assume that the system evolves on an open set; however, the same proofs apply when the state space is closed, so part 1 follows immediately. Regarding part 2, one may reduce to case 1 as follows: given $\xi_1 \succ \xi_2$, we know that $x(t/2, \xi_1) \succ x(t/2, \xi_2)$ (monotonicity plus the fact that a flow is always one-to-one); as the interior is forward invariant, one need only consider these states at time $t/2$ as initial states, to conclude that, using the strong monotonicity results for the interior, $x(t,\xi_1) \gg x(t,\xi_2)$. Incidentally, an even weaker condition may be assumed, namely that irreducibility of the Jacobian holds almost everywhere in time along a trajectory.)

3 Feedbacks with Single Steady-States

It is easy to verify that cascade interconnections of monotone systems admitting characteristics are again monotone and admit characteristics. More interesting is the study of feedback, where the output is fed back to the input of (1) via a static or dynamic feedback law. In this section we study feedback interconnections that lead to global attractivity, and in the next section we study multiple steady states.

We restrict attention to SISO systems, as the results are easier to state in that context. Without loss of generality (otherwise, one may consider $-u$ as an input or $-y$ as an output), we will assume that $K^{\mathcal{U}} = K^{\mathcal{Y}} = \mathbb{R}_{\geq 0}$. Our interest will be in the effect of either "positive" or "activating" feedback (which preserves the orders on inputs and outputs) or "negative" or "inhibitory" feedback (which inverts the orders). The study of general dynamic feedbacks can be reduced to the study of static feedback, simply by viewing a feedback interconnection

$$\begin{aligned} \dot{x} &= f_x(x,u)\,, \quad & y &= h_x(x) \\ \dot{z} &= f_z(z,y)\,, \quad & u &= h_z(z) \end{aligned}$$

as the closure under the static feedback $u = y$ of the composite system

$$\begin{aligned} \dot{x} &= f_x(x,u) \\ \dot{z} &= f_z(z,h_x(x))\,, \quad y = h_z(z) \end{aligned}$$

so we will only state a result for static feedback. (The theorem is proved directly for arbitrary dynamic feedback in [2], and the proof is virtually the same.)

Theorem 3. *Suppose that the system (1) is monotone, with $K^{\mathcal{U}} = K^{\mathcal{Y}} = \mathbb{R}_{\geq 0}$, and that it has a well-defined I/O characteristic $k : \mathcal{U} \to \mathcal{Y}$. Let $\ell : \mathcal{Y} \to \mathcal{U}$ be a monotone (increasing or decreasing) map, and denote $F := \ell \circ k : \mathcal{U} \to \mathcal{U}$. Suppose that every trajectory of the closed-loop system*

$$\dot{x} = f(x,(\ell \circ h)(x)) \tag{4}$$

is bounded. Then, system (4) has a globally attractive equilibrium provided that the scalar discrete time iteration on $\mathcal{U}$

$$u_{k+1} = F(u_k) \tag{5}$$

has a unique globally attractive equilibrium $\bar{u}$.

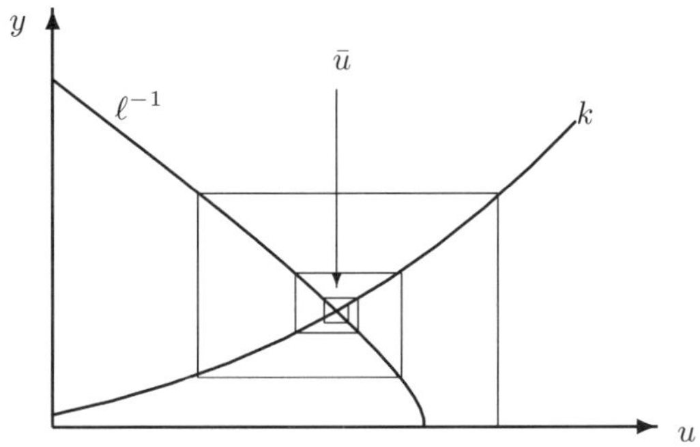

Fig. 1. I/O characteristics in (u, y) plane: negative feedback

For a graphical interpretation of condition (5), for the special case in which ℓ is decreasing ("negative feedback"), see Fig. 1. The figure shows the "convergent spiderweb" diagram that establishes global convergence to the equilibrium of the discrete-time iteration.

The proof of this theorem is not difficult; let us sketch it here. Consider an arbitrary (bounded, by assumption) solution of the closed-loop system, and the signals $u(t)$ and $y(t)$ that appear as input and output of the original system (1). Note that $u(t) = \ell(y(t))$ for all t. The key is to consider the quantities

$$u_+ := \limsup_{t\to+\infty} u(t),\ u_- := \liminf_{t\to+\infty} u(t),\ y_+ := \limsup_{t\to+\infty} y(t),\ y_- := \liminf_{t\to+\infty} y(t),$$

and to use the fact that monotonicity of (1) implies that

$$k(u_-) \le y_- \le y_+ \le k(u_+) \tag{6}$$

and the fact that (from monotonicity of ℓ) either $\ell(u_-) = y_-$ and $\ell(u_+) = y_+$, or $\ell(u_-) = y_+$ and $\ell(u_+) = y_-$. In the first case, we apply the increasing function ℓ to (6) and obtain $F(u_-) \le u_- \le u_+ \le F(u_+)$, and, inductively using that F is increasing,

$$F^i(u_-) \le u_- \le u_+ \le F^i(u_+)$$

for all $i \ge 1$, so taking limits and using $F^i(u_-) \to \bar{u}$ and $F^i(u_+) \to \bar{u}$, we obtain that $u_- = u_+ = \bar{u}$, and hence $u(t) \to \bar{u}$. Then, from the "convergent input convergent state" property for monotone systems with well-defined characteristics (cf. [2]), one concludes that the state $x(t) \to k^X(\bar{u})$. In the second case, we obtain $F(u_-) \ge u_+ \ge u_- \ge F(u_+)$, and, applying F (now a decreasing function) we have that $F^2(u_-) \le u_- \le u_+ \le F^2(u_+)$, and, inductively using that F^2 is increasing,

$$F^i(u_-) \le u_- \le u_+ \le F^i(u_+)$$

for all *even* i; the proof then is finished as before.

It is worth noting that, in the case of increasing ℓ (positive feedback), the global attractivity condition on (5) is satisfied automatically whenever F has a unique fixed point $\bar{u}$ and it holds that $\bar{u} \leq F(u) \leq u$ for all $u \geq \bar{u}$ and $\bar{u} \geq F(u) \geq u$ for all $u \leq \bar{u}$. This is in turn satisfied whenever the graphs of k and ℓ^{-1} intersect at only one point $\bar{u}$ and there $k'(\bar{u}) < (\ell^{-1})'(\bar{u})$.

Boundedness is assumed in our statement of the theorem. Often in biological applications, boundedness is automatic from conservation of mass and similar assumptions; in any event, we show in [2] how the assumption can be weakened provided that some more information is available about the characteristic of the system.

We also remark that arbitrary delays in the feedback loop do not affect the argument. That is, suppose that $u(t) = \ell(y(t - \sigma(t)))$, for some possibly time-varying delay $\sigma(t)$, where σ may be fairly arbitrary, for instance any continuous function such that $t - \sigma(t) \to \infty$ as $t \to \infty$, so that $\limsup y(t - \sigma(t)) = \limsup y(t)$ and similarly for $\liminf$. The same proof then applies. This means that the condition that we gave is a condition for stability under arbitrary delays, as is typically the case with "small-gain" arguments.

A partial converse to this result holds as well: if the feedback ℓ induces a periodic orbit for the discrete iteration, then, for an appropriate delay $\sigma(t)$, an approximate periodic orbit appears. The idea, whose precise statement and proof we leave for a future work, is roughly as follows. Suppose that $k(a) = b$, $\ell(b) = c$, $k(c) = d$, and $\ell(d) = a$. We consider the initial state of the output as $y(t) \equiv d$ for a time long enough that the input $u(t) = \ell(d) = a$ forces the output $y(t)$ to converge to b. Now, for a long enough interval so that convergence to steady state ensues, the input will converge to $\ell(b) = c$, and after a suitable time, the output will converge to $k(c) = d$, after which the pattern will repeat. There results an oscillation of the input between (approximately) the constant values a and c. By adjusting the delay periods, one can achieve sustained oscillations for as long as desired.

4 Multiple Steady States

We now turn to the more general case when several steady states may appear in closed-loop; our result will provide a global analysis tool for systems obtained by positive feedback loops involving monotone systems. In [33], Thomas conjectured that the existence of at least one positive loop in the incidence graph is a necessary condition for the existence of multiple steady states. Proofs of this conjecture were given in [14], [23], [30], and [7], under different assumptions on the system. (The last reference provides the most general result, using a degree theory argument.) However, the existence of positive loops is not sufficient, and our next theorem deals precisely with this question.

Theorem 4. *Assume that the system (1) is monotone, with $K^{\mathcal{U}} = K^{\mathcal{Y}} = \mathbb{R}_{\geq 0}$, and that it has a well-defined nondegenerate characteristic k. Let $\ell : \mathcal{Y} \to$*

$\mathcal{U}$ be a monotone increasing map, and set $F := \ell \circ k : \mathcal{U} \to \mathcal{U}$. Assume that every fixed point of F is non-degenerate ($F'(x) \neq 1$) and that the system (4), namely,

$$\dot{x} = f(x, (\ell \circ h)(x)),$$

is strongly monotone and all of its solutions are bounded. Then, the steady-states of (4) are in 1-1 correspondence with the fixed points of F, and for almost all initial conditions, solutions converge to the set of equilibria of (4) corresponding to inputs for which $F'(u) < 1$.

The proof of this theorem is given in [3]. It relies upon linear control theory arguments for analyzing local stability of the closed loop system, together with the use of a theorem due to Hirsch on almost-everywhere convergence for strongly monotone systems.

Note that the condition $F'(u) < 1$ amounts to local (exponential) stability of the discrete iteration (5). Strong monotonicity of the closed-loop system (4) can be checked in several graph-theoretic ways, based upon conditions on the open-loop system (1); see [2]. Monotonicity of (1) with respect to some orthant and orders $K^{\mathcal{U}} = K^{\mathcal{Y}} = \mathbb{R}_{\geq 0}$ amounts to the incidence graph (defined earlier) having no negative cycles plus the requirement that all paths from the input to the output node must be positive.

5 Example of Multiple Steady States

A typical situation for the application of Theorem 4 is where a monotone system with a well-defined I/O characteristic of sigmoidal shape is closed under unitary feedback: if the sigmoidal function is sufficiently steep, this configuration is known to yield 3 equilibria, 2 stable and 1 unstable. In biological examples, this might arise when a feedback loop comprising any number of positive interactions and an even number of inhibitions is present (but no inhibition at all is also a situation which might lead to the same type of behavior). This is a well-known principle in biology. One of its simplest manifestations is the so called "competitive exclusion" principle, in which one of two competing species completely eliminates the other, or more generally, for appropriate parameters the bistable case in which they coexist but the only possible equilibria are those where either one of the species is strongly inhibited. Of course, the interest of our results is in the high-dimensional case in which phase-plane techniques cannot provide the result, and we turn to such an example next. However, let us note that, for the special case of two-dimensional systems, our techniques are very close to those of [6]. In fact, even the 4-dimensional example of a two-repressor system with RNA dynamics, treated in [6] (Appendix I) in an ad-hoc manner, can be shown to be globally bistable as an immediate application of our techniques.

We now turn to a less trivial example where our tools may be applied. (A different example, involving cascades of systems of this type, and with comparisons with experimental data, is treated in [1].) Consider the following chemical reaction, involving various forms of a protein E:

$$\mathrm{E}_1 \underset{\longleftarrow}{\overset{\mathrm{U}}{\rightleftharpoons}} \mathrm{E}_2 \underset{\longleftarrow}{\overset{\mathrm{U}}{\rightleftharpoons}} \dots \underset{\longleftarrow}{\overset{\mathrm{U}}{\rightleftharpoons}} \mathrm{E}_{n-1} \underset{\longleftarrow}{\overset{\mathrm{U}}{\rightleftharpoons}} \mathrm{E}_n$$

being driven forward by an enzyme U, with the different subscripts indicating an additional degree of phosphorylation, and with constitutive dephosphorylation. This is a very general type of "module" that appears in cell signaling. We will be interested in positive feedback from E_n to U.

A typical way to model such a reaction is as follows. We introduce variables $x_i(t)$, $i = 1, \dots, n$ to indicate the fractional concentrations of the various forms of the enzyme E (so that $x_1 + \dots + x_n \equiv 1$, and $x_i \geq 0$, for the solutions of physical interest), and $u(t) \geq 0$ to indicate the concentration of U. The differential equations are then as follows:

$$\begin{aligned}
\dot{x}_1 &= -\sigma_1(u)\alpha_1(x_1) + \beta_2(x_2) \\
\dot{x}_2 &= \sigma_1(u)\alpha_1(x_1) - \beta_2(x_2) - \sigma_2(u)\alpha_2(x_2) + \beta_3(x_3) \\
&\vdots \\
\dot{x}_{n-1} &= \sigma_{n-2}(u)\alpha_{n-2}(x_{n-2}) - \beta_{n-1}(x_{n-1}) - \sigma_{n-1}(u)\alpha_{n-1}(x_{n-1}) + \beta_n(x_n) \\
\dot{x}_n &= \sigma_{n-1}(u)\alpha_{n-1}(x_{n-1}) - \beta_n(x_n) \,.
\end{aligned}$$

We make the assumptions that α_i and β_i (respectively, σ_i) are differentiable functions $[0, \infty) \to [0, \infty)$ with positive (respectively, either positive or identically zero) derivatives, and $\alpha_i(0) = \beta_i(0) = 0$ and $\sigma_i(0) > 0$ for each i. (We allow some of the σ_i to be constant, and in this manner represent steps that are not controlled by U.) Since we are interested in studying the effect of feeding back E_n, we pick $y = x_n$.

We first prove that the characteristic is well-defined. Recall that we are only interested in those solutions that lie in the intersection X of the plane $x_1 + \dots + x_n \equiv 1$ and the nonnegative orthant in $\mathbb{R}^n$. This set is easily seen to be invariant for the dynamics, and it is convex, so the Brower fixed point theorem guarantees the existence of an equilibrium in X, for any constant input $u(t) \equiv a$. We next prove that this steady-state is unique. Redefining if necessary the functions α_i, we will assume without loss of generality that $\sigma_i(a) = 1$ for all i. Let us introduce the nondecreasing functions

$$G_k = \beta_k^{-1} \circ \alpha_{k-1} \circ \beta_{k-1}^{-1} \circ \dots \circ \beta_2^{-1} \circ \alpha_1$$

for each $k = 2, \dots, n$ and $G(r) := r + G_2(r) + \dots + G_n(r)$. This function is defined on some maximal interval $[0, M]$, consisting of those r such that $\alpha_1(r)$ belongs to the range of β_2, $\alpha_2(\beta_2^{-1}(\alpha_1(r)))$ belongs to the range of β_3, and so forth, and it is strictly increasing. Moreover, for each equilibrium $x = (x_1, \dots, x_n)$, it holds that $x_k = G_k(x_1)$, and therefore, recalling that $x_1 + \dots + x_n = 1$, $G(x_1) = 1$. Thus, if x and $\widetilde{x}$ are two steady states, we have

$G(x_1) = G(\widetilde{x}_1)$. Since G is strictly increasing, it follows that $x_1 = \widetilde{x}_1$, and therefore that $x_k = G_k(x_1) = G_k(\widetilde{x}_1) = \widetilde{x}_k$ for all k, so uniqueness is shown.

We must prove stability. For that, we first perform a change of coordinates:

$$z_1 = x_1,\ z_2 = x_1 + x_2,\ \ldots,\ z_{n-1} = x_1 + \ldots + x_{n-1},\ z_n = x_1 + \ldots + x_n$$

so that the equations in these new variables become (using that $\dot{z}_k = (d/dt)(x_1 + \ldots + x_k)$ and $x_k = z_k - z_{k-1}$ for $k > 1$):

$$\begin{aligned} \dot{z}_1 &= -\sigma_1(u)\alpha_1(z_1) + \beta_2(z_2 - z_1) \\ &\vdots \\ \dot{z}_k &= -\sigma_k(u)\alpha_k(z_k - z_{k-1}) + \beta_{k+1}(z_{k+1} - z_k) \\ &\vdots \\ \dot{z}_{n-1} &= -\sigma_{n-1}(u)\alpha_{n-1}(z_{n-1} - z_{n-2}) + \beta_n(1 - z_{n-1}) \end{aligned}$$

(and $z_n \equiv 1$). When the input $u(t)$ is equal to any given constant, the system described by the first $n-1$ differential equations, seen as evolving in the subset of $\mathbb{R}^{n-1}$ where $0 \le z_1 \le z_2 \le \ldots \le z_{n-1} \le 1$, is a *tridiagonal strongly cooperative system*, and thus a theorem due to Smillie (see [27]) insures that all trajectories converge to the set of equilibria. (The proof given in [28] is also valid when the state-space is closed, as here.) Moreover, linearizing at the equilibrium preserves the structure, so applying the same result to the linearized system we know that we have in fact an exponentially stable equilibrium. Thus, characteristics are well defined and nondegenerate.

It is easy to verify from our graph conditions that the system (in the new coordinates) is monotone, since $df_i/dz_j > 0$ for all pairs $i \neq j$, $df_i/du \le 0$ for all i, and $dh/dz_i = 0$ for all $i < n-1$ and $dh/dz_{n-1} < 0$ (the output is $y = x_n = 1 - z_{n-1}$).

The closed-loop system (4) is obtained in this case (unity feedback) by setting $u = \ell(y) = y$ (and changing to the new coordinates z_i), so, since the output is $1 - z_{n-1}$, we have $u = (\ell \circ h)(z) = 1 - z_{n-1}$. We need to prove strong monotonicity of (4). Closure under positive feedback preserves monotonicity (see [2]). By Proposition 3, it is enough to see that the Jacobian is irreducible in the interior of the state space (where $z_i < z_{i+1}$ strictly) and that solutions exit the boundary instantaneously. In order to prove the irreducibility property, it is enough to note that the size $(n-1) \times (n-1)$ Jacobian matrix A at any interior point has the properties that $a_{i,i-1} = \sigma_i(1 - z_{n-1})\alpha_i'(z_i - z_{i-1}) > 0$ for all $2 \le i \le n-1$, $a_{i,i+1} = \beta_i'(z_{i+1} - z_i) > 0$ for all $1 \le i \le n-3$, and $a_{n-2,n-1} = \sigma_{n-1}'(1 - z_{n-1})\alpha_{n-2}(z_{n-2} - z_{n-3}) + \beta_{n-1}'(z_{n-1} - z_{n-2}) > 0$. To show that every trajectory lies in the interior of X for all $t > 0$, as the interior of X is itself forward invariant (see e.g. [2]), it is sufficient to prove: for any $T > 0$, if Φ is the set of $t \in [0, T]$ such that $x(t)$ is in the boundary of X (relative to the linear space $x_1 + \ldots + x_n = 1$), then $\Phi \neq [0, T]$. (It is more convenient to use x coordinates to show this property.) Assume otherwise. For each i, consider the closed set $\Phi_i = \{t \in [0,T] \mid x_i(t) = 0\}$, and note that $\bigcup_i \Phi_i = \Phi$. If Φ_i would be nowhere dense for every i, then their union Φ would be nowhere dense, contradicting $\Phi = [0,T]$. Thus there is some i so that Φ_i contains an open interval $(a,b) \subseteq [0,T]$. It follows that, for this i, $\dot{x}_i \equiv x_i \equiv 0$ on (a,b), and (looking at the equations) this implies that $x_{i\pm1} \equiv 0$ and, recursively, we obtain $x_j \equiv 0$ for all j, contradicting $x_1 + \ldots + x_n = 1$.

As a numerical example, let us pick $\sigma_i(r) = (0.01+r)/(1+r)$, $\alpha_i(r) = 10\,r/(1+r)$, and $\beta_i(r) = r/(1+r)$ for all i, and $n = 7$. (The constants have no biological significance, but the functional forms are standard models of saturation kinetics.) A plot of the characteristic is shown in Fig. 2(a). Since the intersection with the diagonal has three points as shown, we know that *the closed-loop system (with $u = x_n$) will have two stable and one unstable equilibrium, and almost all trajectories converge to one of these two stable equilibria.* To illustrate this convergence, we simulated six initial conditions, in each case with $x_2(0) = \ldots = x_6(0) = 0$ and with the following choices of $x_7(0)$: 0.1, 0.2, 0.3, 0.4, 0.5, and 0.8 (and $x_1(0) = 1 - x_7(0)$). A plot of $x_7(t)$ for each of these initial conditions is shown in Fig. 2(b); note the convergence to the two predicted steady states.

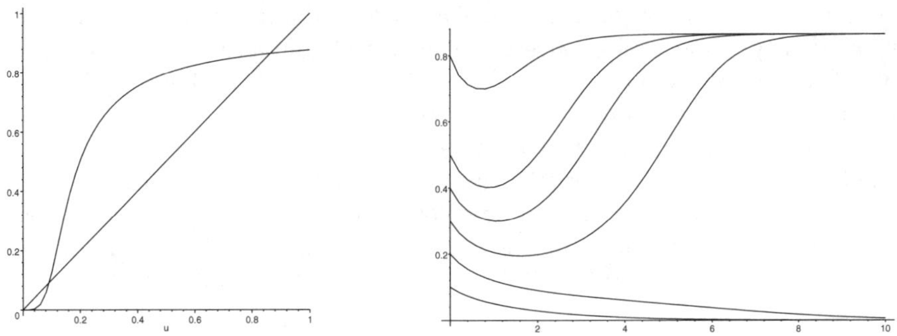

Fig. 2. Enzyme example: (a) Characteristic and (b) Simulations

6 An Example of Negative Feedback

A large variety of eukaryotic cell signal transduction processes operate through "Mitogen-activated protein kinase (MAPK) cascades," which play a role in some of the most fundamental processes of life (cell proliferation and growth, responses to hormones, etc.). A MAPK cascade is a cascade connection of three SISO systems, each of which is a system of the type studied in Section 5, the first with with $n = 2$ and the next two with $n = 3$. We already proved that such systems admit I/O characteristics and are monotone, so cascades have the same property. Alternatively, a proof can be given directly for this example; see [2]. Thus the complete theory may be applied. We pick now a numerical example in order to illustrate the application of Theorem 3.

Instead of the change of variables used in Section 5, we use the change of variables in [2], in which we eliminate x_2 in the case $n = 3$. The cascade then becomes, in terms of the reduced variables, a system of dimension five. As a concrete illustration, let us consider the open-loop system with these equations:

$$
\begin{aligned}
\dot{x}_1 &= \frac{v_2\,(100 - x_1)}{k_2 + (100 - x_1)} - \frac{g_1 x_1}{k_1 + x_1}\frac{g_2 + u}{g_4 + u} \\
\dot{y}_1 &= \frac{v_6\,(300 - y_1 - y_3)}{k_6 + (300 - y_1 - y_3)} - \frac{\kappa_3\,(100 - x_1)\,y_1}{k_3 + y_1} \\
\dot{y}_3 &= \frac{\kappa_4\,(100 - x_1)\,(300 - y_1 - y_3)}{k_4 + (300 - y_1 - y_3)} - \frac{v_5\,y_3}{k_5 + y_3} \\
\dot{z}_1 &= \frac{v_{10}\,(300 - z_1 - z_3)}{k_{10} + (300 - z_1 - z_3)} - \frac{\kappa_7\,y_3\,z_1}{k_7 + z_1} \\
\dot{z}_3 &= \frac{\kappa_8\,y_3\,(300 - z_1 - z_3)}{k_8 + (300 - z_1 - z_3)} - \frac{v_9\,z_3}{k_9 + z_3}.
\end{aligned}
$$

This is the model studied in [20], from which we also borrow the values of constants (with a couple of exceptions indicated below): $g_1 = 0.22$, $g_2 = 45$, $g_4 = 50$, $k_1 = 10$, $v_2 = 0.25$, $k_2 = 8$, $\kappa_3 = 0.025$, $k_3 = 15$, $\kappa_4 = 0.025$ $k_4 = 15$, $v_5 = 0.75$, $k_5 = 15$, $v_6 = 0.75$, $k_6 = 15$, $\kappa_7 = 0.025$, $k_7 = 15$, $\kappa_8 = 0.025$, $k_8 = 15$, $v_9 = 0.5$, $k_9 = 15$, $v_{10} = 0.5$, $k_{10} = 15$. Units are as follows: concentrations and Michaelis constants (k's) are expressed in nM, catalytic rate constants (κ's) in s^{-1}, and maximal enzyme rates (v's) in $nM.s^{-1}$. The paper [20] showed that oscillations may arise in this system for appropriate values of negative feedback gains. (We have slightly changed the input term, using coefficients g_1, g_2, g_4, because we wish to emphasize the open-loop system before considering the effect of negative feedback.) Figure 3 shows the I/O characteristic of this system, as well as the characteristic corresponding to a feedback $u = K/(1 + y)$, with the gain $K = 30000$. It is evident from this planar plot that the small-gain condition is satisfied - a 'spiderweb' diagram shows convergence. Our theorem then guarantees global attraction to a unique equilibrium. Indeed, Figure 4 shows a typical state trajectory.

7 Another Example: Circadian Oscillator

The molecular biology underlying the circadian rhythm in *Drosophila* is the focus of a large amount of both experimental and theoretical work. Goldbeter proposed a simple model for circadian oscillations in [12] (see also his book [13]) using a mechanism based on the negative feedback loop of the protein PER inhibiting its own transcription. This (somewhat oversimplified) model is described by the following equations:

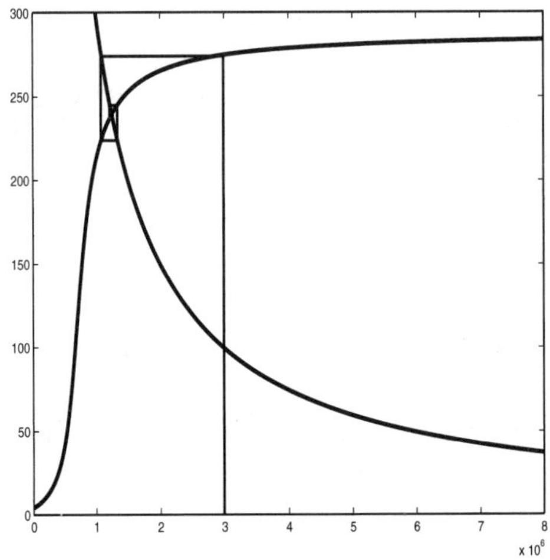

Fig. 3. I/O characteristic and small-gain for MAPK example

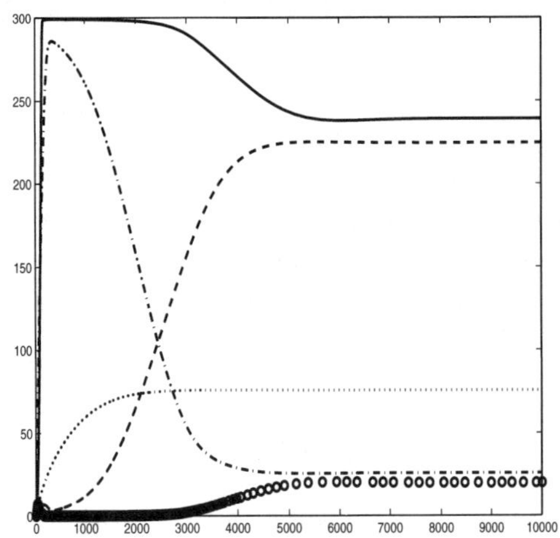

Fig. 4. Simulation of MAPK system under negative feedback satisfying small-gain conditions. Key: x_1 dots, y_1 dashes, y_2 dash-dot, z_1 circles, z_3 solid

$$\begin{aligned}
\dot{M} &= v_s K_I^n/(K_I^n + P_N^n) - v_m M/(k_m + M) \\
\dot{P}_0 &= k_s M - V_1 P_0/(K_1 + P_0) + V_2 P_1/(K_2 + P_1) \\
\dot{P}_1 &= V_1 P_0/(K_1 + P_0) - V_2 P_1/(K_2 + P_1) - V_3 P_1/(K_3 + P_1) + V_4 P_2/(K_4 + P_2)
\end{aligned}$$

$$\dot{P}_2 = V_3 P_1/(K_3+P_1) - V_4 P_2/(K_4+P_2) - k_1 P_2 + k_2 P_N - v_d P_2/(k_d+P_2)$$
$$\dot{P}_N = k_1 P_2 - k_2 P_N$$

where the subscript $i = 0, 1, 2$ in the concentration P_i indicates the degree of phosphorylation of PER protein, P_N is used to indicate the concentration of PER in the nucleus, and M indicates the concentration of *per* mRNA. The parameters (in suitable units μM or h^{-1}) are: $k_2 = 1.3$, $k_1 = 1.9$, $V_1 = 3.2$, $V_2 = 1.58$, $V_3 = 5$, $V_4 = 2.5$, $v_s = 0.76$, $v_m = 0.65$, $k_m = 0.5$, $k_s = 0.38$, $v_d = 0.95$, $k_d = 0.2$, $n = 4$, $K_1 = 2$, $K_2 = 2$, $K_3 = 2$, $K_4 = 2$, $K_I = 1$. The whole point of the model, of course, is that limit cycle oscillations appear. It is important, therefore, to understand to what extent these parameters affect the existence of periodic orbits. We choose to view the system as the feedback closure of the system having M (mRNA) as input and P_N as output (P equations), with the system where P_N negatively regulates the production of M (first equation). The P-subsystem is a monotone tridiagonal system, and characteristics can be analyzed, for small enough inputs M, using techniques as in the previous examples (see [4] for details), while the M subsystem is scalar. We view v_s as a bifurcation parameter. It is remarkable that the conditions for Theorem 3 are satisfied with e.g. $v_s = 0.4$: no oscillations can happen in that case, even under arbitrary delays in the feedback from P_N to M. See Figure 5 for the 'spiderweb diagram' that shows convergence of the discrete iteration. (the dotted and dashed curves are the characteristics). On the other hand,

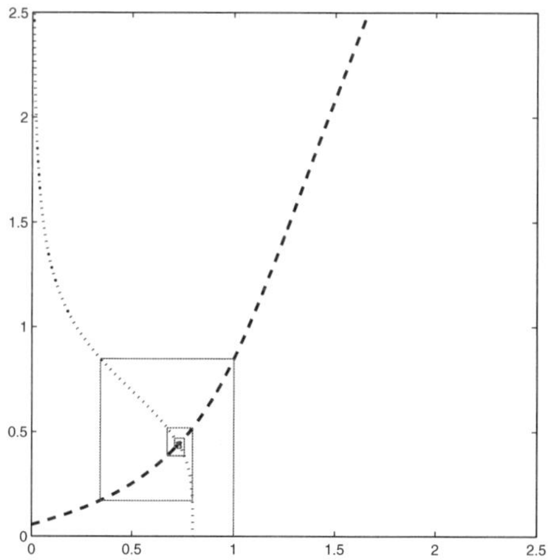

Fig. 5. Stability of spiderweb ($v_s = 0.4$)

for e.g. $v_s = 0.5$, the conditions are violated; see Figure 6 for the 'spiderweb diagram' that shows divergence of the discrete iteration. Thus, and one can expect periodic orbits; and indeed, simulations show that, for large enough delays, such periodic orbits arise, see Figure 7.

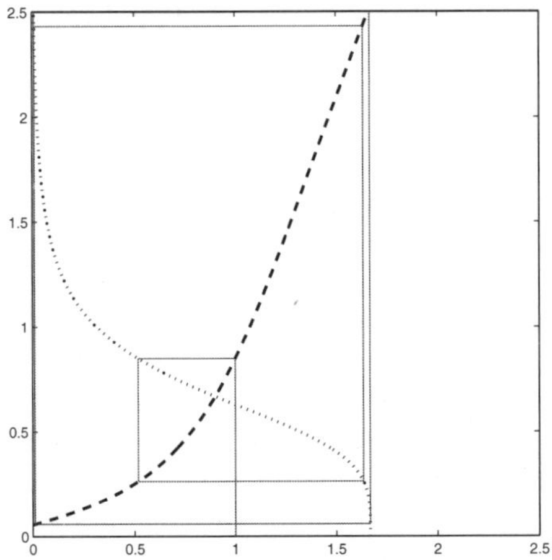

Fig. 6. Instability of spiderweb ($v_s = 0.5$)

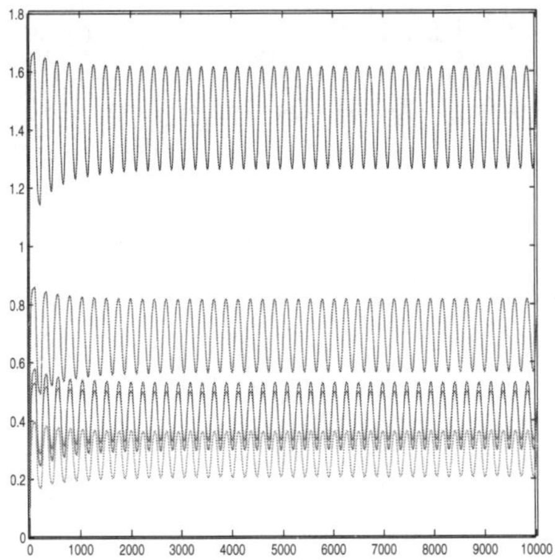

Fig. 7. Oscillations seen in simulations ($v_s = 0.5$, delay of 100, initial conditions all at 0.2), using MATLAB's dde23 package

8 Final Remarks

The theory of monotone systems with well-defined characteristics provide a very powerful method for analyzing both positive and negative feedback interconnections. The theory has only recently started to be developed, but already a large number of nontrivial applications have been handled by it.

For reasons of space, we have omitted a discussion of how complete bifurcation diagrams, with respect to magnitudes of gains, can be immediately derived from the forms of characteristics; see [1, 3] for details.

Another interesting issue, not included here, regards the necessity of the monotonicity assumption in Theorem 4. We show by examples in [2] how relying upon a well-defined characteristic may lead to erroneous conclusions, if monotonicity is not checked first.

References

1. Angeli D, Ferrell J, Sontag E (2003) Detection of multi-stability, bifurcations, and hysteresis in a large class of biological positive-feedback systems, submitted
2. Angeli D, Sontag E (2003) Monotone control systems, IEEE Trans Autom Control 48:1684–1698 (Summarized version: A remark on monotone control systems. In: Proc 41st IEEE Conf Decision and Control Las Vegas 1876–1881)
3. Angeli D, Sontag E (2003) Multi-stability in monotone Input/Output systems, Systems and Control Letters, in press (Summarized version: A note on multi-stability and monotone I/O systems. In: Proc 42nd IEEE Conf Decision Control Maui HI to appear)
4. Angeli D, Sontag E (2003) An analysis of a circadian model using the small-gain approach to monotone systems, arXiv.OC/0312062
5. Bhalla U, Iyengar R (1999) Emergent properties of networks of biological signaling pathways, Science 283:381–387
6. Cherry J, Adler F (2000) How to make a biological switch, J Theor Biol 203:117–133
7. Cinquin O, Demongeot J (2002) Positive and negative feedback: striking a balance between necessary antagonists, J Theor Biol 216:229–241
8. Delbrück M (1949) Génétique du bactériophage. In: Unités Biologiques Douées de Continuité Génétique (Colloques Internationaux du Centre National de la Recherche Scientifique vol 8). CNRS Paris 91–103
9. Ferrell J, Machleder E, (1988) The biochemical basis of an all-r-none cell fate switch in Xenopus Oocytes, Science Reports 280:895-898
10. Ferrell J, Xiong W (2001) Bistability in cell signaling how to make continuous processes discontinuous, and reversible processes irreversible, Chaos 11:227-236
11. Gardner T, Cantor C, Collins J (2000) Construction of a genetic toggle switch in Escherichia coli, Nature 403:339–342
12. Goldbeter A (1995) A model for circadian oscillations in the Drosophila period protein (PER), Proc Royal Soc Lond B 261:319–324
13. Goldbeter A (1996) Biochemical Oscillations and Cellular Rhythms. Cambridge Univ Press Cambridge
14. Gouzé J (1998) Positive and negative circuits in dynamical systems, J Biol Sys 6:11-15
15. Hartwell L, Hopfield J, Leibler S, Murray A (1999) From molecular to modular cell biology, Nature 402(suppl.):C47-C52
16. Hirsch M (1985) Systems of differential equations that are competitive or cooperative II convergence almost everywhere, SIAM J Mathematical Analysis 16:423-439
17. Hirsch M (1988) Stability and convergence in strongly monotone dynamical systems, Reine Angew Math 383:1-53
18. Huang C-Y, Ferrell J (1996) Ultrasensitivity in the mitogen-activated protein kinase cascade, Proc Natl Acad Sci USA 93:10078–10083

19. Hunding A, Engelhardt R (1995) Early biological morphogenesis and nonlinear dynamics, Journal of Theoretical Biology 173:401-413
20. Kholodenko B (2000) Negative feedback and ultrasensitivity can bring about oscillations in the mitogen-activated protein kinase cascades, Eur J Biochem 267:1583–1588
21. Kunze H, Siegel D (1994), A graph theoretical approach to monotonicity with respect to initial conditions. In: Liu X, Siegel D (eds) Comparison Methods and Stability Theory. Marcel Dekker New York
22. Laurent M, Kellershohn N (1999) Multistability a major means of differentiation and evolution in biological systems, Trends in Biochemical Sciences 24:418-422
23. Plahte E, Mestl T, Omholt W (1995) Feedback circuits, stability and multistationarity in dynamical systems, J Biol Syst 3:409-413
24. Pomerening J, Sontag E, Ferrell J (2003) Building a cell cycle oscillator hysteresis and bistability in the activation of Cdc2, Nature Cell Biology 5:346–351
25. Rouche N, Habets P, Laloy M (1977) Stability Theory by Liapunov's Direct Method. Springer-Verlag New York
26. Shvartsman S, Wiley H, Lauffenburger D (2000) Autocrine loop as a module for bidirectional and context-dependent cell signaling, MIT Chemical Engineering Department preprint
27. Smillie J (1984) Competitive and cooperative tridiagonal systems of differential equations, SIAM J Math Anal 15:530–534
28. Smith H (1991) Periodic tridiagonal competitive and cooperative systems of differential equations, SIAM J Math Anal 22:1102-1109
29. Smith H (1995) Monotone Dynamical Systems An Introduction to the Theory of Competitive and Cooperative systems (Mathematical Surveys and Monographs Vol 41). American Mathematical Society Ann Arbor
30. Snoussi E (1998) Necessary conditions for multistationarity and stable periodicity, J Biol Sys 6:3-9
31. Sontag E (1998) Mathematical Control Theory: Deterministic Finite Dimensional Systems Second Edition. Springer-Verlag New York
32. Sontag E (2002) Asymptotic amplitudes and Cauchy gains: A small-gain principle and an application to inhibitory biological feedback, Systems and Control Letters 47:167-179
33. Thomas R (1981) On the relation between the logical structure of systems and their ability to generate multiple steady states or sustained oscillations. In: Springer Series in Synergetics 9. Springer-Verlag Berlin 180-193

Asymptotic Controllability and Input-to-State Stabilization: The Effect of Actuator Errors*

Michael Malisoff[1] and Eduardo Sontag[2]

[1] Department of Mathematics, Louisiana State University and A. & M. College, Baton Rouge LA 70803-4918. malisoff@lsu.edu
[2] Department of Mathematics, Rutgers University, New Brunswick NJ 08901. sontag@control.rutgers.edu

Summary. We discuss several issues related to the stabilizability of nonlinear systems. For a given continuously stabilizable system, we review some constructions of feedbacks that render the system input-to-state stable with respect to actuator errors. We also announce a new feedback design that makes globally asymptotically controllable systems input-to-state stable to actuator errors and small observation noise. We illustrate our constructions using the nonholonomic integrator, and we discuss a related feedback design for systems with disturbances.

1 Introduction

The theory of input-to-state stable (ISS) systems forms the basis for much modern nonlinear feedback design and analysis (cf. [9, 10, 11, 12, 14, 26]). In this survey, we deal with several issues related to the input-to-state stabilizability of nonlinear finite-dimensional continuous-time control systems. We focus on some relatively new feedback designs for asymptotically controllable systems that render the corresponding closed-loop systems ISS with respect to actuator errors. We cover the following topics:

1. Basic controllability and stability concepts
2. Continuously stabilizable systems with actuator errors
3. Asymptotic controllability implies input-to-state stabilization
4. Feedback stabilization of the nonholonomic integrator
5. Systems with large observation noise
6. Integral-input-to-state stabilization

One of our main tools will be the existence theory for smooth control-Lyapunov functions (CLF's) for continuously stabilizable systems. We also use a recent extension of this theory that provides semiconcave CLF's for

*Sponsors: Louisiana Board of Regents Contract LEQSF(2002-04)-ENH-TR-26 (Malisoff); USAF Grant F49620-01-1-0063, NSF Grant CCR-0206789 (Sontag).

M. de Queiroz et al. (Eds.): Optimal Control, Stabilization and Nonsmooth Analysis, LNCIS 301, pp. 155–171, 2004.

systems which are merely asymptotically controllable. Our emphasis will be on relatively new areas of research, including the integral-input-to-state stabilization of systems with disturbances (cf. [1, 13]). While our discussion will be mainly conceptual, we will refer the reader to the relevant literature, where detailed statements and proofs of all our results are found. For a survey on the underlying theory of CLF's, stability, and stabilization, see [25]. As a general reference, we refer the reader to the second author's preprints at `http://www.math.rutgers.edu/~sontag/papers.html`.

2 Basic Controllability and Stability Concepts

In much of what follows, we deal with a general nonlinear finite-dimensional continuous-time deterministic system of the form

$$\dot{x} = f(x, u) \tag{1}$$

evolving over Euclidean space, where $x(t) \in \mathbb{R}^n$ for all $t \geq 0$ and the *controls* u (which are also called *inputs*) are measurable essentially bounded functions $u : [0, \infty) \to \mathcal{U} = \mathbb{R}^m$ taking their values in the control set $\mathcal{U}$ which we have taken to be $\mathbb{R}^m$. We denote the set of all inputs for (1) by $\mathcal{M}^m$. When we wish to restrict the size of controls $u \in \mathcal{M}^m$ in the essential supremum $|\cdot|_\infty$, we will also use the sets $\mathcal{M}^m_N = \{u \in \mathcal{M}^m : |u|_\infty \leq N\}$, defined for each $N > 0$. Many of our results easily generalize to locally essentially bounded inputs mapping into more general control sets $\mathcal{U} \subseteq \mathbb{R}^m$. When we refer to (1), we always assume that f is locally Lipschitz and that $f(0,0) = 0$. We also study systems with no inputs

$$\dot{x} = f(x) \tag{2}$$

for continuous f; all definitions for such systems are implicitly applied to (1) by setting $u = 0$, e.g., we define global asymptotic stability (GAS) for (2), but we say (1) is GAS if the corresponding *zero-input system* $\dot{x} = f(x, 0)$ is GAS.

We denote the solution of (1) starting at any initial state x_o for any given control $u \in \mathcal{M}^m$, defined on its maximal interval, by $x(\cdot, x_o, u)$. We let $|\cdot|$ denote the Euclidean norm, and we set $r\mathcal{B}_k := \{x \in \mathbb{R}^k : |x| < r\}$ for each $k \in \mathbb{N}$ and $r > 0$. We denote the closure of $r\mathcal{B}_k$ by $r\bar{\mathcal{B}}_k$. When we say that a function is smooth, we mean that it is C^1 (i.e., continuously differentiable). We call a function $\alpha : \mathbb{R}^k \to [0, \infty)$ *proper* (a.k.a. *radially unbounded*) if $\alpha(x) \to +\infty$ as $|x| \to +\infty$. We say that a function $\alpha : \mathbb{R}^k \to [0, \infty)$ is *positive definite* provided $\alpha(x) = 0$ iff $x = 0$.

We will frequently use the following types of comparison functions. We let $\mathcal{K}_\infty$ denote the set of all continuous functions $\rho : [0, \infty) \to [0, \infty)$ for which (i) $\rho(0) = 0$ and (ii) ρ is strictly increasing and unbounded. Also, we let $\mathcal{KL}$ denote the set of all continuous functions $\beta : [0, \infty) \times [0, \infty) \to [0, \infty)$ for which (1) $\beta(\cdot, t) \in \mathcal{K}_\infty$ for each $t \geq 0$, (2) $\beta(s, \cdot)$ is nonincreasing for each

$s \geq 0$, and (3) $\beta(s,t) \to 0$ as $t \to +\infty$ for each $s \geq 0$. We also use $\mathcal{N}$ to denote the set of all nondecreasing functions $\sigma : [0,\infty) \to [0,\infty)$.

In terms of comparison functions, the property of *global asymptotic stability* (GAS) of the system (2) without inputs is as follows:

$$(\exists \beta \in \mathcal{KL}) \quad |x(t)| \leq \beta(|x_o|, t)$$

for all $x_o \in \mathbb{R}^n$, all trajectories $x(\cdot)$ of the system (2) starting at x_o, and all $t \geq 0$. More generally, we say that the system (1) with inputs is *globally asymptotically controllable* (GAC) provided:

$$\begin{array}{l}(\exists \beta \in \mathcal{KL})(\exists \sigma \in \mathcal{N}) \ \ \forall x_o \in \mathbb{R}^n, \ \exists u(\cdot) \\ |u|_\infty \leq \sigma(|x_o|), \ |x(t,x_o,u)| \leq \beta(|x_o|,t) \ \ \forall t \geq 0.\end{array}$$

Roughly speaking, our definition of GAC amounts to requiring that for each initial state, there exists a control such that the corresponding solution is defined on $[0,\infty)$ and converges to zero with 'small overshoot' and also that the input remains bounded for x near zero.

The notion of input-to-state stable (ISS) systems provides an alternative generalization of GAS, in which all trajectories of (1) converge to zero with an overshoot depending on the sup norm of the input. The ISS property was introduced in [20], and has become a fundamental concept on which much modern nonlinear feedback analysis and design are based (cf. [9, 10, 11, 12, 14]). By definition, a dynamics $\dot{x} = f(x,u)$ is ISS (with respect to u) provided

$$(\exists \beta \in \mathcal{KL}) \ (\exists \gamma \in \mathcal{K}_\infty) \ \ |x(t)| \leq \beta(|x_o|,t) + \gamma(|u|_\infty) \tag{3}$$

for all $x_o \in \mathbb{R}^n$, all $u \in \mathcal{M}^m$, all trajectories $x(\cdot)$ of the dynamics for u starting at x_o, and all $t \geq 0$. In much of what follows, we will be studying systems

$$\dot{x} = f(x, K(x) + u) \tag{4}$$

where K is a specified feedback map for (1) and the input $u \in \mathcal{M}^m$ represents an actuator error. By a *(memoryless state) feedback*, we mean a locally bounded function $K : \mathbb{R}^n \to \mathcal{U}$ satisfying $K(0) = 0$. For the special case of the dynamics (4) and continuous K, the ISS property (3) implies that K is a *stabilizing* feedback (meaning that the closed-loop system $\dot{x} = f(x, K(x))$ is GAS) and also that the perturbed system (4) exhibits bounded behavior for arbitrary (essentially) bounded actuator errors $u \in \mathcal{M}^m$ (but see also §4 for the definition of trajectories for ISS systems with discontinuous controllers). We say that (1) is *input-to-state stabilizable* (a.k.a. *ISSable)* provided there exists a feedback K such that (4) is ISS. We say that (1) is *continuously stabilizable* (a.k.a. *C^o-stabilizable*) provided there exists a continuous feedback K such that $\dot{x} = f(x, K(x))$ is GAS. Clearly, C^o-stabilizable systems are GAC, but not conversely (cf. §§4-5).

In general, saying that a continuous dynamics $\dot{x} = f(x,u)$ is ISS implies that the corresponding zero-input system $\dot{x} = f(x,0)$ is GAS. This suggests the following natural question:

Question 1. If (1) is C^o-stabilizable, is it also ISSable?

We will address this question in §3. More generally, it is natural to ask:

Question 2. If (1) is GAC, is it also ISSable?

We will address Question 2 in §4. We will see that the answers to these questions are yes, provided the given system (1) is control-affine, i.e., if it has the form $\dot{x} = h(x) + G(x)u$. For the general fully nonlinear system (1), we will see that the answers to these questions are still yes if the system is transformed using a feedback equivalence (cf. §3 for the relevant definitions).

However, if the system is merely GAC, then there may be obstructions to *continuous* (time-invariant) feedback stabilization (cf. [3, 14, 25]). Therefore, we will need to allow discontinuous feedbacks. This produces the technical problem of defining precisely what is meant by a solution of the dynamics when $\dot{x} = f(x, K(x) + u)$ is not continuous, since the standard Carathéodory existence theorems for solutions would not apply. We will resolve this problem by interpreting the trajectories of (4) for discontinuous K as Euler solutions. The preceding issues are central to the rest of this survey.

3 C^o-Stabilizable Systems with Actuator Errors

In this section, we review the main results from [20, 22] on the input-to-state stabilizability of the system (1). Throughout this section, we assume the system (1) is continuously stabilizable (i.e., C^o-stabilizable), meaning, there exists a continuous feedback K_1 for which

$$\dot{x} = f(x, K_1(x)) \tag{5}$$

is GAS. We wish to design a continuous feedback function K (which we allow to be different from the feedback used in (5)) rendering the perturbed system $\dot{x} = f(x, K(x) + u)$ ISS with respect to the actuator error u. This design problem plays an important role in control applications, where one wishes to have a feedback that is robust to noise in the input channel.

For a fully nonlinear system, such a feedback fails to exist in general. For example, consider the scalar system $\dot{x} = -x + x^2u^2$. By considering closed-loop trajectories starting at $x_o = 4$ for the input $u \equiv 1$ (cf. [22]), one can check that there is no continuous feedback K for which the closed-loop system

$$\dot{x} = -x + x^2(K(x) + u)^2 \tag{6}$$

is ISS, even though $\dot{x} = -x$ is obviously GAS (i.e., using $K_1(x) \equiv 0$ as the stabilizer in (5)). In fact, if we interpret the trajectories of (6) in the more general sense of sampling and Euler solutions (cf. §4 for the definitions), then one can show that there is no *dis*continuous feedback rendering (6) ISS either.

On the other hand, if we assume the given system (1) has the control-affine form

$$\dot{x} = h(x) + G(x)u, \tag{7}$$

then we have the following positive result shown in [20]:

Theorem 1. *If (7) is C^o-stabilizable, then there exists a continuous feedback $K : \mathbb{R}^n \to \mathbb{R}^m$ for which*

$$\dot{x} = h(x) + G(x)(K(x) + u) \tag{8}$$

is ISS.

Theorem 1 follows by choosing $K(x) = K_1(x) - \nabla V(x)G(x)$, where V is a C^1 control-Lyapunov function (CLF) for (7) and K_1 is a continuous feedback rendering (7) GAS. The existence of such a CLF was established in [2]. (See also [17, 18] for the existence of semiconcave CLF's for GAC systems, and [25] or §4 below for the definition of CLF's, in terms of set-valued differentials.)

While the example (6) illustrates that C^o-stabilizability does not in general imply input-to-state stabilizability, we can still extend Theorem 1 to fully nonlinear systems by using the following weaker notion of ISS that was introduced in [22]: We say that the system (1) is *ISSable in the weak sense* (a.k.a. *weak ISSable*) provided there exist a continuous feedback K, and an $m \times m$ matrix Θ of continuously differentiable functions which is invertible at each point, such that $\dot{x} = f(x, K(x) + \Theta(x)u)$ is ISS. We then have the following extension of Theorem 1 for fully nonlinear systems shown in [22]:

Theorem 2. *If (1) is C^o-stabilizable, then it is ISSable in the weak sense.*

Theorem 2 can be restated in terms of feedback equivalence, as follows. Recall that two systems $\dot{x} = f(x, u)$ and $\dot{x} = \hat{f}(x, u)$ evolving on $\mathbb{R}^n \times \mathbb{R}^m$ are said to be *feedback equivalent* provided that there exist a feedback $K : \mathbb{R}^n \to \mathbb{R}^m$ and an everywhere invertible function $\Theta : \mathbb{R}^n \to \mathbb{R}^{m \times m}$ for which $\hat{f}(x, u) = f(x, K(x) + \Theta(x)u)$ for all $x \in \mathbb{R}^n$ and $u \in \mathbb{R}^m$; the systems are said to be *continuously feedback equivalent* if, in addition, K can be taken to be continuous. (See also [24] for a more general definition of feedback equivalence, also involving a diffeomorphic transformation of the state variable.) The following is then a direct consequence of Theorem 2:

Corollary 1. *The fully nonlinear control system (1) is C^o-stabilizable if and only if it is continuously feedback equivalent to an ISS system.*

See also [5, 7] for a construction of *time-varying* ISS stabilizing feedback.

4 Asymptotic Controllability Implies ISS Stabilization

In the last section, we saw how C^o-stabilizable systems can be stabilized with respect to actuator errors. However, in many applications, the given system is not C^o-stabilizable. For example, if $m < n$ and $\text{rank}[g_1(0), g_2(0), \ldots, g_m(0)] = m$, then it is impossible to continuously stabilize the drift-free system $\dot{x} = u_1 g_1(x) + \ldots + u_m g_m(x)$, $x \in \mathbb{R}^n$; thus, no totally nonholonomic mechanical system is C^o-stabilizable (cf. [25], p.560). On the other hand, drift-free systems are typically GAC (cf. §5 for an example).

This motivates our study of the more general situation where the given control-affine system $\dot{x} = h(x) + G(x)u$ is GAC, but not necessarily C^o-stabilizable. We wish to design a feedback K so that the closed-loop system

$$\dot{x} = h(x) + G(x)(K(x) + u) \tag{9}$$

is ISS. In fact, we will design our feedback for (9) so that the more general closed-loop system

$$\dot{x} = h(x) + G(x)(K(x + e) + u) \tag{10}$$

is ISS with respect to the actuator error u when the observation error e is sufficiently small (cf. definitions below). In this context, the precise values of $e(t)$ are unknown to the controller, but information about upper bounds on $|e(t)|$ can be used to design the feedback. For the case of *continuous* feedbacks K, small errors in the controller in (10) can be tolerated. However, if we allow *discontinuous* K, then small observation errors can have a substantial effect on the dynamics, in which case the magnitude of e will need to be constrained in terms of the frequency of the sampling (cf. below for the precise restrictions on e). For a construction of a stabilizing feedback K for (10) under the stronger assumption that the given system is C^o-stabilizable, see [25].

Since GAC systems are not in general C^o-stabilizable, we will in fact need to consider discontinuous feedbacks for (10), which produces the technical problem of defining precisely what is meant by a solution for a system that is discontinuous in the state, since the usual existence theorems for solutions of differential equations would not apply. Our solutions will therefore be taken in the more general sense of sampling and Euler solutions. By an Euler solution, we mean a uniform limit of sampling solutions, taken as the frequency of sampling becomes infinite. The following definitions from [14] make these ideas precise.

We say that $\pi = \{t_o, t_1, t_2, \ldots\} \subset [0, \infty)$ is a *partition* provided $t_o = 0$, $t_i < t_{i+1}$ for all $i \geq 0$, and $t_i \to +\infty$ as $i \to +\infty$. The set of all partitions is denoted by Par. Let $F : \mathbb{R}^n \times \mathbb{R}^m \times \mathbb{R}^m \to \mathbb{R}^n : (x, p, u) \mapsto F(x, p, u)$ be a continuous function which is locally Lipschitz in x uniformly on compact subsets of $\mathbb{R}^n \times \mathbb{R}^m \times \mathbb{R}^m$. Recall that a *feedback* (for F) is defined to be any locally bounded function $K : \mathbb{R}^n \to \mathbb{R}^m$ for which $K(0) = 0$. In particular, we allow discontinuous feedbacks. The arguments x, p, and u in F represent the state, feedback value, and actuator error, respectively.

We set $\mathcal{O} := \{e : [0, \infty) \to \mathbb{R}^n\}$ and $\sup(e) := \sup\{|e(t)| : t \geq 0\}$ for all $e \in \mathcal{O}$, and $\mathcal{O}_\eta := \{e \in \mathcal{O} : \sup(e) \leq \eta\}$ for each $\eta > 0$. Given a feedback $K : \mathbb{R}^n \to \mathbb{R}^m$, an initial value $x_o \in \mathbb{R}^n$, a partition $\pi = \{t_o, t_1, t_2, \ldots\} \in \text{Par}$, $e \in \mathcal{O}$, and $u \in \mathcal{M}^m$, the *sampling solution* for the initial value problem

$$\dot{x}(t) = F(x(t), K(x(t) + e(t)), u(t)) \tag{11}$$
$$x(0) = x_o \tag{12}$$

is the continuous function defined by recursively solving

$$\dot{x}(t) \quad = \quad F(x(t), K(x(t_i) + e(t_i)), u(t))$$

from the initial time t_i up until time

$$s_i = t_i \vee \sup\{s \in [t_i, t_{i+1}] : x(\cdot) \text{ is defined on } [t_i, s)\},$$

where $x(0) = x_o$ (cf. [4, 14]). The sampling solution of (11)-(12) is defined from time zero up to the maximal time $\bar{t} = \inf\{s_i : s_i < t_{i+1}\}$. This sampling solution will be denoted by $t \mapsto x_\pi(t; x_o, u, e)$ to exhibit its dependence on $\pi \in \text{Par}$, $x_o \in \mathbb{R}^n$, $u \in \mathcal{M}^m$, and $e \in \mathcal{O}$, or simply by x_π when the dependence is clear from the context. In particular, if $s_i = t_{i+1}$ for all i, then $\bar{t} = +\infty$, so in that case, the sampling solution $t \mapsto x_\pi(t; x_o, u, e)$ is defined on $[0, \infty)$.

We also define the *upper diameter* and *lower diameter* of a given partition $\pi = \{t_o, t_1, t_2, \ldots\} \in \text{Par}$ by

$$\overline{\mathbf{d}}(\pi) = \sup_{i \geq 0}(t_{i+1} - t_i), \quad \underline{\mathbf{d}}(\pi) = \inf_{i \geq 0}(t_{i+1} - t_i)$$

respectively. We let $\text{Par}(\delta) := \left\{\pi \in \text{Par} : \overline{\mathbf{d}}(\pi) < \delta\right\}$ for each $\delta > 0$. We say that a function $y : [0, \infty) \to \mathbb{R}^n$ is an *Euler solution* of (11)-(12) (*robust to small observation errors*) for $u \in \mathcal{M}^m$ provided there are sequences $\pi_r \in \text{Par}$ and $e_r \in \mathcal{O}$ such that

(a) $\overline{\mathbf{d}}(\pi_r) \to 0$; (b) $\sup(e_r)/\underline{\mathbf{d}}(\pi_r) \to 0$; and
(c) $t \mapsto x_{\pi_r}(t; x_o, u, e_r)$ converges uniformly to y

as $r \to +\infty$. In terms of sampling solutions, we have the following analog of ISS from [14]:

Definition 1. *We say that (11) is* ISS for sampling solutions *provided there exist $\beta \in \mathcal{KL}$ and $\gamma \in \mathcal{K}_\infty$ satisfying: For each $\varepsilon, M, N > 0$ with $0 < \varepsilon < M$, there exist positive $\delta = \delta(\varepsilon, M, N)$ and $\kappa = \kappa(\varepsilon, M, N)$ such that for each $\pi \in \text{Par}(\delta)$, $x_o \in M\bar{\mathcal{B}}_n$, $u \in \mathcal{M}_N^m$, and $e \in \mathcal{O}$ for which* $\sup(e) \leq \kappa\underline{\mathbf{d}}(\pi)$,

$$|x_\pi(t; x_o, u, e)| \leq \beta(M, t) + \gamma(N) + \varepsilon \tag{13}$$

for all $t \geq 0$.

Notice that condition (13) is defined to hold in particular for $M = |x_o|$ and $N = |u|_\infty$. Conceptually, condition (13) says that the system is ISS, modulo small overflows, if the sampling is done 'quickly enough' to satisfy $\pi \in \text{Par}(\delta)$, but 'not too quickly', as determined by the additional requirement that $\underline{\mathbf{d}}(\pi) \geq (1/\kappa)\sup(e)$. Moreover, if we restrict to the case where $e = 0$, then the condition on $\underline{\mathbf{d}}(\pi)$ in Definition 1 is no longer needed. Notice that the bounds on e are in the supremum, not the essential supremum.

We also use the following analog of Definition 1 for Euler solutions:

Definition 2. *We say that the system (11) is* ISS for Euler solutions *provided there are $\beta \in \mathcal{KL}$ and $\gamma \in \mathcal{K}_\infty$ satisfying: If $u \in \mathcal{M}^m$ and $x_o \in \mathbb{R}^n$ and $t \mapsto x(t)$ is an Euler solution of (11)-(12), then*

$$|x(t)| \leq \beta(|x_o|, t) + \gamma(|u|_\infty) \tag{14}$$

for all $t \geq 0$.

The following result on control-affine systems is shown in [14]:

Theorem 3. *If (7) is GAC, then there exists a feedback K for which the corresponding closed-loop system (10) is ISS for sampling and Euler solutions.*

The proof of Theorem 3 is based on the existence theory for semiconcave CLF's from [17]-[18], which we review next. Let $\Omega \subseteq \mathbb{R}^n$ be open. A continuous function $g : \Omega \to \mathbb{R}$ is called *semiconcave on Ω* provided for any point $x_o \in \Omega$, there exist $\rho, C > 0$ such that

$$g(x) + g(y) - 2g\left(\frac{x+y}{2}\right) \leq C||x-y||^2$$

for all $x, y \in x_o + \rho\mathcal{B}_n$. The *proximal superdifferential* (resp., *proximal subdifferential*) of a function $V : \Omega \to \mathbb{R}$ at $x \in \Omega$, which is denoted by $\partial^P V(x)$ (resp., $\partial_P V(x)$), is the set of all $\zeta \in \mathbb{R}^n$ for which there exist $\sigma, \eta > 0$ such that $V(y) - V(x) - \sigma|y-x|^2 \leq \langle \zeta, y-x \rangle$ (resp., $V(y) - V(x) - \sigma|y-x|^2 \geq \langle \zeta, y-x \rangle$) for all $y \in x + \eta\mathcal{B}_n$. The *limiting subdifferential* of a continuous function $V : \Omega \to \mathbb{R}$ at $x \in \Omega$ is

$$\partial_L V(x) := \left\{ \begin{array}{l} q \in \mathbb{R}^n : \text{ there exist } x_n \to x \text{ and} \\ q_n \in \partial_P V(x_n) \text{ such that } q_n \to q. \end{array} \right\}.$$

In terms of proximal subdifferentials, a CLF is defined as follows (cf. [25]):

Definition 3. *A* control-Lyapunov function (CLF) *for (1) is a continuous, positive definite, proper function $V : \mathbb{R}^n \to \mathbb{R}$ for which there exist a continuous, positive definite function $W : \mathbb{R}^n \to \mathbb{R}$ and $\alpha \in \mathcal{N}$ satisfying*

$$\forall \zeta \in \partial_P V(x), \quad \inf_{|u| \leq \alpha(|x|)} \langle \zeta, f(x,u) \rangle \leq -W(x)$$

for all $x \in \mathbb{R}^n$.

The existence of *continuous* CLF's for GAC systems was established in [19]. On the other hand, the papers [17]-[18] establish the following stronger existence result: If (1) is GAC, then there exists a CLF V for (1) that is *semiconcave* on $\mathbb{R}^n \setminus \{0\}$ and $\alpha \in \mathcal{N}$ that satisfy

$$\forall \zeta \in \partial_L V(x), \quad \min_{|u| \leq \alpha(|x|)} \langle \zeta, f(x,u) \rangle \leq -V(x) \tag{15}$$

for all $x \in \mathbb{R}^n$. Theorem 3 follows from this stronger existence result by first taking an arbitrary selection $\zeta(x) \in \partial_L V(x)$ for $x \neq 0$, and then using the combined feedback $K = K_1 + K_2$, where for each $x \neq 0$, $K_1(x) \in \alpha(|x|)\mathcal{B}_m$ is any vector u satisfying the inequality in (15) for $\zeta = \zeta(x)$, and

$$K_2(x) = -V(x) \begin{pmatrix} \operatorname{sgn}(\langle \zeta(x), g_1(x) \rangle) \\ \operatorname{sgn}(\langle \zeta(x), g_2(x) \rangle) \\ \vdots \\ \operatorname{sgn}(\langle \zeta(x), g_m(x) \rangle) \end{pmatrix}$$

where g_i is the ith column of the matrix G and

$$\operatorname{sgn}(s) \quad = \quad \begin{cases} 1, & s > 0 \\ -1, & s < 0 \\ 0, & s = 0 \end{cases} .$$

It then follows that the closed-loop system (10) is ISS for sampling and Euler solutions, using the combined feedback $K = K_1 + K_2$ (with $K(0) = 0$).

Here is a sketch of the proof that this combined feedback renders the system ISS for sampling solutions when the observation error $e = 0$. (See [14] for the proof of the general case, and for the proof that this feedback also gives ISS for Euler solutions.) In what follows, $||u(s)||_J$ denotes the essential supremum of the restriction of a function u to an interval J. Let $M, N > 0$ be given, and V be a semiconcave CLF for the given dynamics, as above. Define the functions $\underline{\alpha}, \overline{\alpha} \in \mathcal{K}_\infty$ by

$$\underline{\alpha}(s) := \min\{s, \min\{|x| : V(x) \geq s\}\} \;\; \overline{\alpha}(s) := \max\{|x| : V(x) \leq s\}. \tag{16}$$

Let $x \mapsto \zeta(x)$ be any selection as above and $\zeta(0) \in \mathbb{R}^n$ be arbitrary.

For each $x \in \mathbb{R}^n$, we can choose $u = u_x \in \alpha(|x|)\mathcal{B}_m$ that satisfies the inequality in (15) for the given dynamics. Set $F(x,p,u) := f(x) + G(x)(p+u)$. Choose $S := \{x \in \mathbb{R}^n : V(x) \leq \underline{\alpha}^{-1}(N)\}$ and $\varepsilon \in (0, \min\{1, M\})$ for which $(2\varepsilon)\mathcal{B}_n \subseteq S$. Set

$$\begin{aligned} &Q := \left\{ \left[\overline{\alpha} \circ \underline{\alpha}^{-1}(N+M) + 1 \right] \bar{\mathcal{B}}_n \right\} \setminus \varepsilon \mathcal{B}_n, \\ &\lambda_- := \min \left\{ V(p) : p \in Q^{\varepsilon/2} \right\}, \quad \lambda_+ := \max \left\{ V(p) : p \in Q^{\varepsilon} \right\}. \end{aligned}$$

It follows that $S \subseteq Q^\varepsilon$. We can then choose $\tilde{\varepsilon} \in (0, \varepsilon)$ for which

$$\overline{\alpha}\left(p + \frac{\mathcal{L}_\varepsilon}{4}\tilde{\varepsilon}\right) \leq \overline{\alpha}(p) + \frac{\varepsilon}{8} \quad \forall p \in \left[0, \underline{\alpha}^{-1}(N) + \lambda_+\right] \tag{17}$$

where $\mathcal{L}_\varepsilon > 1$ is a Lipschitz constant for V on $Q^{\varepsilon/2}$. Using local uniformity properties for semiconcave functions (cf. [14]), we can find $\sigma, \mu > 0$ such that

$$V(y) - V(x) \leq \langle \zeta(x), y - x \rangle + \sigma |y - x|^2 \tag{18}$$

for all $y \in x + \mu\mathcal{B}_n$ and $x \in Q^{\varepsilon/2}$. We can then choose

$$\delta = \delta(\varepsilon, M, N) \in \left(0, \frac{\tilde{\varepsilon}}{16 + \lambda_+ + 16\lambda_+}\right) \text{ such that} \tag{19}$$

$$|x_\pi(t; x_o, u, 0) - x_i| \;\leq\; \min\left\{\mu, \frac{\tilde{\varepsilon}}{16(1 + \mathcal{L}_\varepsilon)}, \sqrt{\frac{\lambda_-}{8\sigma}(t - t_i)}\right\} \tag{20}$$

(where $x_i = x_\pi(t_i; x_o, u, 0)$) and

$$\|\zeta(x_i) \cdot (F(x_i, K(x_i), u(s)) - f(x_\pi(s)) - G(x_\pi(s))[u(s) + K(x_i)])\,\|_{J_i} \leq \frac{\lambda_-}{8} \tag{21}$$

where $J_i := [t_i, t_{i+1}]$, for all $u \in \mathcal{M}_N^m$, $t \in [t_i, t_{i+1}]$, $\pi \in \text{Par}(\delta)$, and all i such that $x_i \in Q^{\varepsilon/2}$. Defining $J(t) := 16/(16 + t)$ and

$$\beta(s, t) := \overline{\alpha}\left(\underline{\alpha}^{-1}(s) J(t)\right), \quad \gamma(s) := \overline{\alpha} \circ \underline{\alpha}^{-1}(s), \tag{22}$$

we can then use the estimates (20)-(21) to conclude that the sampling ISS estimate (13) holds for all $x_o \in M\bar{\mathcal{B}}_n$, $u \in \mathcal{M}_N^m$, $\pi \in \text{Par}(\delta)$, and $e = 0$.

The preceding construction uses the control-affine structure of the given GAC system (7) in an essential way. In fact, it is not difficult to construct examples of *fully nonlinear* GAC systems $\dot{x} = f(x, u)$ for which there is no feedback K rendering the closed-loop system $\dot{x} = f(x, K(x) + u)$ ISS for sampling and Euler solutions. One such example was provided by (6), p.158.

On the other hand, it is possible to extend Theorem 3 to fully nonlinear systems if we interpret ISS in terms of feedback equivalence, as follows. In what follows, we assume for simplicity that the observation error in the controller $e \equiv 0$. Recall (cf. [14]) that if f and $\hat{f}$ are feedback equivalent, and if K and Θ satisfy the requirements of our definition of feedback equivalence, then we also say that $\dot{x} = f(x, u)$ is feedback equivalent to (11) with the choice $F(x, p, u) = f(x, p + \Theta(x)u)$. We then have the following corollary from [14]:

Corollary 2. *The fully nonlinear control system (1) is GAC if and only if it is feedback equivalent to a system which is ISS for sampling and Euler solutions.*

For a proof of this corollary, see [14].

5 Feedback Stabilization of the Nonholonomic Integrator

In this section, we use the feedback construction from §4 to ISS stabilize Brockett's nonholonomic integrator control system. The nonholonomic integrator was introduced in [3] as an example of a system that cannot be stabilized using continuous state feedback. It is well-known that if the state space of a system contains topological obstacles (e.g., if the state space is $\mathbb{R}^2 \setminus (-1, 1)^2$, and therefore has a topological obstacle around the origin), then it is impossible to stabilize the system using continuous state feedback. This follows from a theorem of Milnor, which asserts that the domain of attraction of an asymptotically stable vector field must be diffeomorphic to Euclidean space.

Brockett's example illustrates how, even if we assume that the state evolves in Euclidean space, obstructions to continuous stabilization may still occur. These obstructions are not due to the topology of the state space, but instead arise from the form of the control system. Such obstacles occur when it is impossible to move *instantly* in some directions, even though it is possible to move *eventually* in every direction; when this is the case, we call the dynamics nonholonomic (cf. [25]). This gives rise to Brockett's criterion, which was first announced in [3]. Brockett's criterion is a necessary condition for the existence of a continuous (time-invariant) feedback stabilizer for (1); it requires that the mapping $(x, u) \mapsto f(x, u)$ be open at zero. For linear systems $\dot{x} = Ax + Bu$, this says $\text{rank}[A, B] = n$, which is the Hautus controllability condition at the zero mode (cf. [25]). The nonholonomic integrator does not satisfy Brockett's criterion, and therefore cannot be stabilized by continuous state feedbacks.

The physical model for Brockett's example is as follows. Consider a three-wheeled shopping cart whose front wheel acts as a castor. The state variable is $(x_1, x_2, \theta)^T$, where $(x_1, x_2)^T$ is the midpoint of the rear axle of the cart, and θ is the cart's orientation. The front wheel is free to rotate, but there is a "non-slipping" constraint that $(\dot{x}_1, \dot{x}_2)^T$ must always be parallel to $(\cos(\theta), \sin(\theta))^T$. This gives the equations $\dot{x}_1 = v_1 \cos(\theta)$, $\dot{x}_2 = v_1 \sin(\theta)$, and $\dot{\theta} = v_2$, where v_1 is a 'drive' command and v_2 is a steering command. Using a feedback transformation (cf. [25], or §4.3 in [24]) brings the equations into the form

$$\dot{x}_1 = u_1, \quad \dot{x}_2 = u_2, \quad \dot{x}_3 = x_1 u_2 - x_2 u_1, \tag{23}$$

which is called the nonholonomic integrator control system.

One can show that (23) is a GAC system. However, since Brockett's condition is not satisfied for (23), the system has no continuous state feedback stabilizer. While there does not exist a C^1 CLF for the system (23), it is now well-known that every GAC system admits a continuous CLF (see [19]). In fact, it was shown in [16] that the dynamics (23) has the continuous CLF

$$V(x) = \max\left\{\sqrt{x_1^2 + x_2^2}, |x_3| - \sqrt{x_1^2 + x_2^2}\right\}, \tag{24}$$

which is semiconcave outside the cone $x_3^2 = 4(x_1^2 + x_2^2)$. On the other hand, it is not difficult to show (cf. [14]) that the system (23) also has the CLF

$$\hat{V}(x) = \left(\sqrt{x_1^2 + x_2^2} - |x_3|\right)^2 + x_3^2, \tag{25}$$

which is semiconcave on $\mathbb{R}^3 \setminus \{0\}$. This allows us to apply the theory of the previous section to the nonholonomic integrator with the CLF (25). To check the semiconcavity of $\hat{V}$, it suffices to verify that

$$S(x) = -|x_3|\sqrt{x_1^2 + x_2^2}$$

is semiconcave on $\mathbb{R}^3 \setminus \{0\}$, since C^2 functions are semiconcave and the sum of two semiconcave functions is semiconcave. On the other hand, the semiconcavity of S follows easily because $\phi(r,s) = -|rs| = \min\{\pm rs\}$ is semiconcave, as a minimum of two C^2 functions. The proof that $\hat{V}$ is a CLF follows from a slight variant of the change of coordinate arguments used in [16] to show that (24) is a CLF; we leave the proof to the reader as an exercise.

For the nonholonomic integrator dynamics and the CLF (25), the stabilizing K from §4 is constructed as follows. The system vector fields are

$$h(x) = 0, \quad G(x) = \begin{pmatrix} 1 & 0 \\ 0 & 1 \\ -x_2 & x_1 \end{pmatrix},$$

and we can define $\zeta(\cdot)$ by $\zeta(0) = 0$ and

$$\zeta(x) = \begin{cases} 2(|x_3| - r(x))\left(-x_1/r(x), -x_2/r(x), \mathrm{sgn}(x_3)\right) + 2x_3\mathbf{e}_3, & x_3 r(x) \neq 0, \\ -2r(x)\left(-x_1/r(x), -x_2/r(x), 1\right), & x_3 = 0, \\ 2|x_3|(0, -1, \mathrm{sgn}(x_3)) + 2x_3\mathbf{e}_3, & r(x) = 0 \end{cases}$$

for all $x \neq 0$, where $r(x) := \sqrt{x_1^2 + x_2^2}$ and $\mathbf{e}_3 = (0,0,1)^T$. In terms of

$$b(x) = \zeta(x)G(x) = \begin{cases} \begin{pmatrix} \mu(x)(-x_1/r(x) - x_2\mathrm{sgn}(x_3)) - 2x_2x_3 \\ \mu(x)(-x_2/r(x) + x_1\mathrm{sgn}(x_3)) + 2x_1x_3 \end{pmatrix}, & r(x)x_3 \neq 0 \\ -2r(x)\begin{pmatrix} -x_1/r(x) - x_2 \\ -x_2/r(x) + x_1 \end{pmatrix}, & x_3 = 0 \\ \begin{pmatrix} 0 \\ -2|x_3| \end{pmatrix}, & r(x) = 0 \end{cases}$$

where $\mu(x) := 2(|x_3| - r(x))$, the stabilizing feedback is then

$$K(x) = -V(x)\left(\frac{b(x)}{|b(x)|^2} + (\mathrm{sgn}(b_1(x)), \mathrm{sgn}(b_2(x)))\right) \tag{26}$$

for $x \neq 0$ and $K(0) = 0$. It follows that the corresponding closed-loop system $\dot{x} = h(x) + G(x)(K(x+e) + u)$ is ISS for sampling and Euler solutions, using the feedback (26).

6 Systems with Large Observation Noise

In the preceding section, we studied the problem of constructing a state feedback K such that the closed loop system

$$\dot{x} = h(x) + G(x)(K(x+e) + u) \tag{27}$$

is ISS with respect to the actuator error u, where e is an observation noise. We assumed that the given system $\dot{x} = h(x) + G(x)u$ was GAC, and we also used the fact that the observation noise e in the controller was *sufficiently small*. It is natural to ask whether our analysis can be extended to the case where the observation error e is an arbitrary bounded measurable function. More generally, one could consider the following conjecture (where we use w to denote the input to distinguish it from the actuator error):

Conjecture L. If $h : \mathbb{R}^n \to \mathbb{R}^n$ and $G : \mathbb{R}^n \to \mathbb{R}^{n\times m}$ are locally Lipschitz with $h(0) = 0$, and if $\dot{x} = h(x) + G(x)w$ is C^o-stabilizable, then there exists a continuous feedback $K : \mathbb{R}^n \to \mathbb{R}^m$ satisfying the following: There exist $\beta \in \mathcal{KL}$ and $\gamma_1, \gamma_2 \in \mathcal{K}_\infty$ such that for each initial value $x_o \in \mathbb{R}^n$ and each $e \in \mathcal{M}^n$ and each $u \in \mathcal{M}^m$, each solution $x(t)$ of the closed-loop system (27) starting at x_o satisfies $|x(t)| \leq \beta(|x_o|, t) + \gamma_1(|u|_\infty) + \gamma_2(|e|_\infty)$ for all $t \geq 0$.

An important difference between this problem and the previous one is that the "small" observation noise e is now allowed to be large, but we have relaxed the stabilization objective to an ISS one. This problem is of interest in the context of observer design. For details, see [25].

It turns out that Conjecture L is false, even if the effect of actuator errors u is ignored. A counterexample to the conjecture for a single input plant was provided in [6]. The counterexample is based on the C^o-stabilizable system

$$\dot{x} = \left(I + 2\Theta\left(\frac{\pi}{2}\right) x x^T\right) \Theta(x^T x) \cdot \left(\begin{bmatrix} -1 & 0 \\ 0 & x^T x \end{bmatrix} \Theta\left(-x^T x\right) x + \begin{bmatrix} 0 \\ 1 \end{bmatrix} w \right), \quad (28)$$

where $x \in \mathbb{R}^2$ and Θ is the rotation matrix

$$\Theta(\theta) = \begin{bmatrix} \cos\theta & -\sin\theta \\ \sin\theta & \cos\theta \end{bmatrix}.$$

One can show that there is no continuous feedback K for this system that can satisfy the requirements of Conjecture L.

This can be seen by choosing the actuator error $u \equiv 0$ and expressing the system in terms of the variable

$$z := \Theta(-x^T x)x,$$

and then considering only initial conditions that lie on the invariant set

$$S = \{x \in \mathbb{R}^2 : z_1 = 0\}.$$

One then constructs a sensor disturbance e with the property that the corresponding value for the perturbed state $x(t) + e(t)$ would lie on $S \cap \{z_2 < 0\}$ when the actual state $x(t)$ lies on $S \cap \{z_2 > 0\}$ (cf. [6] for details). See [8] for a construction of continuous feedback rendering $\dot{x} = f(x) + G(x)K(x + e)$ ISS with respect to e, which applies to a *restricted* class of C^o-stabilizable single input nonlinear systems in lower triangular form. See also [7] for *time-varying* feedbacks that make control-affine systems ISS to observation errors; the feedbacks in [7] apply for scalar systems, and for systems that satisfy a feedback passivity property.

7 Integral-Input-to-State Stabilization

In the previous sections, we constructed feedbacks that render systems ISS to actuator errors. A system that is ISS exhibits low total energy or low overshoot when excited by energy bounded or uniformly bounded signals, respectively. While these are highly desirable characteristics, it is sometimes the case that feedback design does not render ISS behavior, or that a property weaker than ISS is verified in a step in a recursive design. One such weaker property, which was introduced in [23], is integral-input-to-state stability (iISS). The iISS condition reflects the qualitative property of having small overshoots when the disturbances have finite energy. It provides a nonlinear analog of "finite H^2 norm" for linear systems, and as such has obvious physical relevance and significance.

In this section, we study the integral-input-to-state stabilizability of

$$\dot{x} = h(x,d) + G(x)u, \tag{29}$$

where $u \in \mathcal{M}^m$ is an input, $d \in \mathcal{M}^m$ is a disturbance, and the functions h and G are locally Lipschitz with $h(0,0) = 0$. This includes control-affine systems with no disturbances, as a special case. We wish to design a state feedback k for which the closed-loop system

$$\dot{x} = h(x,d) + G(x)k(x), \tag{30}$$

exhibits iISS. This design problem is related to the problems in the previous sections, because k will be used to mitigate the effect of the error term d. We will require that k be *almost smooth*, i.e., smooth on $\mathbb{R}^n \setminus \{0\}$ and continuous on $\mathbb{R}^n$. Recall from [13] that the iISS condition for (30) states that there exist $\alpha_o, \gamma \in \mathcal{K}_\infty$ and $\beta \in \mathcal{KL}$ such that for each disturbance $d \in \mathcal{M}^m$ and initial value $x_o \in \mathbb{R}^n$, each solution $x(t)$ of (30) starting at x_o satisfies

$$\alpha_o(|x(t)|) \leq \beta(|x_o|, t) + \int_0^t \gamma(|d(s)|)\mathrm{d}s$$

for all $t \geq 0$. As pointed out in [13, 23], iISS is a weaker requirement than ISS; for example, the scalar system $\dot{x} = -x + xd$ is iISS but not ISS.

The feedback constructions in the preceding sections are based on the existence of CLF's. In parallel to those results, the feedback k for (30) can be constructed using an iISS-CLF for (29), which is defined as follows (cf. [13]):

Definition 4. *We say that a positive definite radially unbounded smooth function $V : \mathbb{R}^n \to \mathbb{R}$ is an* iISS-CLF *for the system (29) provided there exist a positive definite function $\alpha : \mathbb{R} \to [0, \infty)$ and a class $\mathcal{K}_\infty$ function χ such that*

$$\inf_{u \in \mathbb{R}^m} \{a(x,d) + b(x)u\} \leq -\alpha(|x|) + \chi(|d|) \quad \forall x \in \mathbb{R}^n,\ d \in \mathbb{R}^m, \tag{31}$$

where $a(x,d) = \nabla V(x)h(x,d)$ and $b(x) = \nabla V(x)G(x)$.

Our construction of the feedback k is based on the well-known 'universal formula' for feedback stabilization, which was introduced in [21] and later generalized to p-norm bounded controls in [15]. We will use the functions

$$K_u(a,b) := -\begin{cases} \dfrac{a+\sqrt{a^2+|b|^4}}{|b|^2}\, b, & b \neq 0 \\ 0, & b = 0 \end{cases} \tag{32}$$

and

$$\omega(x) := \max_d \{a(x,d) - \chi(|d|)\}$$

where $a(x,d)$ and χ are from the iISS-CLF definition, and where we assume without loss of generality that $\chi(r)/r \to +\infty$ as $r \to +\infty$, which is necessary for ω to be well-defined. The function K_u comes from the 'universal' stabilizing formula from [21] (cf. Remark 1 for details). In terms of the functions α and b from Definition 4, we say that an iISS-CLF V has the *small control property (scp)* provided: for each $\varepsilon > 0$, there exists $\delta > 0$ such that if $0 < |x| < \delta$, then there exists u with $|u| < \varepsilon$ for which $\omega(x) + b(x)u \leq -\alpha(|x|)$. We also choose an almost smooth function $\bar{\omega}$ that satisfies

$$\omega(x) + \alpha(|x|)/3 \leq \bar{\omega}(x) \leq \omega(x) + 2\alpha(|x|)/3 \quad \forall x \in \mathrm{I\!R}^n,$$

where α is the positive definite function from the definition of an iISS-CLF. The existence of $\bar{\omega}$ is standard (cf. [13]). In terms of the functions we have defined, we have the following feedback construction shown in [13]:

Theorem 4. *If V is an iISS-CLF for (29) satisfying the small control property, then the feedback $k(x) := K_u(\bar{\omega}(x), b(x))$ is almost smooth and renders the closed-loop system (30) integral-input-to-state stable.*

The proof of Theorem 4 is based on the iISS-Lyapunov function characterizations from [1].

Remark 1. In addition to the preceding construction, formula (32) can also be used to give an explicit expression for our feedback $K = K_1 + K_2$ from §3, under the additional assumption that the CLF V for the given system $\dot{x} = h(x) + G(x)u$ satisfies the small control property (cf. [21]). Indeed, in this case, it suffices to use the feedback $K_1(x) = K_u(a(x), b(x))$ and K_2 as before, where $a(x) = \nabla V(x)h(x)$ and $b(x) = \nabla V(x)G(x)$ are the Lie derivatives in the direction of V. When defined in this way, K_1 is the so-called universal formula for feedback stabilization from [21], and is almost smooth. A similar construction can be made when the given system is merely GAC (cf. §4); in this case, we replace the Lie derivatives $a(x)$ and $b(x)$ in the formula for K_1 with $\tilde{a}(x) = \zeta(x)h(x)$ and $\tilde{b}(x) = \zeta(x)G(x)$, respectively, for a selection $\zeta(x) \in \partial_L V(x)$ for the (possibly nonsmooth, but semiconcave) CLF.

Remark 2. The novelty of Theorem 4 is that it uses a universal formula, rather than the more commonly used pointwise min norm control laws (which are in general just continuous) or partition of unity arguments (which are nonconstructive). For extensions of Theorem 4 to systems with outputs, see [13]. The proof of Theorem 4 is based on the fact that the disturbance d and the input u in (29) are decoupled. This allows the interchange of the order of the max from the definition of ω and the inf in the decay condition (31) (cf. [13], §5). Consequently, the proof breaks down for systems where u and d are coupled. The search for extensions of the theorem to more general systems is an important question that is of considerable ongoing research interest.

References

1. Angeli D, Sontag E, Wang Y (2000) A characterization of integral input-to-state stability, IEEE Trans Automat Control 45:1082-1097
2. Artstein Z (1983) Stabilization with relaxed controls, Nonlinear Analysis 11:1163-1173
3. Brockett R (1983) Asymptotic stability and feedback stabilization. In: Brockett R, Millman R, Sussmann H (eds) Differential Geometric Control Theory. Birkhäuser Boston
4. Clarke F, Ledyaev Y, Sontag E, Subbotin A (1997) Asymptotic controllability implies feedback stabilization, IEEE Trans Automat Control 42:1394-1407
5. Fah N (1999) Input-to-state stability with respect to measurement disturbances for one-dimensional systems, ESAIM Control Optim Calc Var 4:99-121
6. Freeman R (1995) Global internal stabilizability does not imply global external stabilizability for small sensor disturbances, IEEE Trans Automat Control 40:2119-2122
7. Freeman R (1997) Time-varying feedback for the global stabilization of nonlinear systems with measurement errors. In: Proceedings of the European Conference on Control Brussels
8. Freeman R, Kokotovic P (1993) Global robustness of nonlinear systems to state measurement disturbances. In: Proceedings of the 32nd IEEE Conference on Decision and Control San Antonio TX
9. Khalil H (1996) Nonlinear systems Second Edition. Prentice-Hall Upper Saddle River NJ
10. Kokotović P, Arcak M (1999) Constructive nonlinear control progress in the 90's. In: Proceedings of the 14th IFAC World Congress, Plenary and Index Volume Beijing
11. Krstic M, Deng H (1998) Stabilization of uncertain nonlinear systems. Springer London
12. Krstic M, Kanellakopoulos I, Kokotović P (1995) Nonlinear and adaptive control design. John Wiley and Sons New York
13. Liberzon D, Sontag E, Wang Y (2002) Universal construction of feedback laws achieving ISS and integral-ISS disturbance attenuation, Systems Control Lett 46:111-127
14. Malisoff M, Rifford L, Sontag E (2003) Global asymptotic controllability implies input-to-state stabilization, SIAM J Control Optim to appear
15. Malisoff M, Sontag E (2000) Universal formulas for feedback stabilization with respect to Minkowski balls, Systems Control Lett 40:247-260

16. Rifford L (2000) Problèmes de stabilisation en théorie du contrôle, Thèse Université Claude Bernard Lyon I
17. Rifford L (2000) Existence of Lipschitz and semiconcave control-Lyapunov functions, SIAM J Control Optim 39:1043-1064
18. Rifford L (2002) Semiconcave control-Lyapunov functions and stabilizing feedbacks, SIAM J Control Optim 41:659-681
19. Sontag E (1983) A Lyapunov-like characterization of asymptotic controllability, SIAM J Control Optim 21:462-471
20. Sontag E (1989) Smooth stabilization implies coprime factorization, IEEE Trans Automat Control 34:435-443
21. Sontag E (1989) A 'universal' construction of Artstein's theorem on nonlinear stabilization, Systems Control Lett 13:117-123
22. Sontag E (1990) Further facts about input to state stabilization, IEEE Trans Automat Control 35:473-476
23. Sontag E (1998) Comments on integral variants of ISS, Systems Control Lett 34:93-100
24. Sontag E (1998) Mathematical Control Theory Second Edition. Springer New York
25. Sontag E (1999) Stability and stabilization discontinuities and the effect of disturbances. In: Clarke F, Stern R (eds) Nonlinear Analysis Differential Equations and Control. Kluwer Dordrecht
26. Sontag E (2000) The ISS philosophy as a unifying framework for stability-like behavior. In: Isidori A, Lamnabhi-Lagarrigue F, Respondek W (eds) Nonlinear Control in the Year 2000 Volume 2. Springer-Verlag Berlin

The Stabilization Problem: AGAS and SRS Feedbacks

Ludovic Rifford

Institut Girard Desargues, Université Claude Bernard, 43, Bd du 11 Novembre 1918, 69622 Villeurbanne Cedex, France. rifford@igd.univ-lyon1.fr

1 The Problem

Throughout this paper, M denotes a smooth manifold of dimension n. We are given a control system on M of the form

$$\dot{x} = f(x,u) := \sum_{i=1}^{m} u_i f_i(x), \tag{1}$$

where $f_1, \cdots, f_m$ are smooth vector fields on M and where the control

$$u = (u_1, \cdots, u_m)$$

belongs to the closed unit ball $\overline{B_m}$ in $\mathbb{R}^m$. Throughout the paper, "smooth" means "of class C^∞". We let $x(\cdot;\bar{x},u(\cdot))$ denote the trajectory for (1) for the control u starting at $\bar{x} \in M$. Such a control system is said to be *globally asymptotically controllable* at the point $O \in M$ (abbreviated GAC in the sequel) provided the following two properties are satisfied:

1. Attractivity: For each $\bar{x} \in M$ there exists a control $u(\cdot) : \mathbb{R}_{\geq 0} \to \overline{B_m}$ such that the corresponding trajectory $x(\cdot;\bar{x},u(\cdot))$ of (1) tends to O as t tends to infinity.
2. Lyapunov stability: For each neighborhood $\mathcal{V}$ of O, there exists some neighborhood $\mathcal{U}$ of O such that if $\bar{x} \in \mathcal{U}$ then the control $u(\cdot)$ above can be chosen such that $x(t;\bar{x},u(\cdot)) \in \mathcal{V}, \forall t \geq 0$.

Example 1. The control system in the plane defined by

$$\begin{aligned} \dot{x} &= u(x^2 - y^2) \\ \dot{y} &= u(2xy), u \in [-1,1], \end{aligned}$$

is GAC at the point $(0,0)$. In fact, as shown in Figure 1, for $(x,y) \neq (0,0)$ in the plane, the set $\{u(x^2-y^2, 2xy) : u \in [-1,1]\}$ is a subinterval of the tangent space to the circle passing through (x,y) and $(0,0)$ with center on the y-axis. The GAC property becomes obvious.

M. de Queiroz et al. (Eds.): Optimal Control, Stabilization and Nonsmooth Analysis, LNCIS 301, pp. 173–183, 2004.

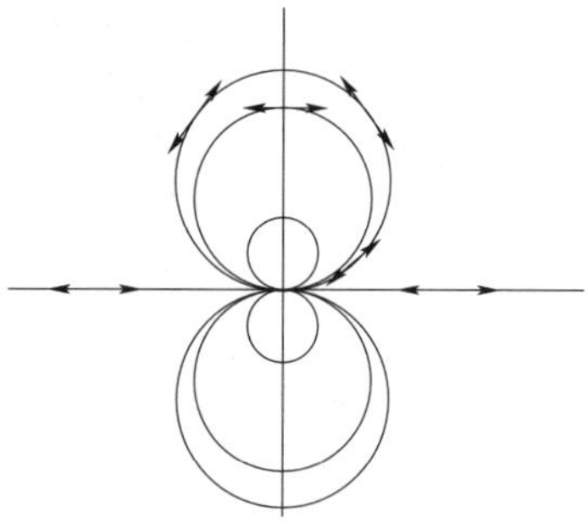

Fig. 1.

Example 2. In $\mathbb{R}^3$, the nonholonomic integrator defined by

$$\begin{aligned} \dot{x}_1 &= u_1 \\ \dot{x}_2 &= u_2 \\ \dot{x}_3 &= x_1 u_2 - x_2 u_1 \end{aligned}$$

with $u_1^2 + u_2^2 \leq 1$, is a famous example of GAC control system (at any point of the space).

Example 3. More generally, if M is a connected manifold and if the vector fields $f_1, \cdots, f_m$ satisfy the Hörmander's bracket generating condition

$$\forall x \in M, \quad \text{Lie}\,\{f_1, \cdots, f_m\}(x) = T_x M,$$

then a classical result of Chow says that every two points can be joined by a trajectory of the control system (1). Hence (1) is GAC at any point of the manifold.

Given a GAC control system of the form (1), the purpose of the stabilization problem is to study the possible existence of a feedback $k(\cdot) : M \mapsto \overline{B_m}$ which makes the closed-loop system

$$\dot{x} = f(x, k(x)) = \sum_{i=1}^{m} k_i(x) f_i(x) \tag{2}$$

globally asymptotically stable at the point O (abbreviated GAS in the sequel), i.e., such that all the trajectories of the closed-loop system converge asymptotically to the point O, and in addition such that the local property of Lyapunov stability is satisfied.

Example 4. As shown in [19, pp. 561-562], the control system given in Example 1 above does not admit a continuous stabilizing feedback. One proof is by noticing that the circles defined in Figure 1 are invariant under the closed-loop system, and that on these circles the only way for a closed-loop system to go to O would be to have $k < 0$ or $k > 0$ along its trajectories. Since by Lyapunov stability, the closed-loop control should be negative (resp., positive) on the right side (resp., on the left side) of the origin, we conclude by connectedness of the circle that a continuous stabilizing feedback should have another equilibrium on each invariant circle!

As we shall see in the sequel, the absence of continuous stabilizing feedbacks for general GAC control systems motivated many authors to define new kinds of stabilizing feedbacks. The main contributions in that direction have been Sussmann [21], Artstein [3], Brockett [4], Sontag [18], Coron [7, 8], Clarke, Ledyaev, Sontag and Subbotin [6], and Ancona and Bressan [2].

2 The Kurzweil Theorem

Given a continuous vector field X on the manifold M which has an equilibrium at $O \in M$, the classical Lyapunov function method asserts that if there exists some function $V : M \to \mathbb{R}$ that satisfies the following properties:

(i) V is smooth on $M \setminus \{0\}$ and continuous at the origin,
(ii) $V \geq 0$ and $V(x) = 0 \Longleftrightarrow x = O$,
(iii) $\forall x \in M \setminus \{O\}, (L_g V)_x < 0$,

then the dynamical system given by

$$\dot{x} = X(x(t)) \tag{3}$$

is GAS at the point O. Such a function is called a *Lyapunov function* for the dynamical system (3). The Kurzweil Theorem establishes the converse result; it asserts that if the dynamical system is GAS (at O) on the manifold M, then it admits a Lyapunov function on M.

This result has an important consequence. If the control system (1) admits a continuous stabilizing feedback $k : M \to \overline{B_m}$ then the dynamical system given by the closed-loop system (2) is GAS, and as a consequence it admits a Lyapunov function. Thus there exists some function $V : M \to \mathbb{R}$ which satisfies the following properties:

(i) V is smooth on $M \setminus \{0\}$ and continuous at the origin,
(ii) $V \geq 0$ and $V(x) = 0 \Longleftrightarrow x = O$,
(iii) $\forall x \in M \setminus \{O\}$, there exists $u \in \overline{B_m}$ such that $(L_{f(\cdot,u)}V)_x < 0$.

Such a function is called a *smooth control-Lyapunov function* for (1). To summarize, using the Kurzweil Theorem, we prove that if a control system has a continuous stabilizing feedback, then it admits a smooth control Lyapunov function.

Example 5. This result allows us to give another proof of the nonexistence of continuous stabilizing feedback for the control system given in Example 1.

If such a stabilizing feedback existed then the control system would admit a smooth control-Lyapunov function. But such a function would attain its maximum over any compact circle of Figure 1, that is, there would be some point where the gradient of V and the tangent space to the circle are orthogonal, which contradicts (iii).

The converse of this result is true (for the type of control systems that we consider, that is, affine in the control and without drift). Artstein proved in [3] that if the control system admits a smooth control-Lyapunov function then there exists a continuous stabilizing feedback (and the feedback can indeed be taken to be smooth outside O). This gives the following characterization of control systems that admit continuous stabilizing feedbacks:

Theorem 1. *The control system (1) has a continuous stabilizing feedback on M if and only if it admits a smooth control-Lyapunov function on M.*

This classical result leads us to the consideration of two major obstructions to the existence of continuous stabilizing feedbacks. We consider these obstructions next.

3 Two Obstructions

3.1 A Local Obstruction: The Brockett Condition

Since the obstruction we are talking about is local, let us assume that we work in $\mathbb{R}^n$. If there exists a continuous vector field X defined in some neighborhood $\mathcal{U}$ of O which is locally GAS, then for all ϵ small enough,

$$\exists \delta > 0 \text{ such that } \delta B \subset X(\epsilon B),$$

where B is the open unit ball in $\mathbb{R}^n$. This result, of topological nature, can indeed be seen as a consequence of Kurzweil Theorem. We refer the reader to Sontag [19] for its proof based on Kurzweil and Brouwer theorems. This property enables us to deduce Brockett's necessary condition.

If there exists a continuous feedback $k : \mathcal{U} \to \mathbb{R}^m$ ($k(O) = 0$) such that the closed-loop system (2) is locally GAS, then the result above applies to the dynamics $\dot{x} = f(x, k(x))$, and then we deduce that for all ϵ small enough,

$$\exists \delta > 0 \text{ such that } \delta B \subset f(\epsilon B, \overline{B_m}),$$

which is Brockett's necessary condition. This necessary condition gives us an easy way to verify if a given GAC control system possesses continuous stabilizing feedbacks.

Example 6. The Brockett necessary condition is not satisfied for the nonholonomic integrator defined by (2). It is easy to see that the vector $(0, 0, \epsilon)$ is not in the set $f(x, \overline{B_2})$ for x in a neighborhood of the origin.

Example 7. More generally, if the tangent vectors $f_1(O), \cdots, f_m(O)$ are independent in $T_O M$ (with $m < n$), then the control system (1) does not satisfy Brockett's necessary condition.

3.2 A Global Obstruction

If (1) has a continuous stabilizing feedback, then it admits a smooth control-Lyapunov function. In fact, this function can be seen as a Morse function on M with a unique (possibly degenerate) critical point. Indeed, by a result of Milnor (see [10]), if a manifold M admits such a Morse function, then it is diffeomorphic to $\mathbb{R}^n$. Therefore, we proved that if a control system has a continuous stabilizing feedback on M, then the manifold M must be diffeomorphic to $\mathbb{R}^n$.

3.3 Consequences

The obstructions above make it impossible to prove the existence of continuous stabilizing feedbacks for general GAC control systems. Actually, they motivate the design of new kinds of stabilizing feedbacks and moreover, as we shall see, it turns out to be essential to consider nonsmooth control-Lyapunov functions.

In order to simplify the statements of the results, we assume from now that the manifold M is the Euclidean space $\mathbb{R}^n$. Of course, all the results remain on smooth manifolds.

4 Semiconcave Control-Lyapunov Functions

Let Ω be an open set in $\mathbb{R}^n$. A function $g : \Omega \to \mathbb{R}$ is said to be *semiconcave* on Ω provided it is continuous on Ω and for any $x_0 \in \Omega$ there are constants $\rho, C > 0$ such that

$$\frac{1}{2}(g(x) + g(y)) - g\left(\frac{x+y}{2}\right) \leq C\|x - y\|^2, \quad \forall x, y \in x_0 + \rho B_n. \tag{4}$$

Equivalently, this means that the function g can be written locally as the sum of a concave function and a smooth (quadratic) function:

$$g(x) = [g(x) - 4C\|x\|^2] + 4C\|x\|^2.$$

Observe that any function of class C^2 is semiconcave. Also, any semiconcave function is locally Lipschitz, since both concave functions and smooth functions have that property. Other classical types of semiconcave functions are given by the following examples:

Example 8. If S denotes some closed subset of $\mathbb{R}^n$, then the distance function to the set S is semiconcave on $\mathbb{R}^n \setminus S$.

Example 9. If $\phi_1, \cdots, \phi_p : \Omega \to \mathbb{R}$ is a finite family of smooth functions, then the function defined by $g(x) := \min\{\phi_1(x), \cdots, \phi_p(x)\}$ is semiconcave on Ω.

We next consider the notion of a *nonsmooth control-Lyapunov function.* A continuous function $V : \mathbb{R}^n \to \mathbb{R}$ is said to be a *control-Lyapunov function* for the control system (1) provided it is a positive definite proper viscosity supersolution of the Hamilton-Jacobi equation

$$\max_{u \in \overline{B_m}} \{-\langle f(x,u), DV(x)\rangle\} - V(x) \geq 0. \tag{5}$$

In [13] we proved the following result:

Theorem 2. *If the control system (1) is GAC, then there exists a control-Lyapunov function $V : M \to \mathbb{R}$ which is continuous on M and semiconcave on $M \setminus \{O\}$.*

Sontag introduced a similar notion of nonsmooth control-Lyapunov function in his seminal paper [18], and proved the equivalence of global asymptotic controllability and the existence of continuous control-Lyapunov function. The converse of this result is true. Furthermore, we notice that whenever a continuous function $V : M \to \mathbb{R}$ is semiconcave on $M \setminus \{O\}$, it is a control-Lyapunov function if and only if it is positive definite, proper and such that for every $x \in M \setminus \{0\}$ where V is differentiable, we have

$$\min_{u \in \overline{B_m}} \langle \nabla V(x), f(x,u)\rangle \leq -V(x).$$

(Note that since V is locally Lipschitz on $M \setminus \{O\}$, it is differentiable almost everywhere in M.) Let us show, in the following sections, the relevance of the semiconcave property to the construction of stabilizing feedbacks.

5 Stabilizing Feedbacks

Clarke *et al.* [6] proved that any GAC control system can be stabilized by means of a discontinuous stabilizing feedback. Later in [13], we showed that a semiconcave control-Lyapunov function generates this kind of stabilizing feedback in a very simple and natural way. (We refer the reader to the Clarke's article elsewhere in this volume for the presentation of the construction.) Recall that the closed-loop system (2) is said to be GAS in the sense of Carathéodory if for every $x \in \mathbb{R}^N$ the solutions (which are called Carathéodory solutions) of

$$\dot{x}(t) = f(x(t), \alpha(x(t)) \text{ a.e. }, x(0) = x_0,$$

exist, converge to the origin as $t \to \infty$ and satisfy the property of Lyapunov stability. In the case of control systems that are affine in the control, we proved that in fact the method that we developed leads to the following result:

Theorem 3. *If the control system (1) is GAC, then there exists an open dense set $\mathcal{D}$ of full measure in $M \setminus \{O\}$ and a feedback $k : M \to \mathbb{R}^m$ such that*

(i) *k is smooth on $\mathcal{D}$; and*
(ii) *the closed-loop system (2) is GAS in the sense of Carathéodory.*

This result has been initially proven by Ancona and Bressan (see [2]) who do not make use of control-Lyapunov functions but paste together different stabilizing open-loops into what they call a patchy feedback. Although Theorem 3 provides an easy way to design stabilizing feedbacks (in the sense of Carathéodory), it does not explicitly detail the behavior of the closed-loop system near the singularities (*i.e.* the points where the feedback is not continuous). From now on, our aim is to explain how further regularity of the control-Lyapunov function can improve the above result.

6 A Classification of Singularities in the Plane

In [14], we proved that the nature of the singularities of the stabilizing feedback given by Theorem 3 can be described very precisely, whenever we work on surfaces. In this case, we demonstrated that a part of the singular set of a semiconcave function (that is, the set of points where the function is not differentiable) can be "stratified" by Lipschitz submanifolds of dimension 1 and by isolated points. This permitted us to achieve the following result (and then to answer a conjecture of Bressan):

Theorem 4. *If M is a smooth manifold of dimension 2 and if the control system (1) is GAC, then there exists a Carathéodory stabilizing feedback $k : M \to \mathbb{R}^m$ with singularities as shown in Figure 2.*

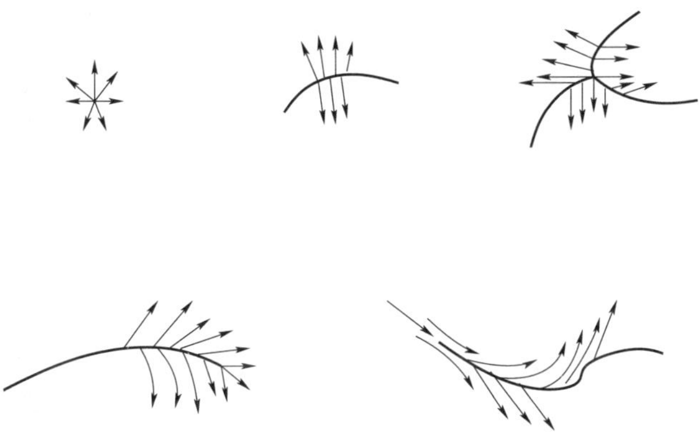

Fig. 2. Different types of singularities on surfaces

Unfortunately, the "natural" stratification that appeared in dimension two cannot be achieved in higher dimensions. Hence, if we want to better understand the behavior of the closed-loop system near its singularities, we have to consider nonsmooth control-Lyapunov functions with a stronger property of semiconcavity.

7 Stratified Semiconcave Control-Lyapunov Functions

Let Ω be an open set of $\mathbb{R}^n$ and let $g : \Omega \to \mathbb{R}$ be a semiconcave function. By Rademacher's theorem, we know that g is differentiable almost everywhere in Ω. Let us denote by $\Sigma(g)$ the singular set of g, *i.e.* the set of points of Ω where g is not differentiable. We can also view $\Sigma(g)$ as the set of $x \in \Omega$ such that $\dim(\partial g(x)) \geq 1$; this point of view leads to a natural partition of the singular set. As a matter of fact, following more or less the seminal work of Alberti, Ambrosio and Cannarsa [1], $\Sigma(g)$ can be written as the disjoint union of N sets $\Sigma^k(g)$ (for $k \in \{1, \cdots, N\}$) defined by

$$\Sigma^k(g) := \{x \in \Omega : \dim(\partial g(x)) = k\}.$$

Alberti *et al.* proved that for any $k \in \{1, \cdots, N\}$, the set $\Sigma^k(u)$ is countably $\mathcal{H}^{n-k}$-rectifiable, *i.e.* it is contained (up to a $\mathcal{H}^{n-k}$-negligible set) in a countable union of C^1 hypersurfaces of dimension $N - k$. But each set $\Sigma^k(u)$ is certainly not an exact hypersurface (or submanifold) of Ω. That is why we introduce the concept of stratified semiconcave function.

The semiconcave function $g : \Omega \to \mathbb{R}$ is said to be *stratified semiconcave (in Ω)* provided the following conditions are satisfied:

(1) The set $\Sigma(g)$ is a Whitney stratification[1] such that the strata of dimension $N - k$ are the connected components of $\Sigma^k(g)$;
(2) For every stratum S of $\Sigma(g)$, the set $\overline{S}$ is a smooth submanifold with boundary;
(3) For every stratum S of $\Sigma(g)$, the function g is smooth on $\overline{S}$; and
(4) For every $x \in \Sigma^k(g)$, the set $\partial g(x)$ is a compact convex set of dimension k with exactly $k+1$ extreme points $\zeta_1(x), \cdots, \zeta_{k+1}(x)$. In addition, the maps $\zeta_1(\cdot), \cdots, \zeta_{k+1}(\cdot)$ are smooth on $\Sigma^k(g)$ and for any stratum S of $\Sigma^k(g)$, they can be smoothly extended to $\overline{S}$.

Example 10. Assume we are given a finite family $(h_i)_{i \in I}$ of affine functions in $\mathbb{R}^N$. If $u : \mathbb{R}^N \to \mathbb{R}$ is defined by

$$u := \min_{i \in I}\{h_i\},$$

then it is a stratified semiconcave function and moreover it satisfies:

[1]We refer the reader to our paper [15] for the precise definition of a Whitney stratification.

(i) For every $k \in \{1, \cdots, N\}$, the set $\Sigma^k(u)$ is a finite disjoint union of open polyhedra of dimension $N - k$.
(ii) For every $k \in \{1, \cdots, N\}$, the multivalued map $x \mapsto \partial u(x) = \partial^P u(x)$ is constant on each connected component of $\Sigma^k(u)$.

In [15], we showed that every semiconcave function can be approximated locally by a stratified semiconcave function, in such a way that the proximal subdifferentials of both functions are very close. Furthermore, by transversality arguments, we managed to paste together different approximations to get the following:

Theorem 5. *If the control system (1) is GAC, then there exists a control-Lyapunov function which is stratified semiconcave on* $M \setminus \{O\}$.

Theorem 5 has its own importance, but in fact it appeared to be the key tool in the construction of AGAS feedbacks. We turn to this feedback construction next.

8 AGAS Feedbacks

A smooth dynamical system of the form $\dot{x} = X(x)$ is said to be *almost globally asymptotically stable* at the point O (abreviated AGAS) if the two following properties are satisfied:

1. Almost attractivity: For almost every $x \in M$, the solution of $\dot{x} = X(x)$ starting at x converges to O;
2. Lyapunov stability: For each neighborhood $\mathcal{V}$ of O, there exists some neighborhood $\mathcal{U}$ of O such that if $x \in \mathcal{U}$ then the solution of $\dot{x} = X(x)$ starting at x satisfies $x(t) \in \mathcal{V}, \forall t \geq 0$.

This particular property of stability has been introduced by Rantzer in [11]. The "very" regular control-Lyapunov function given by Theorem 5 permits us to prove the following result:

Theorem 6. *If the control system (1) is GAC, then there exists a continuous feedback* $k : M \to \mathbb{R}^m$ *which is smooth outside the origin, and such that the closed-loop system (2) is AGAS.*

Knowing this theorem, we can wonder if it is possible to detail the nature of the set (of measure zero) of points which are not stabilized by our smooth feedback. This set can be proved to be closed and repulsive in some cases.

9 SRS Feedbacks

The feedback $k : M \to \mathbb{R}^m$ is said to be a *smooth repulsive stabilizing* feedback (called SRS feedback) if the following properties are satisfied:

1. There exists a set $\mathcal{S} \subset M \setminus \{O\}$ which is closed in $M \setminus \{O\}$ and of measure zero;
2. The feedback k is smooth outside the set $\mathcal{S} \cup \{O\}$; and
3. The closed-loop system (2) is GAS in the sense of Carathéodory;
4. At all positive times, the trajectories of the closed-loop system do not belong to the set $\mathcal{S}$.

The first result of existence of SRS feedbacks is a classical fact in sub-Riemannian geometry. We refer the reader to [9] and [20] for its proof and for the definition of fat distribution.

Theorem 7. *If the distribution defined by the control system (1) is fat, then the control system admits a SRS feedback.*

This theorem applies for instance in the case of the nonholonomic integrator. The second result that we want to highlight is concerned with control system in dimension three. In [17] we prove the following:

Theorem 8. *Assume that $n = 3$. If the control system (1) satisfies the Hörmander's condition at the origin,*

$$Lie\,\{f_1, \cdots, f_m\}(0) = \mathbb{R}^3,$$

then there exists a local SRS feedback.

The proof of this result is based on the classification of the singularities given in Section 6; it does not appear to extendable in higher dimension. Actually, we do not know if this result hold in dimension greater than three (or even globally).

References

1. Alberti G, Ambrosio L, Cannarsa P (1992) On the singularities of convex functions, Manuscripta Math 76:421–435
2. Ancona F, Bressan (1999) Patchy vector fields and asymptotic stabilization, ESAIM Control Optim Calc Var 4:445–471
3. Artstein Z (1983) Stabilization with relaxed controls, Nonlinear Analysis TMA 7:1163–1173
4. Brockett R (1983) Asymptotic stability and feedback stabilization. In: Brockett R, Millman R, Sussmann H (eds) Differential Geometric Control Theory. Birkhäuser Boston
5. Clarke F (2004) Lyapunov functions and feedback in nonlinear control, this collection
6. Clarke F, Ledyaev Y, Sontag E, Subbotin A (1997) Asymptotic controllability implies feedback stabilization, IEEE Trans Aut Control 42:1394–1407
7. Coron J-M (1992) Global Asymptotic stabilization of controllable systems without drift, Math Cont Sign Sys 5:295–312

8. Coron J-M (1994) Linearized control systems and application to smooth stabilization, SIAM J Control Optim 32:358–386
9. Ge Z (1993) Horizontal path spaces and Carnot-Carathéodory metrics, Pacific J Math 161:255–286
10. Milnor J (1964) Differential topology. In: Lectures on Modern Mathematics Vol II. Wiley New York 165–183
11. Rantzer A (2000) A dual to Lyapunov stability theorem, Systems Control Lett 42:161–168
12. Rifford L (2000) Existence of Lipschitz and semiconcave control-Lyapunov functions, SIAM J Control Optim 39:1043–1064
13. Rifford L (2002) Semiconcave control-Lyapunov functions and stabilizing feedbacks, SIAM J Control Optim 41:659–681
14. Rifford L (2003) Singularities of viscosity solutions and the stabilization problem in the plane, Indiana Univ Math J 52:1373-1396
15. Rifford L (2004) Stratified semiconcave control-Lyapunov functions and the stabilization problem, submitted
16. Rifford L (2004) The stabilization problem on surfaces, in preparation
17. Rifford L (2004) On the existence of smooth repulsive stabilizing feedbacks in dimension three, in preparation
18. Sontag E (1983) A Lyapunov-like characterization of asymptotic controllability, SIAM J Control Optim 21:462–471
19. Sontag E (1999) Stability and stabilization discontinuities and the effect of disturbances. In: Nonlinear Analysis Differential Equations and Control Montreal QC 1998. Kluwer Dordrecht 307–367
20. Strichartz R (1986) Sub-Riemannian geometry, J Differential Geom 24:221–263
21. Sussmann H (1979) Subanalytic sets and feedback control, J Differential Equations 31:31–52

Stabilization by Patchy Feedbacks and Robustness Properties

Fabio Ancona[1] and Alberto Bressan[2]

[1] Dipartimento di Matematica and C.I.R.A.M., Università di Bologna, P.zza Porta S. Donato 5, Bologna 40127, Italy. ancona@ciram.unibo.it.
[2] S.I.S.S.A., Via Beirut 4, Trieste 34014, Italy. bressan@sissa.it

Summary. This note is concerned with the problem of robust stabilization of control systems via discontinuous feedback laws. A particular class of discontinuous, piecewise smooth vector fields ("patchy vector fields") was introduced by the authors in a previous paper to construct discontinuous stabilizing feedbacks for globally asymptotically controllable systems. Here we study the stability of the corresponding solution set with respect to impulsive perturbations with small total variation. These results yield the robustness of "patchy feedback controls" in the presence of measurement errors and external disturbances.

1 Introduction

The construction of discontinuous stabilizing feedback laws for nonlinear control systems has been the focus of vast research in recent years. Indeed, the presence of certain topological obstructions can prevent the existence of continuous stabilizing feedbacks even for globally asymptotically controllable systems. This fact was first pointed out by Sussmann [19] for a two-dimensional system, and by Sontag and Sussmann [18] for one–dimensional systems, and then it was further analyzed by Brockett [5] and Coron [8] for general nonlinear systems. Unfortunately, allowing non-regular feedbacks immediately leads to a major technical difficulty since then one has to analyze the stability of a discontinuous O.D.E. that may not even possess any solution (in the classical Carathéodory sense). To overcome this problem, suitable definitions of "generalized solution" (sampling and Euler solutions) originating from the theory of positional differential games [10] were proposed in [7, 14, 16]. In these works it is shown that any asymptotically controllable system can be stabilized in the sense of such solutions by means of feedback laws that are constructed using Lyapunov-like techniques in connection with nonsmooth control-Lyapunov functions.

An alternative strategy was followed in [1], where the authors introduced a particular class of discontinuous, piecewise smooth vector fields ("patchy vector fields") which always admit Carathéodory solutions. It was shown that by

M. de Queiroz et al. (Eds.): Optimal Control, Stabilization and Nonsmooth Analysis, LNCIS 301, pp. 185–199, 2004.

patching together open loop controls of an asymptotically controllable system one can always construct a stabilizing feedback that generates a discontinuous vector field belonging to this class, i.e., a "patchy feedback". In particular, one can develop the whole theory by studying the corresponding discontinuous O.D.E's within the classical framework of Carathéodory solutions.

In this note we are concerned with the robustness properties of a discontinuous feedback, which are particularly relevant in many practical situations since one of the main reasons for using a state feedback is precisely the fact that open loop controls are usually very sensitive to disturbances. The feedback laws constructed in [7, 14, 16] are robust with respect to actuator errors as well as to perturbation of the external dynamics, while they may be highly sensitive to errors in the measurement of the state vectors. In particular, it is shown in [6, 16] that such feedbacks enjoy a "relative robustness" with respect to observation errors, meaning that the magnitude of the error must not exceed the maximum step size of the underlying discretization of the sampling solution taken in consideration.

On the contrary, we have shown in [2] that a patchy feedback exhibits a "full robustness" with respect to measurement errors having sufficiently small total variation so to avoid the possible chattering behavior that may arise at discontinuity points. This result is obtained by establishing the stability of the set of trajectories of a patchy vector field with respect to a wide class of impulsive perturbations and by applying the patchy feedback construction developed in [1]. For a generic class of "polygonal" patchy vector fields in the plane, we have also shown in [3] that the distance between a perturbed trajectory and an unperturbed one is of the same order of magnitude as the impulsive forcing term. In the present note we will survey the results in [2, 3] and outline some of their proofs.

1.1 Patchy Vector Fields

We start by recalling the main definitions introduced in [1]:

Definition 1. *By a* patch *we mean a pair* (Ω, g) *where* $\Omega \subset \mathbb{R}^n$ *is an open domain with smooth boundary* $\partial\Omega$, *and* g *is a smooth vector field defined on a neighborhood of the closure* $\overline{\Omega}$, *which points strictly inward at each boundary point* $x \in \partial\Omega$. *Calling* $\mathbf{n}(x)$ *the outer normal at the boundary point* x, *we thus require*

$$\langle g(x),\ \mathbf{n}(x)\rangle < 0 \qquad \textit{for all} \quad x \in \partial\Omega. \tag{1}$$

Definition 2. *We say that* $g : \Omega \mapsto \mathbb{R}^n$ *is a* patchy vector field *on the open domain* Ω *if there exists a family of patches* $\{(\Omega_\alpha,\ g_\alpha);\ \ \alpha \in \mathcal{A}\}$ *such that*

- *$\mathcal{A}$ is a totally ordered set of indices;*
- *the open sets Ω_α form a locally finite covering of Ω, i.e. $\Omega = \bigcup_{\alpha\in\mathcal{A}}\Omega_\alpha$ and every compact set $K \subset \mathbb{R}^n$ intersects only a finite number of domains Ω_α, $\alpha \in \mathcal{A}$;*

- *the vector field* g *can be written in the form*

$$g(x) = g_\alpha(x) \qquad \textit{if} \qquad x \in \Omega_\alpha \setminus \bigcup_{\beta > \alpha} \Omega_\beta. \tag{2}$$

By setting

$$\alpha^*(x) \doteq \max\{\alpha \in \mathcal{A}\,;\;\; x \in \Omega_\alpha\}, \tag{3}$$

we can write (3) in the equivalent form

$$g(x) = g_{\alpha^*(x)}(x) \qquad \text{for all} \quad x \in \Omega. \tag{4}$$

REMARK 1. Notice that the patches $(\Omega_\alpha,\, g_\alpha)$ are not uniquely determined by the patchy vector field g. Indeed, whenever $\alpha < \beta$, by (2) the values of g_α on the set $\Omega_\alpha \cap \Omega_\beta$ are irrelevant. Therefore, if the open sets Ω_α form a locally finite covering of Ω and we assume that, for each $\alpha \in \mathcal{A}$, the vector field g_α satisfies (1) at every point $x \in \partial\Omega_\alpha \setminus \bigcup_{\beta>\alpha} \Omega_\beta$, then the vector field g defined according with (2) is again a patchy vector field. To see this, it suffices to construct vector fields $\tilde{g}_\alpha$ (defined on a neighborhood of $\overline{\Omega}_\alpha$ as g_α) which satisfy the inward pointing property (1) at every point $x \in \partial\Omega_\alpha$ and such that $\tilde{g}_\alpha = g_\alpha$ on $\Omega_\alpha \setminus \bigcup_{\beta>\alpha} \Omega_\beta$ (cfr. [1, REMARK 2.1]).

REMARK 2. In some situations it is useful to adopt a more general definition of patchy vector field than the one formulated above. Indeed, one can take into consideration patches $(\Omega_\alpha,\;\, g_\alpha)$ where the domain Ω_α has a piecewise smooth boundary (cfr. [3]). In this case, the inward-pointing condition (1) can be expressed by requiring that

$$g(x) \in \operatorname{int}(T_\Omega(x)) \tag{5}$$

where $\operatorname{int}(T_\Omega(x))$ denotes the interior of the tangent cone to Ω at the point x, defined by

$$T_\Omega(x) \doteq \left\{ v \in \mathbb{R}^n \;:\; \liminf_{t \downarrow 0} \frac{d\big(x + tv,\; \Omega\big)}{t} = 0 \right\}. \tag{6}$$

Clearly, at any regular point $x \in \partial\Omega$, the interior of the tangent cone $T_\Omega(x)$ is precisely the set of all vectors $v \in \mathbb{R}^n$ that satisfy $\langle v,\; \mathbf{n}(x)\rangle < 0$ and hence (5) coincides with the inward-pointing condition (1). One can easily see that all the results concerning patchy vector fields established in [1, 2] remain true within this more general formulation. For the sake of simplicity, throughout the paper we shall almost always adopt the original formulation of Definition 1.

We shall occasionally adopt the longer notation $(\Omega,\, g,\, (\Omega_\alpha,\, g_\alpha)_{\alpha \in \mathcal{A}})$ to indicate a patchy vector field, specifying both the domain and the single patches. If g is a patchy vector field, then the differential equation

$$\dot{x} = g(x) \tag{7}$$

has many interesting properties. In particular, we have proved in [1] that the set of Carathéodory solutions of (7) is closed in the topology of uniform convergence, but possibly not connected. Using patchy vector fields, we can thus avoid the topological obstructions that sometimes prevent the existence of a continuous stabilizing feedback (whenever we interpret $g(x)$ as the closed-loop vector field $f(x, U(x))$ generated by a given state feedback $u = U(x)$). Moreover, given an initial condition

$$x(t_0) = x_0, \tag{8}$$

the Cauchy problem (7)-(8) has at least one forward solution, and at most one backward solution. For every Carathéodory solution $x = x(t)$ of (7), the map $t \mapsto \alpha^*(x(t))$ is left continuous and non-decreasing.

1.2 Impulsive Perturbations

We now address the problem of the stability of the solution set of (7) w.r.t. various perturbations. Most of our analysis will be concerned with impulsive perturbations, described by

$$\dot{y} = g(y) + \dot{w}. \tag{9}$$

Here $w = w(t)$ is any left continuous function with bounded variation. By a solution of the perturbed system (9) with an initial condition $y(t_0) = y_0$ we mean a measurable function $t \mapsto y(t)$ such that

$$y(t) = y_0 + \int_{t_0}^{t} g\big(y(s)\big)\, ds + \big[w(t) - w(t_0)\big]. \tag{10}$$

If $w(\cdot)$ is discontinuous, the system (9) has impulsive behavior and the solution $y(\cdot)$ will be discontinuous as well. We choose to work with (9) because it provides a simple and general framework to study robustness properties. Indeed, consider a system with both inner and outer perturbations, of the form

$$\dot{x} = g\big(x + e_I(t)\big) + e_O(t). \tag{11}$$

The map $t \mapsto y(t) \doteq x(t) + e_I(t)$ then satisfies the impulsive equation

$$\dot{y} = g(y) + e_O(t) + \dot{e}_I(t) = g(y) + \dot{w},$$

where

$$w(t) = e_I(t) + \int_{t_0}^{t} e_O(s)\, ds.$$

Therefore, from the stability of solutions of (9) w. r. t. perturbations w that have small total variation, one can immediately deduce a result on the stability of solutions of (11), when Tot.Var.$\{e_I\}$ and $\|e_O\|_{\mathbf{L}^1}$ are suitably small.

Here, Tot.Var.$\{e_I\}$ denotes the total variation of the function e_I over the whole interval where it is defined, while Tot.Var.$\{e_I\,;\,J\}$ denotes the total variation of e_I over a subset J. We shall also denote by BV the space of all functions of bounded variation. Any function of bounded variation $w = w(t)$ can be redefined up to $\mathbf{L}^1$ equivalence. For sake of definiteness, throughout the paper we shall always consider left continuous representatives, so that $w(t) = w(t^-) \doteq \lim_{s \to t-} w(s)$ for every t. The Lebesgue measure of a Borel set $J \subset \mathbb{R}$ will be denoted by meas(J). We will use $B(x,r)$ to denote the closed ball centered at x with radius r and, for every given set A, we let $B(A,r) \doteq \cup_{x\in A} B(x,r)$. We write $B_r = B(0,r)$. The closure, the interior and the boundary of a set Ω are written as $\overline{\Omega}$, int(Ω) and $\partial\Omega$, respectively.

2 Stability of Patchy Vector Fields

We observe that, since the Cauchy problem for (7) does not have forward uniqueness and continuous dependence, one clearly cannot expect that a single solution of (7) be stable under small perturbations. What we extablished in [2] is a different stability property, involving not a single trajectory but the whole solution set: if the perturbation w is small in the BV norm, then every solution of (9) is close to some solution of (7). This is essentially an upper semicontinuity property of the solution set, namely, the following results hold.

Theorem 1. *Let g be a patchy vector field on an open domain $\Omega \subset \mathbb{R}^n$. Consider a sequence of solutions $y_\nu(\cdot)$ of the perturbed system*

$$\dot{y}_\nu = g(y_\nu) + \dot{w}_\nu \qquad\qquad t \in [0,T], \tag{12}$$

with Tot.Var.$\{w_\nu\} \to 0$ *as* $\nu \to \infty$. *If the $y_\nu : [0,T] \mapsto \Omega$ converge to a function $y : [0,T] \mapsto \Omega$, uniformly on $[0,T]$, then $y(\cdot)$ is a Carathéodory solution of (7) on $[0,\,T]$.*

Corollary 1. *Let g be a patchy vector field on an open domain $\Omega \subset \mathbb{R}^n$. Given any compact set $K \subset \Omega$, and any $\varepsilon > 0$, there exists $\delta = \delta(K,\varepsilon) > 0$ such that the following holds: If $y : [0,T] \mapsto K$ is a solution of the perturbed system (9), with* Tot.Var.$\{w\} < \delta$, *then there exists a solution $x : [0,T] \mapsto \Omega$ of the unperturbed equation (7) with*

$$\big\|x - y\big\|_{\mathbf{L}^\infty} < \varepsilon\,. \tag{13}$$

We sketch the main arguments in the proofs of Theorem 1 and Corollary 1. For details see [2].

Proof of Theorem 1.
1. By the basic properties of a patchy vector field, for every solution $t \mapsto x(t)$ of (7) the corresponding map $t \mapsto \alpha^*(x(t))$ in (3) is non-decreasing. Roughly speaking, a trajectory can move from a patch Ω_α to another patch Ω_β only

if $\alpha < \beta$. This property no longer holds in the presence of an impulsive perturbation. However, one can show that, for a solution $y : [0,T] \mapsto \Omega$ of (9) taking values in a given compact set $K \subset \Omega$, one can slightly modify the impulsive perturbation $\dot{w}$, say replacing it by another perturbation $\dot{w}^\diamond$, so that the map $t \mapsto \alpha^*(y^\diamond(t))$ is still non-decreasing along the corresponding trajectory $t \mapsto y^\diamond(t)$, and so that there holds

$$\begin{aligned} \text{Tot.Var.}\{w^\diamond\} &\leq C \cdot \text{Tot.Var.}\{w\}\,, \\ \big\|y^\diamond - y\big\|_{\mathbf{L}^\infty} &\leq C \cdot \text{Tot.Var.}\{w\}\,, \end{aligned} \tag{14}$$

for some constant C (depending only on the compact set K).

2. Thanks to the above observation, in order to extablish Theorem 1 it is sufficient to consider a sequence of solutions $y_\nu^\diamond : [0,T] \mapsto \Omega$ of the perturbed system (12), corresponding to impulsive perturbations $\dot{w}_\nu^\diamond$, with the following properties:

(i) Tot.Var.$\{w_\nu^\diamond\} \to 0$ as $\nu \to \infty$;
(ii) $t \mapsto \alpha^*\big(y_\nu^\diamond(t)\big)$ is non-decreasing and left continuous, for every ν;
(iii) $\big(y_\nu^\diamond\big)_\nu$ converges to some function $y : [0,T] \mapsto \Omega$ uniformly on $[0,T]$.

Because of (ii), there exist finitely many indices $\alpha_1 < \cdots < \alpha_N$ and partitions $0 = \tau_{1,\nu} \leq \tau_{2,\nu} \leq \ldots \leq \tau_{N+1,\nu} = T$ of $[0,\,T]$, such that

$$\alpha^*(y_\nu^\diamond(t)) = \alpha_i \qquad \forall\, t \in\,]\tau_{i,\nu},\, \tau_{i+1,\nu}], \quad i = 1, \ldots, N. \tag{15}$$

Then, recalling (4) and (10), we have

$$\begin{aligned} y_\nu^\diamond(t) = y_\nu^\diamond(0) \;+\; &\sum_{\ell=1}^{i-1} \int_{\tau_{\ell,\nu}}^{\tau_{\ell+1,\nu}} g_{\alpha_\ell}(y_\nu^\diamond(s))\, ds \;+ \\ &+ \int_{\tau_{i,\nu}}^{t} g_{\alpha_i}(y_\nu^\diamond(s))\, ds \;+\; [w_\nu^\diamond(t) - w_\nu^\diamond(0)], \\ &\forall\, t \in [\tau_{i,\nu},\, \tau_{i+1,\nu}], \qquad i = 1, \ldots, N. \end{aligned} \tag{16}$$

By taking a subsequence, we can assume that every sequence $(\tau_{i,\nu})_\nu$ converges to some limit point $\overline{\tau}_i$ as $\nu \to \infty$. From (16) and because of (iii) it thus follows that on each interval $[\overline{\tau}_i,\, \overline{\tau}_{i+1}]$, the function $y(\cdot)$ is a classical solution of $\dot{y} = g_{\alpha_i}(y)$ and satisfies

$$\left.\begin{aligned} y(t) &\in \overline{\Omega}_{\alpha_i} \setminus \textstyle\bigcup_{\beta > \alpha_i} \Omega_\beta, \\ \dot{y}(t) &= g_{\alpha_i}(y(t)), \end{aligned}\right. \qquad \forall\, t \in\,]\overline{\tau}_i,\, \overline{\tau}_{i+1}]. \tag{17}$$

Moreover, the inward-pointing condition (1) guarantees that the set $\{t \in [\overline{\tau}_i,\, \overline{\tau}_{i+1}] : y(t) \in \partial\,\Omega_{\alpha_i}\}$ is nowhere dense in $[\overline{\tau}_i,\, \overline{\tau}_{i+1}]$. Therefore, recalling Definition 2, from (16) we conclude

$$
\begin{aligned}
& y(t) \in \Omega_{\alpha_i} \setminus \bigcup_{\beta > \alpha_i} \Omega_\beta \qquad \forall\, t \in]\overline{\tau}_i,\, \overline{\tau}_{i+1}], \quad i = 1, \dots, N, \\
& y(t) = y(0) + \int_0^t g\big(y(s)\big)\, ds \qquad\qquad \forall\, t \in [0,\, T],
\end{aligned}
\tag{18}
$$

proving that $y : [0,\, T] \mapsto \Omega$ is a Carathéodory solution of (7) on $[0,\, T]$.

Proof of Corollary 1. We shall argue by contradiction assuming that the statement is false. Suppose that, for some compact set K, and some $\varepsilon > 0$, there exists a sequence of solutions $y_\nu : [0, T] \mapsto K$ of the perturbed system (9), corresponding to impulsive perturbations $\dot{w}_\nu$, that enjoy the following properties:

(i) $\delta_\nu \doteq \text{Tot.Var.}\{w_\nu\} \to 0 \quad \text{as} \quad \nu \to \infty$;

(ii) for every solution $x : [0, T] \mapsto \Omega$ of the unperturbed equation (7) there holds

$$
\big\| x - y_\nu \big\|_{\mathbf{L}^\infty} \geq \varepsilon \qquad \forall\, \nu\,. \tag{19}
$$

Then, for each ν, consider the polygonal curve $y_\nu^\diamond$ with vertices at the points $y_\nu(\ell\delta_\nu)$, $\ell \geq 0$, defined by setting

$$
\begin{aligned}
& y_\nu^\diamond(t) \doteq y_\nu\big(\ell\delta_\nu\big) + \tfrac{t - \ell\delta_\nu}{\delta_\nu} \cdot \Big(y_\nu\big((\ell+1)\delta_\nu\big) - y_\nu\big(\ell\delta_\nu\big)\Big) \\
& \qquad \forall\, t \in [\ell\delta_\nu,\, (\ell+1)\delta_\nu] \cap [0, T], \qquad 0 \leq \ell \leq \lfloor T/\delta_\nu \rfloor\,,
\end{aligned}
\tag{20}
$$

where $\lfloor T/\delta_\nu \rfloor$ denotes the integer part of T/δ_ν. Observe that, letting

$$
M \doteq \sup_{y \in K} |g(y)|,
$$

we get

$$
\big| y_\nu\big((\ell+1)\delta_\nu\big) - y_\nu\big(\ell\delta_\nu\big) \big| \leq M\delta_\nu + \text{Tot.Var.}\{w_\nu\} \quad \forall \ell\,, \tag{21}
$$

and hence we deduce

$$
\big| y_\nu^\diamond(t') - y_\nu^\diamond(t) \big| \leq (t' - t) \bigg(\frac{\text{Tot.Var.}\{w_\nu;\, [t, t']\,\}}{\delta_\nu} + M \bigg) \tag{22}
$$

for all $0 \leq t < t' \leq T$. It follows that $y_\nu^\diamond(\cdot)$ is a uniformly bounded sequence of Lipschitz maps, having Lipschitz constant $\text{Lip}(y_\nu^\diamond) \leq (1 + M)$. Hence, applying Ascoli-Arzelà Theorem, we can find a subsequence, still denoted by $y_\nu^\diamond(\cdot)$, which converges to some function $y(\cdot)$ uniformly on $[0,\, T]$. On the other hand, by construction and thanks to *(i)* and (21), we deduce that $\big\| y_\nu - y_\nu^\diamond \big\|_{\mathbf{L}^\infty} \to 0$ as $\nu \to \infty$. It is then also clear that the sequence $y_\nu(\cdot)$ converges to $y(\cdot)$ uniformly on $[0,\, T]$, and hence we can apply Theorem 1 to $y_\nu(\cdot)$ concluding that the function $y(\cdot)$ is a Carathéodory solution of the unperturbed equation (7) that satisfies $|y - y_\nu|_{\mathbf{L}^\infty} < \varepsilon$ for all ν sufficiently large. We thus obtain a contradiction with (19), concluding the proof.

3 Stability Rates for Patchy Vector Fields

We address in this section the problem of deriving an estimate of the distance in the $\mathbf{L}^\infty$ norm, between solutions of the Cauchy problem for the perturbed system (9) and the unperturbed on (7).

It is well-known that, in the case of a Lipschitz continuous vector field g, the Cauchy problems

$$\begin{cases} \dot{x} = g(x), \\ x(0) = x_0, \end{cases} \qquad \begin{cases} \dot{y} = g(y) + \dot{w}, \\ y(0) = x_0, \end{cases} \tag{23}$$

have unique solutions, obtained by a fixed point argument (see [4]). Their distance grows linearly with Tot.Var.$\{w\}$ and can be estimated by

$$\big|y(t) - x(t)\big| \le \int_0^t e^{L(t-s)} \big|dw(s)\big| \le e^{Lt} \cdot \text{Tot.Var.}\{w\}, \tag{24}$$

denoting with L the Lipschitz constant of g. On the contrary, one cannot expect, in general, a linear dependence in the case of a patchy vector fields. To see this, consider the following

Example 1. Consider a patchy vector field on $\mathbb{R}^2$, as in Figure 1. Assume $g = (1, 0)$ is below the curve γ_1 and to the right of the curve γ_2, while $g = (0, 1)$ above the curve γ_1. Observe that there exists a Carathéodory solution $x(\cdot)$ of (7) going through the points A and B. Next, consider a perturbed solution x^ε, following the vector field horizontally up to P, jumping from P to P', then moving vertically to Q and horizontally afterwards. To fix the ideas, assume that

$$A = (0, 0), \quad B = (0, 1), \quad P = (\varepsilon, -\varepsilon^\alpha), \quad P' = (\varepsilon, \varepsilon^\alpha), \tag{25}$$

$$\gamma_1 = \{x_2 = |x_1|^\alpha\}, \qquad \gamma_2 = \{x_1 = |x_2 - 1|^\beta\}, \tag{26}$$

In this case the trajectory x^ε is a solution of a perturbed system where $\dot{w}$ is a single Dirac mass of strength $|P' - P| = 2|\varepsilon|^\alpha$. On the other hand, after both trajectories have switched to the right of the curve γ_2 their distance is $\|x^\varepsilon - x\| = \varepsilon^{1/\beta}$. In this example, the distance between solutions grows much worse than linearly w.r.t. the strength of the perturbation. Indeed, the only estimate available is

$$\big\|y - x\big\|_{\mathbf{L}^\infty} = \mathcal{O}(1) \cdot \big(\text{Tot.Var.}\{w\}\big)^{1/\alpha\beta}. \tag{27}$$

The situation is better when the patches are in "generic" position. Observe that in (27) the numbers α and β are determined by the order of tangency of the curves γ_1, γ_2 with the vector field g. By an arbitrarily small displacement of the curves γ_1, γ_2 we can arrange so that there is no trajectory connecting the two point of tangency A and B (fig. 2). Moreover, we can assume that the tangency is only of first order. For generic patchy vector fields on $\mathbb{R}^2$, in Corollary 1 one thus expects an estimate of the form

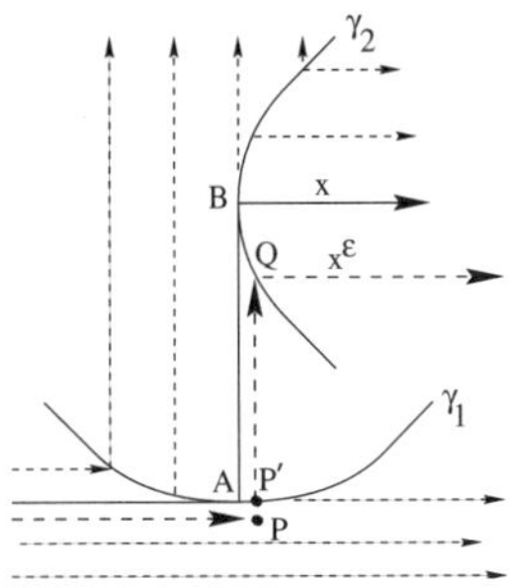

Fig. 1.

$$\|y-x\|_{\mathbf{L}}^{\infty} = \mathcal{O}(1)\cdot\big(\text{Tot.Var.}\{w\}\big)^{1/2}. \tag{28}$$

Here the exponent 1/2 is due to the fact that first order tangencies are not removable by small perturbations. In higher space dimensions, an even lower exponent is expected. To obtain an error estimate which is linear with respect to the strength of the perturbation, one thus needs to remove all these tangencies. This cannot be achieved if the patches have smooth boundary, but is quite possible if we allow "polyhedral" patches (fig. 3), i.e., patches (Ω, g) for which the boundary of the domain Ω is contained in a finite set of hyperplanes. Replacing "patches" with "polyhedral patches" in Definition 2

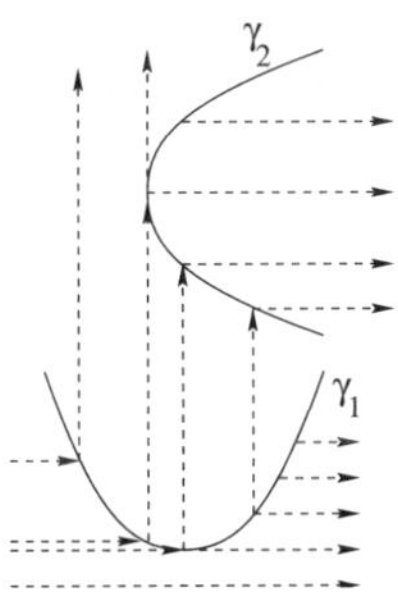

Fig. 2.

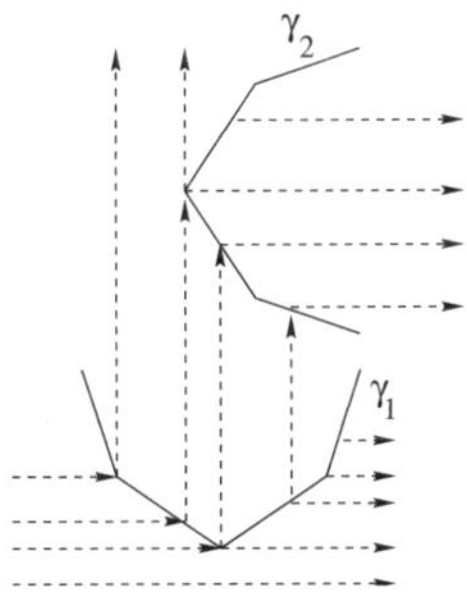

Fig. 3.

(cfr. Remark 2) we obtain the notion of *polyhedral patchy vector field.* For such fields, it is expected that impulsive perturbations of the form (9) should generically produce a perturbation on the set of trajectories which is of exactly the same order of magnitude as the strength of the impulse on the right hand side.

To avoid lengthy technicalities, we have considered in [3] only the planar case, i.e. *polygonal patchy vector fields*, for which we have established the following:

Theorem 2. *For a generic polygonal patchy vector field g on $\mathbb{R}^2$, whose values are bounded away from zero, one has the the following stability property:*

Given any compact set $K \subset \mathbb{R}^2$, there exist constants C, $\delta > 0$, such that the following holds. For every solution $y : [0,T] \mapsto K$ of (9), with Tot.Var.$\{w\} < \delta$, *there exists a solution $x : [0,T] \mapsto \mathbb{R}^2$ of (7) such that*

$$\big\|x - y\big\|_{\mathbf{L}^\infty} \leq C \cdot \text{Tot.Var.}\{w\}. \tag{29}$$

The generic condition that guarantees the estimate (29) requires that the boundary of every patch Ω_α be transversal to all fields g_β, with $\beta \leq \alpha$. We conjecture that the same type of result holds true for generic polyhedral patchy vector fields on $\mathbb{R}^n$.

The basic steps of the proof of Theorem 2 are sketched below. Further details can be found in [3].

1. We first show that, given a polygonal patchy vector field g satisfying the above mentioned transversality condition, we can always replace an arbitrary solution $y : [0,T] \mapsto K$ of (9) with a piecewise smooth map $y^\diamond : [0,T] \mapsto K$ that is a (finite) concatenation of solutions of the unperturbed system (7), so that the distance between $y(\cdot)$ and $y^\diamond(\cdot)$ is of the same order of magnitude as the strength of the impulse in (9).

2. Next, we prove that, for any (discontinuous) solution $y^\diamond : [0,T] \mapsto K$ of (9) which is a concatenation of two solutions of the unperturbed system (7), we can find a solution $y^\flat : [0,T] \mapsto K$ of (7) whose distance from $y^\diamond(\cdot)$ is of the same order of magnitude as the jump in $y^\diamond(\cdot)$.

3. Finally, relying on **2.**, we show with an inductive arguments that, given any solution $y^\diamond : [0,T] \mapsto K$ of (9) which is a (finite) concatenation of solutions of the unperturbed system (7), we can construct a finite sequence of solutions $(y_k(\cdot))_{k=1}^{m-1}$ of (9), together with a sequence of points $\tau_1 = 0 < \tau_2, \ldots, \tau_m = T$, having the following properties:

(i) $y_k|_{[0,\tau_{k+1}]}$ is a Carathéodory solution of (7);

(ii) letting $\alpha_k \doteq \max\{\alpha \; ; \; \alpha \in \mathrm{Im}((\alpha^* \circ y_k)|_{[0,\,\hat{\tau}_{k+1}]})\}$, one has $\alpha_k > \alpha_{k-1}$ for all $k > 0$;

(iii) letting $\Delta(y)$ denotes the total amount of jumps in $y(\cdot)$, there holds

$$\Delta(y_k) \leq C^{k-1} \cdot \Delta(y) \qquad \forall\, k > 0\,, \tag{30}$$

for some constant C depending only on the compact set K.

The conclusion then follows from **1.**,**3.** by observing that, by *(i)*, $y_{m-1}(\cdot)$ is a Carathéodory solution of (7) on $[0,T]$, while (30) yields

$$\begin{aligned} \big|y_{m-1}(t) - y(t)\big| &\leq \sum_{k=2}^{m-1} \big|y_k(t) - y_{k-1}(t)\big| \\ &\leq NC^N \cdot \Delta(y) \qquad \forall\, t \in [0,\,T]\,, \end{aligned} \tag{31}$$

where we denote with N the number of indices in $\{\alpha \in \mathcal{A} \; ; \; \Omega_\alpha \cap K \neq \emptyset\}$.

4 Robustness of Patchy Feedbacks

The type of stability described in the previous sections is precisely what is needed in many applications to feedback control. Consider the stabilization problem for a nonlinear control system on $\mathbb{R}^n$ of the form

$$\dot{x} = f(x, u) \qquad\qquad u(t) \in \mathcal{U}, \tag{32}$$

where the control set $\mathcal{U} \subset \mathbb{R}^m$ is compact and the map $f : \mathbb{R}^n \times \mathbb{R}^m \mapsto \mathbb{R}^n$ is smooth. Assume that (32) is globally (null)-asymptotically controllable, i.e. that every initial state $\overline{x} \in \mathbb{R}^n$ can be steered to the origin by an open-loop control $u = u^{\overline{x}}(t)$. We seek a state feedback control $u = U(x) \in \mathcal{U}$ that stabilizes all trajectories of the closed-loop system

$$\dot{x} = f\big(x,\ U(x)\big) \tag{33}$$

at the origin. As observed in the introduction, a continuous stabilizing feedback fails to exist in general, and hence we are usually forced to look for a stabilizing control within a class of discontinuous functions. Let us consider, for the moment, a semi-global "practical" stabilization problem for (32) (steering all states from a given compact set of initial data into a prescribed neighborhood of zero). Given $r > \varepsilon > 0$, and a compact set $K \supset B_r$, assume that we have constructed a piecewise constant feedback $u = U(x)$ so that $g(x) \doteq f(x, U(x))$ is a patchy vector field on $\mathbb{R}^n \setminus \{0\}$, and such that every solution $t \mapsto x(t)$ of (7) starting from a point $x(0) \in B_r$ is steered inside the ball B_ε within a time T, and satisfies $x(t) \in K$ for all $t \in [0, T]$. Our objective is to design the feedback law $u = U(x)$ so to preserve robustness with respect to small internal and external persistent perturbations. Thus, consider the perturbed system

$$\dot{x} = f\big(x,\ U(x + \zeta(t))\big) + d(t), \tag{34}$$

where $\zeta = \zeta(t)$ represents a state measurement error, and $d = d(t)$ represents an external disturbance of the system dynamics (33). Since we are dealing with a discontinuous O.D.E., one cannot expect the full robustness of the feedback $U(x)$ with respect to arbitrary measurement errors because of possible chattering behavior that may arise at discontinuity points (see [9], [11], [15], [16]). Therefore, we shall consider state measurement errors which are small in BV norm, avoiding such phenomena, while the external disturbances will be assumed to be small in the $\mathbf{L}^1$ norm. Then, by Corollary 1, every solution of the perturbed system (9) will be steered inside the ball $B_{2\varepsilon}$ within time T. In other words, the feedback still performs well in the presence of small (internal and external) perturbations.

Before stating our main result in this direction, we recall here some basic definitions and Proposition 4.2 in [1] , which provides the semi-global practical stabilization of an asymptotycally controllable system, by means of a feedback control that generates a patchy vector field. Such a class of piecewise constant feedbacks was introduced in [1] to provide a class of discontinuous feedback laws that always admit Carathéodory solutions.

Definition 3. *Let* $(\Omega,\ g,\ (\Omega_\alpha,\ g_\alpha)_{\alpha\in\mathcal{A}})$ *be a patchy vector field. Assume that there exist control values* $k_\alpha \in \mathcal{U}$ *such that, for each* $\alpha \in \mathcal{A}$,

$$g_\alpha(x) \doteq f(x,\ k_\alpha) \qquad\qquad \forall\, x \in D_\alpha \doteq \Omega_\alpha \setminus \bigcup_{\beta>\alpha} \Omega_\beta. \tag{35}$$

Then, the piecewise constant map

$$U(x) \doteq k_\alpha \qquad \textit{if} \qquad x \in D_\alpha \tag{36}$$

is called a patchy feedback control *on* Ω, *and denoted* $(\Omega,\ U,\ (\Omega_\alpha,\ k_\alpha)_{\alpha\in\mathcal{A}})$.

REMARK 3. From Definitions 2 and 3, it follows that the field

$$g(x) = f\big(x,\ U(x)\big)$$

defined in connection with a given patchy feedback $(\Omega,\ U,\ (\Omega_\alpha,\ k_\alpha)_{\alpha\in\mathcal{A}})$ is precisely the patchy vector field $(\Omega,\ g,\ (\Omega_\alpha,\ g_\alpha)_{\alpha\in\mathcal{A}})$ associated with a family of fields $\{g_\alpha : \alpha \in \mathcal{A}\}$ satisfying (1). Clearly, the patches $(\Omega_\alpha,\ g_\alpha)$ are not uniquely determined by the patchy feedback U. Indeed, whenever $\alpha < \beta$, by (35) the values of g_α on the set $\Omega_\alpha \cap \Omega_\beta$ are irrelevant. Moreover, recalling the notation (3) we have

$$U(x) = k_{\alpha^*(x)} \qquad\qquad \forall\, x \in \Omega. \tag{37}$$

As usual, we consider as (open-loop) *admissible controls* all the measurable functions $u : [0,\ \infty) \to \mathbb{R}^m$ such that $u(t) \in \mathcal{U}$ for a.e. $t \geq 0$.

Definition 4. *The system (32) is globally* asymptotically controllable *to the origin provided the following holds:*

1. **Attractiveness:** *for each* $\overline{x} \in \mathbb{R}^n$ *there exists some admissible (open-loop) control* $u = u^{\overline{x}}(t)$ *such that the corresponding trajectory of*
$$\dot{x}(t) = f\big(x(t),\ u^{\overline{x}}(t)\big), \qquad\qquad x(0) = \overline{x}\,, \tag{38}$$
either reaches the origin in finite time, or tends to the origin as $t \to \infty$.
2. **Lyapunov stability:** *for each* $\varepsilon > 0$ *there exists* $\delta > 0$ *such that the following holds: For every* $\overline{x} \in \mathbb{R}^n$ *with* $|\overline{x}| < \delta$, *there is an admissible control* $u^{\overline{x}}$ *as in 1. steering the system from* $\overline{x}$ *to the origin, so that the corresponding trajectory of (38) satisfies* $|x(t)| < \varepsilon$ *for all* $t \geq 0$.

Proposition 1. *Let system (32) be globally asymptotically controllable to the origin. Then, for every* $0 < r < s$, *one can find* $T > 0$, *and a patchy feedback control* $U : D \mapsto \mathcal{U}$, *defined on some domain*

$$D \supset \big\{x \in \mathbb{R}^n\ ;\ r \leq |x| \leq s\big\} \tag{39}$$

so that the following holds: For any initial state x_0 *with* $r \leq |x_0| \leq s$, *and for any (forward) Carathéodory trajectory* $t \mapsto \gamma(t)$, $t \geq 0$, *starting from* x_0, *one has*

$$\gamma(t) \in D \qquad \forall\, t \geq 0, \tag{40}$$

and there exists $\bar{t}_\gamma < T$ *such that*

$$\big|\gamma(\bar{t}_\gamma)\big| < r. \tag{41}$$

Relying on Corollary 1 of Theorem 1 and on Proposition 1, we obtained in [2] the following result concerning robustness of a stabilizing feedback with respect to both internal and external perturbations.

Theorem 3. *Let system (32) be globally asymptotically controllable to the origin. Then, for every* $0 < r < s$, *one can find* $T' > 0$, $\chi' > 0$, *and a patchy feedback control* $U' : D' \mapsto \mathcal{U}$ *defined on some domain* D' *satisfying (39), so that the following holds: Given any pair of maps* $\zeta \in BV([0, T'])$, $d \in \mathbf{L}^1([0, T'])$, *such that*

$$\text{Tot.Var.}\{\zeta\} \leq \chi', \qquad \big\|d\big\|_{\mathbf{L}^1([0,\, T'])} \leq \chi', \tag{42}$$

and any initial state x_0 *with* $r \leq |x_0| \leq s$, *for every solution* $t \mapsto x(t)$, $t \geq 0$, *of the perturbed system (34), starting from* x_0, *one has*

$$x(t) \in D' \qquad \forall\, t \in [0, T'], \tag{43}$$

and there exists $\bar{t}_x < T'$, *such that*

$$\big|x(\bar{t}_x)\big| < r. \tag{44}$$

5 Robustness for Sampling Solutions

For discontinuous stabilizing feedbacks constructed in terms of sampling solutions, an alternative concept of robustness was introduced in [6, 16, 17]. In this case, one considers a partition of the time interval and applies a constant control between two consecutive sampling times. To preserve stability, the measurement error should be sufficiently small compared to the maximum step size. On the other hand, each step size should be big enough to prevent possible chattering phenomena. The next result shows that the feedback provided by Proposition 1 also enjoys this type of "relative" robustness. Before stating this result we describe now the concept of sampling trajectory associated to the perturbed system (34) that was introduced in [6, 17].

Let an initial condition x_0, and a partition $\pi = \{0 = \tau_0 < \tau_1 < \cdots < \tau_{m+1} = T\}$ of the interval $[0, T]$, be given. A *sampling trajectory* x_π of the perturbed system (34), corresponding to a set of measurement errors $\{e_i\}_{i=0}^{m}$

and an external disturbance $d \in \mathbf{L}^\infty([0, T])$, is defined in a step-by-step fashion as follows. Between τ_0 and τ_1, let $x_\pi(\cdot)$ be a Carathéodory solution of

$$\dot{x} = f\big(x,\ U(x_0 + e_0)\big) + d(t) \qquad t \in [\tau_0,\, \tau_1]\,, \tag{45}$$

with initial condition $x_\pi(0) = x_0$. Then, $x_\pi(\cdot)$ is recursively obtained by solving the system

$$\dot{x} = f\big(x,\ U(x_\pi(\tau_i) + e_i)\big) + d(t) \qquad t \in [\tau_i,\, \tau_{i+1}], \quad i > 0. \tag{46}$$

The sequence $\{x_\pi(\tau_i) + e_i\}_{i=0}^m$ corresponds to the non-exact measurements used to select control values.

Theorem 4. *Let system (32) be globally asymptotically controllable to the origin. Then, for every* $0 < r < s$, *one can find* $T'' > 0$, $\chi'' > 0$, $\overline{\delta} > 0$, $\overline{k} > 0$, *and a patchy feedback control* $U'' : D'' \mapsto \mathcal{U}$ *defined on some domain* D'' *satisfying (39) so that the following holds: Given an initial state* x_0 *with* $r \le |x_0| \le s$, *a partition* $\pi = \{\tau_0 = 0, \tau_1, \ldots, \tau_{m+1} = T''\}$ *of the interval* $[0, T'']$ *having the property*

$$\frac{\delta}{2} \le \tau_{i+1} - \tau_i \le \delta \quad \forall\, i\,, \qquad \text{for some} \qquad \delta \in\ ]0, \overline{\delta}\,]\,, \tag{47}$$

a set of measurement errors $\{e_i\}_{i=0}^m$ *and an external disturbance* $d \in \mathbf{L}^\infty$ *that satisfy*

$$\begin{aligned} \max_i |e_i| &\le \overline{k} \cdot \delta\,, \\ \big\|d\big\|_{\mathbf{L}^\infty} &\le \chi''\,, \end{aligned} \tag{48}$$

the resulting sampling solution $x_\pi(\cdot)$ *starting from* x_0 *has the property*

$$x_\pi(t) \in D'' \qquad \forall\, t \in [0, T''\,]\,. \tag{49}$$

Moreover, there exists $\overline{t}_{x_\pi} < T''$, *such that*

$$\big|x_\pi(\overline{t}_{x_\pi})\big| < r. \tag{50}$$

REMARK 4. Consider a partition $\pi = \{\tau_0 = 0, \tau_1, \ldots, \tau_{m+1} = T\}$ of the interval $[0, T\,]$ having the property (47). If we associate to a set of measurement errors $\{e_i\}_{i=1}^m$ satisfying (48) the piecewise constant function $\zeta : [0, T] \mapsto \mathbb{R}^n$ defined as

$$\zeta(t) = e_i \qquad \forall t \in\ ]\tau_i,\, \tau_{i+1}]\,,$$

then

$$\text{Tot.Var.}\{\zeta\} \le 4\overline{k} \cdot T\,.$$

Thus, taking the constant $\overline{k}$ sufficiently small we may reinterpret the *discrete* internal disturbance allowed for a sampling solution in Theorem 4 as a particular case of the measurement errors with small total variation considered in Theorem 3.

References

1. Ancona F, Bressan A (1999) Patchy vector fields and asymptotic stabilization, ESAIM Control Optim Calc Var 4:445-471
2. Ancona F, Bressan A (2003) Flow stability of patchy vector fields and robust feedback stabilization, SIAM J Control Optim 41:1455-1476
3. Ancona F, Bressan A, (2003) Stability rates for Patchy vector fields, ESAIM Control Optim Calc Var to appear
4. Bressan A (1987) On differential systems with impulsive controls, Rend Sem Mat Univ Padova 78:227-235
5. Brocket R (1983) Asymptotic stability and feedback stabilization. In: Brockett R, Millman R, Sussmann H (eds) Differential Geometric Control Theory. Birkhaüser Boston
6. Clarke F, Ledyaev Y, Rifford L, Stern R (2000) Feedback stabilization and Lyapunov functions, SIAM J Control Optim 39:25-48
7. Clarke F, Ledyaev Y, Sontag E, Subbotin A (1997) Asymptotic controllability implies feedback stabilization, IEEE Trans Autom Control 42:1394-1407
8. Coron J-M (1990) A necessary condition for feedback stabilization, Systems Control Lett 14:227-232
9. Coron J-M, Rosier L, (1994) A relation between continuous time-varying and discontinuous feedback stabilization, J Math Systems Estim and Control 4:67-84
10. Krasovskii N, Subbotin A (1974) Positional Differential Games. Nauka Moscow
11. Hermes H (1967) Discontinuous vector fields and feedback control. In: Hale J, La Salle J (eds) Differential Equations and Dynamical Systems. Academic Press New York
12. Ledyaev Y, Sontag E (1999) A Lyapunov characterization of robust stabilization, J Nonlinear Anal 37:813-840
13. Rifford L (2000) Existence of Lipschitz and semiconcave control-Lyapunov functions, SIAM J Control Optim 39:1043-1064
14. Rifford L (2002) Semiconcave control-Lyapunov functions and stabilizing feedbacks, SIAM J Control Optim 41:659-681
15. Ryan E (1994) On Brockett's condition for smooth stabilizability and its necessity in a context of nonsmooth feedback, SIAM J Control Optim 32:1957-1604
16. Sontag E (1999) Stability and stabilization: discontinuities and the effect of disturbances. In: Clarke F, Stern R (eds) Nonlinear Analysis Differential Equations and Control (Proc NATO Advanced Study Institute Montreal). Kluwer Dordrecht
17. Sontag E (1999) Clocks and insensitivity to small measurement errors, ESAIM Control Optim Calc Var 4:537-557
18. Sontag E, Sussmann H (1980) Remarks on continuous feedback. In: Proc IEEE Conf Decision and Control Aulbuquerque NM
19. Sussmann H (1979) Subanalytic sets and feedback control, J Differential Equations 31:31-52

Control of Interconnected Nonlinear Systems: A Small-Gain Viewpoint

Zhong-Ping Jiang

Department of Electrical and Computer Engineering, Polytechnic University, Six Metrotech Center, Brooklyn, NY 11201. zjiang@control.poly.edu

Summary. The small-gain method for robust nonlinear control is reviewed and is further ameliorated to cover important classes of interconnected systems in dynamically perturbed, decentralized, or discrete time forms. Both partial-state and output feedback control problems are considered. It is shown that assumptions required in related but independent work of others can be significantly relaxed. An application to underactuated ship control is shown as an illustrative practical example.

1 Introduction

The last two decades have witnessed tremendous progress in robust and adaptive control of nonlinear systems. Geometric and analytical methods play a crucial role in identifying important classes of nonlinear systems and in generating constructive nonlinear design tools. The main purpose of this paper is to give a quick review of recent advances in Sontag's ISS ("input-to-state stability") theory and nonlinear small-gain theorems, with a special emphasis on their applications to nonlinear control design. As shown in our recent work and the work of others, ISS small gain methods provide us with a general framework to unify various approaches to stabilization by partial-state and output feedback, robust adaptive nonlinear control with dynamic uncertainties, and decentralized control for large-scale nonlinear systems. The list of references given at the end of the paper is by no means complete, but serves the unique purpose of guiding the reader to find many other references on the related subjects.

As shown in past literature of linear robust control, classical small-gain theorems have laid a solid foundation for robustness analysis and robustification of various linear feedback control laws. Nonlinear small-gain theorems involve the concept of 'nonlinear gain' functions, instead of linear finite-gain, and quantify the effects of initial conditions in system dynamics via the use of class $\mathcal{KL}$ functions. Through a proper decomposition, stability properties of a complex and nonlinear system can be derived from those of simpler low-dimensional subsystems. With these features in mind, it is our firm belief that

M. de Queiroz et al. (Eds.): Optimal Control, Stabilization and Nonsmooth Analysis, LNCIS 301, pp. 201–214, 2004.

ISS and nonlinear small-gain techniques will remain as useful tools for solving challenging problems in robust nonlinear control theory and applications.

In the rest of the paper, we review some of the ingredients in nonlinear ISS small-gain methods, and discuss their applications in global stabilization by partial-state and output feedback, disturbance attenuation with internal global stability, and decentralized control. Among some other practical applications, we show that ISS and small-gain viewpoints play a key role in our first solution to the longstanding problem of simultaneous stabilization and tracking for an underactuated surface ship.

2 Tools for Control of Interconnected Systems

The concept of "input-to-state stability" (for short, ISS) was introduced by Sontag in his seminal paper [49], and is an important concept in a lot of recent work in stability and control of nonlinear systems; see [50] for an excellent tutorial. We first review this definition and recall some slight modifications from [31]. Then, we present nonlinear extensions of the classical small-gain theorem [8] along with gain assignment results.

2.1 Input-to-State Stability

The ISS property can be seen as a nonlinear generalization of the well-known 'Superposition Principle' for a dynamical system with inputs:

$$\dot{x} = f(x,u), \quad x \in \mathbb{R}^n , \quad u \in \mathbb{R}^m \tag{1}$$

More precisely, system (1) is ISS provided there exist a function β of class $\mathcal{KL}$ and a function γ of class $\mathcal{K}$ (see [50]) such that, for any initial condition $x(0)$ and any input $u : \mathbb{R}_+ \to \mathbb{R}^m$, the associated solution $x(t)$ satisfies

$$|x(t)| \leq \max\{\beta(|x(0)|,t)\, ,\, \gamma(\|u_t\|)\} , \quad \forall\, t \geq 0 \tag{2}$$

where u_t is the truncated function of u at t, $|\cdot|$ is the usual Euclidean norm, and $\|\cdot\|$ is the $\mathcal{L}_\infty$-norm. Such a function γ in (2) is referred to as an *ISS-gain* for system (1). Note that we can also use "+" instead of "max" in the bound of (2).

The *local* version of ISS (for short, local ISS) means that inequality (2) holds for small initial conditions $x(0) \in \mathbb{R}^n$ and small inputs $u : [0,\infty) \to \mathbb{R}^m$. Recall that ISS (resp., local ISS) implies GAS (resp., LAS) when $u = 0$ and BIBS-stability when $u \neq 0$ (resp., when u is small in magnitude). However, the converse is not true; see [50] for elementary counterexamples and the definitions of GAS and LAS. This is why ISS is often viewed as an extension of Lyapunov stability to control systems.

To accommodate the practical situation where nonvanishing disturbances may occur and only partial stability is achievable, we have introduced in [31] the notion of "*input-to-output practical stability*" (for short, IOpS). Essentially, system (1) with outputs $y = h(x)$ for a continuous function h is said to be *input-to-output practically stable* (IOpS) if inequality (2) is relaxed into a more general property:

$$|y(t)| \le \max\{\beta(|x(0)|, t), \gamma(\|u_t\|), d\}, \quad \forall t \ge 0 \tag{3}$$

where $y(t) = h(x(t))$ and d is a nonnegative constant. In particular, when $y = x$, IOpS is called ISpS; when $d = 0$ in (3), IOpS is called IOS. Obviously, IOS reduces to ISS when $y = x$.

Clearly, IOpS is a natural extension of "BIBO" (bounded-input bounded-output) stability property. The transient behavior of the system output is quantified by the function β. There is a close connection between IOpS and ISpS, and IOS and ISS. Indeed, it is shown in [31, Prop. 3.1] that, for system (1) with general outputs $y = h(x)$, IOS is equivalent to ISS under the additional condition of "*strong unboundedness observability*" (for short, SUO) –also called IOSS in [56]– defined as

$$|x(t)| \le \max\{\beta^o(|x(0)|, t), \gamma^o(\|(u_t, y_t)\|)\} \tag{4}$$

with functions $\beta^o \in \mathcal{KL}$ and $\gamma^o \in \mathcal{K}$.

Lyapunov, among many other equivalent, characterizations of ISS and its variants can be found in [2, 50, 54] and references cited therein.

2.2 Generalized Nonlinear Small-Gain Theorems

One of the fundamental questions in nonlinear stability is concerned with (sufficient) conditions that guarantee that the interconnection of two stable systems is still stable in some sense. A general interconnected nonlinear system is depicted in Figure 1.

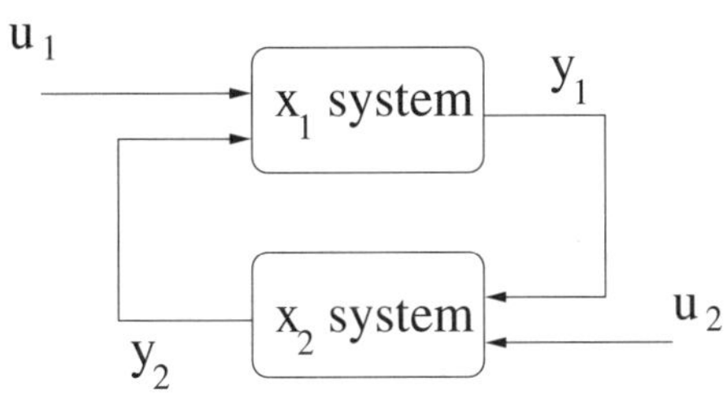

Fig. 1. Configuration of interconnected nonlinear systems

Here, we consider an interconnected system composed of two finite-dimensional nonlinear systems of the general form

$$\dot{x}_1 = f_1(x_1, y_2, u_1), \quad y_1 = h_1(x_1, y_2, u_1) \tag{5}$$

$$\dot{x}_2 = f_2(x_2, y_1, u_2), \quad y_2 = h_2(x_2, y_1, u_2) \tag{6}$$

where, for $i = 1, 2$, $x_i \in \mathbb{R}^{n_i}$, $u_i \in \mathbb{R}^{m_i}$ and $y_i \in \mathbb{R}^{p_i}$. It is assumed that f_1, f_2, h_1 and h_2 are suitable functions guaranteeing existence and uniqueness of solutions of the total system (5)–(6).

A simple adaptation of the proof of Theorem 2.1 of [31] yields a nonlinear generalization of the classical small-gain theorem:

Theorem 1. *Assume that the subsystems (5) and (6) are IOpS in the sense that*

$$\begin{aligned} |y_1(t)| &\leq \max\{\beta_1(|x_1(0)|, t)\,,\ \gamma_1^y(\|y_{2t}\|)\,,\ \gamma_1^u(\|u_{1t}\|)\,,\ d_1\}\,, \\ |y_2(t)| &\leq \max\{\beta_2(|x_2(0)|, t)\,,\ \gamma_2^y(\|y_{1t}\|)\,,\ \gamma_2^u(\|u_{2t}\|)\,,\ d_2\}\,. \end{aligned} \tag{7}$$

It is also assumed that both the subsystems have a weaker property of unboundedness observability *(UO) of the type*

$$|x_i(t)| \leq \alpha_i^o(|x_i(0)| + \|(u_{it}, y_t)\|) + D_i^o \tag{8}$$

for $i = 1, 2$, functions $\alpha_i^o \in \mathcal{K}$ and constants $D_i^o \geq 0$. If there exists a constant $s_\ell \geq 0$ such that

$$\gamma_1^y \circ \gamma_2^y(s) < s\,, \quad \forall\, s > s_\ell\,, \tag{9}$$

then the interconnected system (5)–(6) is IOpS from $u = (u_1, u_2)$ to $y = (y_1, y_2)$ and satisfies the weak UO property. Furthermore, when $d_1 = d_2 = D_1^o = D_2^o = s_\ell = 0$, the interconnected system (5)–(6) is IOS.

Comments

- The condition (9) is referred to as the *small-gain condition.* It is different from the ones in [31, 39] because of the use "max", instead of "+", in the IOpS property. It clearly holds in the case of classical small-gain theorems [8].
- The generalized nonlinear small-gain Theorem 1 complements an earlier monotone gain theorem of Mareels and Hill [39] based on input-output operator theory. More specifically, the effects of the initial conditions on the system dynamics are taken into account explicitly via $\mathcal{KL}$ functions β_i.
- In the special but interesting case where the system (5)–(6) is a cascade interconnection of two ISS systems via the outputs $y_1 = h_1(u_1)$ and $y_2 = h_2(x_2, u_2)$, the small-gain condition (9) obviously holds. Indeed, the ISS-gain of the x_2-subsystem with respect to input x_1 is $\gamma_2^x \equiv 0$. The cascade system is therefore ISS as well.
- A local version of Theorem 1 can be established if the conditions are satisfied for small signals; see, for example, [63]. Also see [23, 32, 38] for extensions of Theorem 1 to discrete-time nonlinear systems and [52, 59] for extensions to infinite-dimensional nonlinear systems. The ISS property for time-varying systems and applications have also been considered in [10, 24, 33, 61].

2.3 Gain Assignment Theorems

Assigning an "appropriate gain" to a system by means of feedback is a key step in the application of small-gain theorems to robust control design. As shown in [31, Theorem 2.2], this is possible if the system in question has a special feedback form. Here, we provide an improved version of the gain assignment Theorem 2.2 of [31].

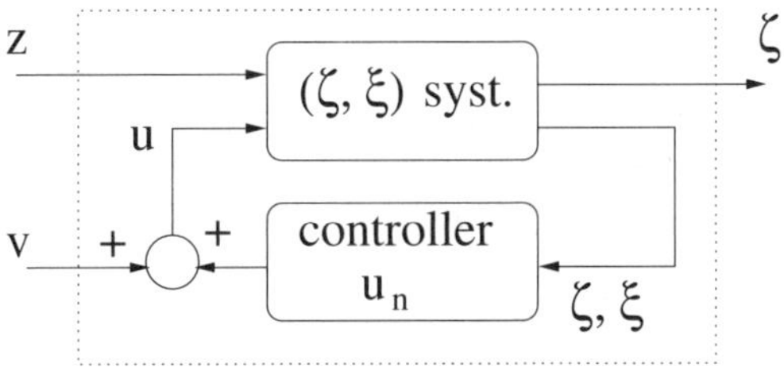

Fig. 2. Gain assignment via feedback

Theorem 2. *Consider the dynamical control system, depicted in Figure 2,*

$$\begin{cases} \dot{\zeta} = A\zeta + B(C_o\xi + \omega_0(\zeta, z)) \\ \dot{\xi} = A_o\xi + B_o u + \omega(\zeta, \xi, z) \end{cases} \tag{10}$$

where $u \in \mathbb{R}$ is the control input and $z \in \mathbb{R}^{n_z}$ is the disturbance input, (A, B) is a pair of (known) stabilizable matrices and (A_o, B_o, C_o) is in the controller and observer canonical forms, i.e., $B_o = \mathrm{col}(0, \ldots, 0, 1)$, $C_o^T = \mathrm{col}(1, 0, \ldots, 0)$, and A_o is the $n \times n$ matrix with ones on the upper-diagonal and zeros elsewhere. It is further assumed that ω_0 and each component ω_i of ω, with $1 \leq i \leq n$, satisfy

$$|\omega_0(\zeta, z)| \leq \psi_{01}(|B^T\zeta|) + \psi_{02}(|z|) \;, \tag{11}$$

$$|\omega_i(\zeta, \xi, z)| \leq \psi_{i1}(|(\zeta, \xi_1, \ldots, \xi_i)|) + \psi_{i2}(|z|) \tag{12}$$

for known functions ψ_{01} and ψ_{i1}, of class C^1, and ψ_{02} and ψ_{i2}, of class $\mathcal{K}$, all vanishing at the origin.

Then, for any function γ of class $\mathcal{K}_\infty$, there exists a smooth function $u_n(\zeta, \xi)$, with $u_n(0,0) = 0$, such that the system (10) in closed-loop with $u = u_n(\zeta, \xi) + v$ is :

P1) ISS with (z, v) as input,
P2) IOpS with (z, v) as input, ζ as output and the function γ as gain.

Moreover, if for every $0 \leq i \leq n$, the composed function $\psi_{i2} \circ \gamma^{-1}$ is linearly bounded on a neighborhood of 0, then the closed-loop system (10) can be rendered IOS with (z, v) as input, ζ as output and the function γ as gain.

Comments

- At first sight, the inequalities (11) and (12) may look restrictive. In fact, when B is a nonzero scalar, (11) is always true for *any* continuous function ω_0, as explicitly pointed out in [42, eqs. (22) and (14)]. Eq. (12) restricts the coupling term ω to satisfy the usual "lower-triangularity" condition introduced in recent literature of robust/adaptive nonlinear control; see, for instance, [15, 35, 36, 40].
- Clearly, that the functions ψ_{i2} and γ^{-1} are C^1 is a sufficient, but not necessary, condition for $\psi_{i2}\circ\gamma^{-1}$ being linearly bounded on a neighborhood of 0.

Proof of Theorem 2. Let K be a vector rendering $A - BK$ asymptotically stable and $P = P^T > 0$ be the solution to the Lyapunov matrix equation

$$(A - BK)^T P + P(A - BK) = -I \tag{13}$$

where I stands for an identity matrix of appropriate dimension.

As in the proof of Theorem 2.2 of [31], for any constant $\eta > 0$ and any function γ_1 of class $\mathcal{K}_\infty$, take a smooth, convex function k of class $\mathcal{K}_\infty$ such that

$$k(s+\eta) \geq \gamma_1(s) \quad , \qquad s\,\frac{dk}{ds}(s) \geq k(s) \qquad \forall\, s \geq 0\,. \tag{14}$$

In (14), η can be chosen as 0 whenever γ_1 is linearly bounded on a neighborhood of 0.

Differentiating $V_0 = k(\zeta^T P \zeta)$ with respect to time yields

$$\begin{aligned}\dot V_0 &= \frac{dk}{ds}(\zeta^T P\zeta)\left[-\zeta^T\zeta + 2\zeta^T PB(\xi_1 + K\zeta + \omega_0(\zeta, z))\right] \qquad (15)\\ &\leq -\lambda_{\max}(P)^{-1}V_0 + \bar\xi_1^2 + \psi_{02}^2(|z|) \qquad (16)\end{aligned}$$

provided we take

$$\bar\xi_1 = \xi_1 - u_0(\zeta)\,, \quad u_0 = -K\zeta - 2\frac{dk}{ds}(\zeta^T P\zeta)\, B^T P\zeta - \frac{B^T P\zeta\hat\psi_{01}(\zeta)}{\lambda_{\min}(P)} \tag{17}$$

where $\hat\psi_{01}$ is a smooth function such that $\psi_{01}(|B^T\zeta|) \leq |B^T\zeta|\hat\psi_{01}(\zeta)$ for all ζ. Such a function $\hat\psi_{01}$ always exists because ψ_{01} is of class C^1 and vanishes at the origin.

By recursive application of backstepping as in [31, Proof of Theorem 2.2] and [17, Section 3.2], it follows that there are a smooth, positive definite and proper function $V_n(\zeta, \bar\xi_1, \ldots, \bar\xi_n)$ and a smooth function $u_n(\zeta, \bar\xi_1, \ldots, \bar\xi_n)$ such that

$$u_n(0, \ldots, 0) = 0\,, \quad V_0(\zeta) \leq V_n(\zeta, \bar\xi_1, \ldots, \bar\xi_n) \quad \forall\,(\zeta, \bar\xi_1, \ldots, \bar\xi_n) \in \mathbb{R}^{n_\zeta + n}\,. \tag{18}$$

In addition, the time derivative of V_n along the solutions of (10) in closed-loop with $u = u_n + v$ satisfies

$$\dot{V}_n \leq -c_1 V_n(\zeta(t), \bar{\xi}_1(t), \ldots, \bar{\xi}_n(t)) + c_2 \max_{i\in\{0,1,\ldots,n\}} \{\psi_{i2}^2(|z(t)|)\ ,\ |v(t)|^2\} \quad (19)$$

for some constants $c_1, c_2 > 0$.

From here, the rest of the proof follows readily using the same reasoning as in the proof of Theorem 2.2 of [31]. **Q.E.D.**

Note that, for systems with *matched* dynamic uncertainties, gain assignment results are available in the independent work [46, 55].

3 Control Applications

3.1 Systems with Dynamic Uncertainties

As an illustration of ISS and small-gain techniques, we examine the problem of partial-state feedback stabilization for the class of nonlinear triangular systems (see [61] for a broader class)

$$\begin{aligned} \dot{z} &= q(z,\zeta) \\ \dot{\zeta} &= A\zeta + B(C_o\xi + \omega_0(\zeta, z)) \\ \dot{\xi} &= A_o\xi + B_o u + \omega(\zeta,\xi,z) \end{aligned} \quad (20)$$

where the ζ and ξ subsystems satisfy the conditions of Theorem 2, and the z-subsystem is ISS with gain γ_z from ζ to z. As opposed to the work of others on the topic, a nice feature of the small gain method is that there is no need to assume the functions q, ω_0 and ω to be C^1 or to have a C^1 upper-bound.

A direct application of Theorems 1 and 2 yields

Theorem 3. *Consider the system (20). If for every $0 \leq i \leq n$, the composed function $\psi_{i2} \circ \epsilon\gamma_z$ is linearly bounded on a neighborhood of 0 for a constant $\epsilon > 1$, then the Problem of Global Stabilization is solvable by a smooth partial-state feedback law of the form $u = \alpha(\zeta, \xi)$.*

It is worth noting that the condition on $\psi_{i2} \circ \epsilon\gamma_z$ is needed only for existence of a *smooth* controller.

Applications of nonlinear small-gain theorems to the problem of global output feedback stabilization are studied in [5, 42, 45]. Small-gain based robust control designs for interconnected nonlinear systems with *input* dynamic uncertainties are proposed in [22, 24, 37, 43, 44] (also see [3, 47]).

As advocated in [36] (also see [35, 40]), there has been much progress in the development of new adaptive schemes for nonlinear plants since the late 1980s. However, relatively less attention has been paid to robustness and robust redesign issues in adaptive nonlinear control. It has been shown in our early work [27] and more recent work [17, 18, 25, 28, 29] that ISS and small-gain related techniques turn out to be helpful in robustifying current adaptive algorithms for popular classes of nonlinear triangular systems with *dynamic uncertainties*.

3.2 Nonlinear Disturbance Attenuation with Internal Stability

In practical applications, the design of a control law should meet both the objectives of internal and external stability. This is the topic of disturbance attenuation with internal stability. There have been many research publications on the subject of state-feedback, L_2-gain, and disturbance attenuation with internal stability; see, for example, [14, 40, 64] and references therein. However, fewer constructive solutions are available for nonlinear uncertain systems when only output measurements are available to the designer. As opposed to restricting ourselves to the L_2-gain property as in [14, 40, 64], we are interested in designing a disturbance attenuating controller in a broader context of *nonlinear*-gain disturbance attenuation with internal stability. The following figure illustrates the problem setting where u is the control input, w is the exogenous disturbance input, y_c is the to-be-controlled output, and y is the measured output:

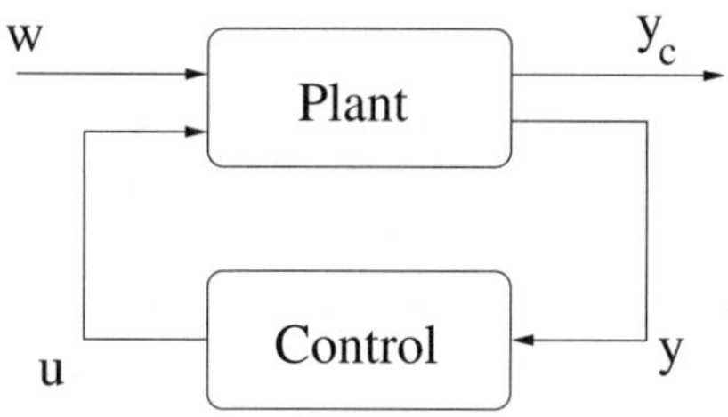

Fig. 3. Nonlinear disturbance attenuation with internal stability.

In our preliminary work [19], we consider $y_c = y$ and address nonlinear disturbance attenuation with stability for a large class of nonlinear systems. A constructive tool is presented to solve the problem of output-feedback disturbance attenuation with internal global stability for minimum-phase nonlinear systems described by

$$\begin{aligned} \dot{z} &= q_0(t,z,y) + q_1(t,z,y)w \\ \dot{x} &= A_o x + B_o u + f(t,z,y) + g(t,z,y)w \\ y &= C_o x \end{aligned} \tag{21}$$

where the triple of matrices (A_o, B_o, C_o), of appropriate dimensions, is defined as in (10). The control problem addressed in [19] is precisely stated as follows: find a smooth, dynamic, output feedback law so that the closed-loop system enjoys the following properties:

P3)When $w = 0$, the closed-loop system is GAS at the origin.
P4)When $w \in L^m_{\infty e}$, the closed-loop system solutions satisfy the inequality

$$\int_{t_0}^{t} |y(\tau)|^2 d\tau \leq \epsilon \int_{t_0}^{t} \gamma(|w(\tau)|) d\tau + \beta \tag{22}$$

for all $t \geq t_0 \geq 0$, some β dependent on the initial conditions, some (fixed) $\mathcal{K}$-function γ, and an arbitrary constant $\epsilon > 0$.

Lyapunov-based ISS and small-gain techniques are applied for controller design and stability analysis. The crucial ideas are: (i) to treat the z-dynamics as a disturbance and to propose a reduced-order observer-like dynamic system to approximate the unmeasured states x; (ii) to transform the output feedback control problem into a corresponding partial-state feedback control issue; finally, (iii) to apply Lyapunov-based ISS and small-gain theorems to generate a desired observer-based output feedback controller.

Due to the importance of decentralized control for many engineering applications ranging from power systems to telecommunications networks to biological systems, an extension of the small-gain method to decentralized control of large-scale nonlinear systems is given in [20, 30] (see, e.g., [21] for a review of recent results). As explained in [21], our earlier work in small gain based decentralized control provides a unifying framework by including as special cases previously published results in the independent work of others on the decentralized control for large-scale nonlinear systems. Specifically, the class of decentralized systems considered in our work does not require any matching condition on parametric and nonparametric uncertainties nor growth conditions of any kind on the nonlinear interconnections. The relative degree of every isolated subsystem is not required to be less than two. The problems of decentralized stabilization with disturbance attenuation and adaptive asymptotic tracking—using state and output feedback—have been studied for an enlarged class of large-scale systems with strong nonlinear interconnections and significant uncertainties.

3.3 Nonlinear Ship Control

Nonlinear control for an underactuated surface ship is a challenging issue because of its involved dynamics and the fact that there is a smaller number of actuators than the number of degrees of freedom. A simplified mathematical model of a ship, studied by several authors in recent literature, is described by the six first-order ordinary differential equations [11]

$$\begin{aligned}
\dot{x} &= u\cos\psi - v\sin\psi \\
\dot{y} &= u\sin\psi + v\cos\psi \\
\dot{\psi} &= r \\
\dot{u} &= \tfrac{m_{22}}{m_{11}} vr - \tfrac{d_{11}}{m_{11}} u + \tfrac{1}{m_{11}} \tau_u \\
\dot{v} &= -\tfrac{m_{11}}{m_{22}} ur - \tfrac{d_{22}}{m_{22}} v \\
\dot{r} &= \tfrac{m_{11}-m_{22}}{m_{33}} uv - \tfrac{d_{33}}{m_{33}} r + \tfrac{1}{m_{33}} \tau_r
\end{aligned} \tag{23}$$

where x, y and ψ are the surge displacement, sway displacement and yaw angle in the earth fixed frame; and u, v and r denote surge, sway and yaw velocities respectively. The positive constants d_{ii}, m_{ii}, with $1 \leq i \leq 3$, denote the hydrodynamic damping and ship inertia including added mass in surge, sway and yaw. The surge force τ_u and the yaw moment τ_r are considered as the controls. Note that there is no control actuator in the direction of sway, leading to the v-variable being not directly controlled by a controller.

As can be directly checked, the ship model does not satisfy the well-known necessary condition of Brockett for asymptotic stabilization (cf. [6]). As a result, there is no continuous time-invariant state-feedback control law that makes the trivial solution (locally) asymptotically stable. In addition, system (23) is not transformable to the standard chained form for which there are various time-varying and/or discontinuous feedback designs developed in the recent literature of control of nonholonomic systems. Therefore, some researchers have developed novel algorithms to solve the global stabilization (or so-called parking or dynamic positioning) problem for the above ship system. While the linearized model of the ship system (23) around the origin is uncontrollable, its linearization around an appropriate "moving" (quantified by a "persistent excitation (PE)" condition) trajectory is controllable. In other words, for the trajectory tracking problem, a (local) solution can be obtained via conventional linear control theory. Nevertheless, a global tracking solution is achievable only via alternative methods such as Lyapunov's direct method.

A longstanding open question is whether or not it is possible to find one single (time-varying, or discontinuous) controller to achieve simultaneously (global) stabilization and (global) tracking, without switching between stabilizers and trackers. In the recent contribution [9], we have obtained the *first* time-varying solution to the problem of simultaneous stabilization and tracking for underactuated ships. The key observation is that the sway v-dynamics in (23) is ISS with respect to (u, r). Appropriate state and input transformations bring the system into a nice cascade-interconnected form which is ready for constructive controller design and stability analysis.

3.4 Other Applications

Last but not the least, we mention some other applications of small-gain techniques. Sontag and his coauthors have developed some new solutions to old problems in mathematical biology, based on small-gain principles [1, 7, 51]. Global output regulation of nonlinear systems has been a hard problem because of the challenges in solving the regulator equation and the lack of appropriate tools for explicit controller design. Some neat results in semiglobal output regulation are proposed in recent literature; see [48] and references therein. Recently, Huang and Chen [13] have linked the global output regulation problem to the global stabilization problem in the context of nonlinear uncertain systems. This is an important step because all techniques proposed in the stabilization field are now ready for direct applications in solving global output regulation problems for important classes of nonlinear systems. In particular, some nice small-gain based solutions are already available in the to-be-published book of Huang [12]. Sampled-data control, and discrete-time control, for nonlinear systems with partial-state and output measurements, have been left largely unexplored because of the daunting challenges on the computational level [23, 41]. More remains to be accomplished in these areas that have been paid growing attention.

Acknowledgement. It is a pleasure to thank I. Mareels, L. Praly, A. Teel and Y. Wang for fruitful collaboration on topics discussed here. This work has been supported partly by the U.S. National Science Foundation under grants ECS-0093176 and INT-9987317.

References

1. Angeli D, Sontag E (2003) Monotone control systems, IEEE Trans Automat Control 48:1684–1698
2. Angeli D, Sontag E, Wang Y (2000) A characterization of integral input to state stability, IEEE Trans Automat Control, 45:1082–1097
3. Arcak M, Seron M, Braslavsky J, Kokotović P (1999) Robustification of backstepping against input unmodeled dynamics. In: Proc 38th IEEE Conf Decision Control 2495-2500
4. Astolfi A, Ortega R (2003) Immersion and invariance: a new tool for stabilization and adaptive control of nonlinear systems, IEEE Trans Automat Control 48:590–606
5. Battilotti S (1998) Robust output feedback stabilization via a small-gain theorem, Int J Robust and Nonlinear Control 9:211–229
6. Brockett R (1983) Asymptotic stability and feedback stabilization. In: Brockett R, Millman R, Sussmann H (eds). Differential Geometric Control Theory 181-191
7. De Leenheer P, Angeli D, Sontag E (2003) On predator-prey systems and small-gain theorems, submitted
8. Desoer C, Vidyasagar M (1975) Feedback Systems: Input-Output Properties. Academic Press New York
9. Do K, Jiang Z-P, Pan J (2002) Universal controllers for stabilization and tracking of underactuated ships, Systems and Control Letters 47:299–317
10. Edwards H, Lin Y, Wang Y (2000) On input-to-state stability for time-varying nonlinear systems. In: Proc 39th IEEE Conf Dec Control 3501–3506
11. Fossen T (1994) Guidance and Control of Ocean Vehicles. Wiley Chichester
12. Huang J (2003) Nonlinear Output Regulation Theory and Applications, in preparation
13. Huang J, Chen Z (2002) A general framework for tackling output regulation problem. In: Proc 2002 American Control Conference 102–109
14. Isidori A (1995) Nonlinear Control Systems 3rd edition. Springer London
15. Isidori A (1999) Nonlinear Control Systems Vol II. Springer London
16. Ito H, Jiang Z-P (2003) Robust disturbance attenuation of nonlinear systems using output feedback and state-dependent scaling, Automatica to appear
17. Jiang Z-P (1999) A robustification tool for adaptive nonlinear control, Int J Adaptive Control and Signal Processing 13:719-732

18. Jiang Z-P (1999) A combined backstepping and small-gain approach to adaptive output feedback control, Automatica 35:1131–1139
19. Jiang Z-P (2000) Global output feedback control with disturbance attenuation for minimum-phase nonlinear systems, Systems and Control Letters 39:155–164
20. Jiang Z-P (2002) Decentralized disturbance attenuating output-feedback trackers for large-scale nonlinear systems, Automatica 38:1407–1415
21. Jiang Z-P (2003) Decentralized control for large-scale nonlinear systems: a review of recent results, Dynamics of Continuous Discrete and Impulsive Systems (Special Issue in honor of Prof Siljak's 70th birthday), to appear
22. Jiang Z-P, Arcak M (2001) Robust global stabilization with ignored input dynamics an input-to-state stability (ISS) small-gain approach, IEEE Trans Automat Control 46:1411–1415
23. Jiang Z-P, Lin Y, Wang Y (2003) Nonlinear small-gain theorems for discrete-time feedback systems and applications, Automatica to appear
24. Jiang Z-P, Mareels I (1997) A small gain control method for nonlinear cascaded systems with dynamic uncertainties, IEEE Trans Automat Control 42:292-308
25. Jiang Z-P Mareels I (2001) Robust nonlinear integral control, IEEE Trans Automat Control 46:1336–1342
26. Jiang Z-P, Mareels I, Wang Y (1996), A Lyapunov formulation of the nonlinear small gain theorem for interconnected ISS systems, Automatica 32:1211-1215
27. Jiang Z-P, Praly P (1992) Preliminary results about robust Lagrange stability in adaptive nonlinear regulation, Int J Adaptive Control Signal Processing 6:285-307
28. Jiang Z-P, Praly L (1998) Design of robust adaptive controllers for nonlinear systems with dynamic uncertainties, Automatica 34:825–840
29. Jiang Z-P, Repperger D (2001) New results in decentralized adaptive nonlinear stabilization using output feedback, Int J Control 74:659–673
30. Jiang Z-P, Repperger D, Hill D (2001) Decentralized nonlinear output-feedback stabilization with disturbance attenuation, IEEE Trans Automatic Control 46:1623–1629
31. Jiang Z-P, Teel A, Praly L (1994) Small-gain theorem for ISS systems and applications, Mathematics of Control Signals and Systems 7:95–120
32. Jiang Z-P, Wang Y (2001) Input-to-state stability for discrete-time nonlinear systems, Automatica 37:857-869
33. Karafyllis I, Tsinias J (2003) Non-uniform in time input-to-state stability and the small-gain theorem, IEEE Trans Automat Control to appear
34. Khalil H (2002) Nonlinear Systems 3rd Edition. Prentice-Hall
35. Kokotović P, Arcak M (2001) Constructive nonlinear control a historical perspective, Automatica 37:637–662
36. Krstić M, Kanellakopoulos I, Kokotović P (1995) Nonlinear and Adaptive Control Design. John Wiley & Sons New York
37. Krstić M, Kokotović P (1994) On extending the Praly-Jiang-Teel design to systems with nonlinear input unmodeled dynamics. Technical Report CCEC 94-0211 University of California
38. Laila D, Nešic D (2002) Lyapunov based small-gain theorem for parameterized discrete-time interconnected ISS systems. In: Proc 41st IEEE Conf Dec Control 2292–2297
39. Mareels I, Hill D (1992) Monotone stability of nonlinear feedback systems, J Math Systems Estimation Control 2:275–291

40. Marino R, Tomei P (1995) Nonlinear Control Design Geometric Adaptive and Robust. Prentice-Hall London
41. Nesic D, Teel A (2001) Sampled-data control of nonlinear systems: an overview of recent results. In: Moheimani R (ed) Perspectives on Robust Control. Springer-Verlag New York 221-239
42. Praly L, Jiang Z-P (1993) Stabilization by output feedback for systems with ISS inverse dynamics, Systems and Control Letters 21:19–33
43. Praly L, Jiang Z-P (1998) Semiglobal stabilization in the presence of minimum-phase dynamic input uncertainties. In: Proc Nonlinear Control System Design Symposium NOLCOS 326-331
44. Praly L, Jiang Z-P (1998). Further results on robust semiglobal stabilization with dynamic input uncertainties. Proc 37th IEEE Conf Dec Control 891–896
45. Praly L, Jiang Z-P (2003), Linear output feedback with dynamic high gain for nonlinear systems, Systems Control Letters to appear
46. Praly L, Wang Y (1996) Stabilization in spite of matched unmodeled dynamics and an equivalent definition of input-to-state stability, Math Control Sign Systems 9:1–33
47. Sepulchre R, Janković M, Kokotović P (1997) Constructive Nonlinear Control. Springer-Verlag New York
48. Serrani A, Isidori A, Marconi L (2001) Semiglobal nonlinear output regulation with adaptive internal model, IEEE Trans Automat Control 46:1178–1194
49. Sontag E (1989) Smooth stabilization implies coprime factorization, IEEE Trans Automat Control 34:435–443
50. Sontag E (2000) The ISS philosophy as a unifying framework for stability-like behavior. In: Isidori A, Lamnabhi-Lagarrigue F, Respondek W (eds) Nonlinear Control in the Year 2000 Vol 2. Springer-Verlag Berlin 443-468
51. Sontag E (2002) Asymptotic amplitudes and Cauchy gains: a small-gain principle and an application to inhibitory biological feedback, Systems and Control Letters 47:167–179
52. Sontag E, Ingalls B (2002) A small-gain theorem with applications to input/output systems incremental stability detectability, and interconnections, J Franklin Inst 339:211–229
53. Sontag E, Teel A (1995) Changing supply rates in input-to-state stable systems, IEEE Trans Automat Control 40:1476–1478
54. Sontag E, Wang Y (1995) On characterizations of the input-to-state stability property, Systems and Control Letters 24:351–359.
55. Sontag E, Wang Y (1995) On characterizations of input-to-state stability with respect to compact sets, Proc IFAC NOLCOS 226-231
56. Sontag E, Wang Y (1997) Output-to-state stability and detectability of nonlinear systems, Systems and Control Letters 29:279–290
57. Spooner J, Maggiore M, Ordonez R, Passino K (2002) Stable Adaptive Control and Estimation for Nonlinear Systems. John Wiley & Sons New York
58. Teel A (1996) A nonlinear small gain theorem for the analysis of control systems with saturation, IEEE Trans Automat Control 41:1256–1270
59. Teel A (1998) Connections between Razumikhin-type theorems and the ISS nonlinear small-gain theorems, IEEE Trans Automat Control 43:960–964
60. Teel A, Praly L (1995) Tools for semiglobal stabilization by partial-state and output feedback, SIAM J Control Optim 33:1443–1488

61. Tsinias J (1997) Control Lyapunov functions, input-to-state stability and applications to global feedback stabilization for composite systems, J Math Syst Est Control 7:1–31
62. Tsinias J, Karafyllis I (1999) ISS property for time-varying systems and application to partial-state feedback stabilization and asymptotic tracking, IEEE Trans Automat Control 44:2179–2184
63. Tzamtzi M, Tzafestas S (1999) A small gain theorem for locally input-to-state stable interconnected systems, J Franklin Inst 336:893-901
64. van der Schaft A (1996) L_2-Gain and Passivity Techniques in Nonlinear Control. Springer-Verlag London

Quantitative Aspects of the Input–to–State–Stability Property

Lars Grüne

Mathematical Institute, University of Bayreuth, 95440 Bayreuth, Germany.
lars.gruene@uni-bayreuth.de

1 Introduction

Since its introduction in Sontag [17] in 1989, the input–to–state stability (ISS) property has become one of the most influential concepts in nonlinear stability theory for perturbed systems. The property generalizes the well known asymptotic stability property by assuming that each trajectory φ of a perturbed nonlinear system with time–varying perturbation $u(t)$ satisfies the inequality

$$\|\varphi(t, x, u)\| \leq \max\{\beta(\|x\|, t), \, \rho(\|u\|_\infty)\}$$

for suitable functions β of class $\mathcal{KL}$ and ρ of class $\mathcal{K}_\infty$ (cf. §3 for the relevant definitions).

The fact that this concept was used by many authors (see, e.g., [1, 4, 9, 10, 12, 14, 16, 22]) is mainly due to the intuitive simplicity of the concept, which captures the qualitative essence of robust asymptotic stability in a truly nonlinear manner; for details and the relation to other nonlinear robust stability concepts see, e.g., [8, 18, 21] and the survey [20]. On the other hand, the use of the comparison functions β and ρ in its formulation immediately leads to the idea of explicitly using the *quantitative* information contained in the ISS inequality, i.e., the rate of convergence β and the robustness gain ρ, with one of the most prominent applications being the nonlinear small gain theorem [10], for which the quantitative information contained in ρ is crucial.

In this quantitative context, however, it turns out that the original ISS formulation has some drawbacks, which are caused by the fact that it does not yield explicit information about what happens for vanishing perturbations, i.e., for perturbations u for which $u(t) \to 0$ as $t \to \infty$. Implicitly, ISS ensures that if $u(t)$ tends to 0 as t tends to infinity then also $\varphi(t, x, u)$ converges to 0 for t tending to infinity, but no explicit rate of convergence can be deduced. The main idea in order to overcome this difficulty is introducing a certain "memory fading" effect into the u–term of the ISS formulation, an idea which was used before by Praly and Wang [14] in their notion of exp–ISS. There the

M. de Queiroz et al. (Eds.): Optimal Control, Stabilization and Nonsmooth Analysis, LNCIS 301, pp. 215–230, 2004.

perturbation is first fed into a one–dimensional control system whose output then enters the right hand side of the ISS estimate. Here, instead, we use the value of the perturbation at each time instant as an initial value of a one–dimensional dynamical system, which leads to the concept of *input–to–state dynamical stability* (ISDS). Proceeding this way, we are in particular able to "synchronize" the effects of past disturbances and large initial values by using the same dynamical system for both terms. It turns out that ISDS is qualitatively equivalent to ISS and, in addition, that we can pass from ISS to ISDS with only slightly larger robustness gains.

One of the most important features of the ISS property is that it can be characterized by a dissipation inequality using a so-called ISS Lyapunov function; see [21]. One of the central properties of the ISDS estimate is that it admits an ISDS Lyapunov function, which not only characterizes ISDS as a qualitative property[1] but also represents the respective decay rate, the overshoot gain and the robustness gain. The respective results are given in Section 4.

Certainly, there are many applications where quantitative robust stability properties are of interest. Particular areas of applications are numerical investigations, where one interprets a numerical approximation as a perturbation of the original system and vice versa. One example is given in Section 5; for a comprehensive treatment of this subject we refer to the monograph [5]. In Section 6, we present two control theoretic applications of the ISDS property, which also illustrate the difference between ISDS and the ISS property.

2 Motivation

In order to explain and motivate our approach, in this section we briefly recall some classical results for systems without input, i.e., for nonlinear autonomous differential equations of the type

$$\dot{x}(t) = f(x(t)) \tag{1}$$

with $x \in \mathbb{R}^n$ and $f : \mathbb{R}^n \to \mathbb{R}^n$ locally Lipschitz. The solutions of (1) for initial value $x \in \mathbb{R}^n$ at initial time $t = 0$ will be denoted by $\varphi(t, x)$. If we assume that the origin is globally asymptotically stable for (1), then it is well known that there exists a Lyapunov function $V : \mathbb{R}^n \to \mathbb{R}$, i.e., a positive definite and proper function, which is C^∞ on $\mathbb{R}^n \setminus \{0\}$ and satisfies

$$DV(x) \cdot f(x) < 0 \text{ for all } x \in \mathbb{R}^n \setminus \{0\}.$$

By a suitabl rescaling of V we may assume that there exists a class $\mathcal{K}_\infty$–function σ (see Section 3 for a definition), such that the inequalities

[1]In fact, the qualitative equivalence ISS $\Leftrightarrow$ ISDS immediately implies that the well known ISS Lyapunov function would be sufficient for this.

$$\|x\| \le V(x) \le \sigma(\|x\|) \tag{2}$$

hold. Furthermore, it is easily seen that there exists a continuous function $g : \mathbb{R}_0^+ \to \mathbb{R}_0^+$ with $g(r) > 0$ for $r > 0$ such that the inequality

$$DV(x) \cdot f(x) < -g(V(x)) \text{ for all } x \in \mathbb{R}^n \setminus \{0\} \tag{3}$$

holds.

Integrating inequality (3) and using (2) then yields the estimate

$$\|\varphi(t, x)\| \le \mu(\sigma(\|x\|), t) \text{ for all } x \in \mathbb{R}^n,\, t \ge 0, \tag{4}$$

where μ is the solution of the 1d differential equation

$$\frac{d}{dt}\mu(r, t) = -g(\mu(r, t)), \quad \mu(r, 0) = r.$$

This means that we get a special type of $\mathcal{KL}$–estimate for the norm of the solution trajectories $\varphi(t, x)$, which in turn implies global asymptotic stability.

Now the nice property of an inequality of type (4) is that it admits a converse Lyapunov theorem using a construction of Yoshisawa [23]. If we assume (4) and set

$$V(x) := \sup_{t \ge 0} \mu(\|\varphi(t, x)\|, -t),$$

then this function satisfies (2) and

$$V(\varphi(t, x)) \le \mu(V(x), t) \text{ for all } x \in \mathbb{R}^n,\, t \ge 0,$$

from which we can in turn conclude (4). This function V, however, may be discontinuous, so we cannot conclude (3). In order to obtain a smooth function we can fix an arbitrary $\varepsilon > 0$ and set

$$V(x) := \sup_{t \ge 0} \mu(\|\varphi(t, x)\|, -(1 - \varepsilon)t).$$

This function is Lipschitz continuous and satisfies (2). It also satisfies (3) with $(1-\varepsilon)g$ instead of g in a nonsmooth sense (more precisely in the sense of viscosity supersolutions). Hence by an appropriate smoothing technique (see, e.g., [13]) we can obtain a smooth function (at least away from 0) satisfying

$$\|x\| \le V(x) \le (1 + \varepsilon)\sigma(\|x\|)$$

and

$$DV(x) \cdot f(x) < -(1 - 2\varepsilon)g(V(x)) \text{ for all } x \in \mathbb{R}^n \setminus \{0\}.$$

Thus, the particular form of the decay estimate (4) allows a converse Lyapunov theorem, which preserves the decay rate $\mu(\sigma(r), t)$ up to an arbitrarily small $\varepsilon > 0$. Our aim in this paper is to generalize this approach to the ISS property, i.e.,

- formulate a suitable variant of ISS similar to (4), which leads to the ISDS property
- find a Lyapunov function which implies ISDS
- prove a converse Lyapunov theorem that preserves the rate and gains at least up to some arbitrarily small parameter $\varepsilon > 0$.

3 Input–to–State Dynamical Stability

We consider nonlinear systems of the form

$$\dot{x}(t) = f(x(t), u(t)), \tag{5}$$

where we assume that $f : \mathbb{R}^n \times \mathbb{R}^m \to \mathbb{R}^n$ is continuous and that for each two compact subsets $K \subset \mathbb{R}^n$ and $W \subset \mathbb{R}^m$ there exists a constant $L = L(K, W)$ such that $\|f(x,u) - f(y,u)\| \leq L\|x - y\|$ for all $x, y \in K$ and all $u \in W$. The perturbation functions u are supposed to lie in the space $\mathcal{U}$ of measurable and locally essentially bounded functions with values in U, where U is an arbitrary subset of $\mathbb{R}^m$. The trajectories of (5) with initial value x at time $t = 0$ are denoted by $\varphi(t, x, u)$.

We recall that a continuous function $\alpha : \mathbb{R}_0^+ \to \mathbb{R}_0^+$ is said to be of *class* $\mathcal{K}$ provided it is strictly increasing with $\alpha(0) = 0$, and is said to be of *class* $\mathcal{K}_\infty$ provided, in addition, it is unbounded. A continuous function $\beta : \mathbb{R}_0^+ \times \mathbb{R}_0^+ \to \mathbb{R}_0^+$ is said to be of *class* $\mathcal{KL}$ provided it is of class $\mathcal{K}_\infty$ in the first and strictly decreasing to 0 in the second argument. We define a continuous function $\mu : \mathbb{R}_0^+ \times \mathbb{R} \to \mathbb{R}_0^+$ to be of *class* $\mathcal{KLD}$ provided its restriction to $\mathbb{R}_0^+ \times \mathbb{R}_0^+$ is of class $\mathcal{KL}$ and, in addition, it is a one dimensional dynamical system, i.e., it satisfies

$$\mu(r, t+s) = \mu(\mu(r,t), s) \text{ for all } t, s \in \mathbb{R}.$$

Observe that this condition implies $\mu(r, 0) = r$.

The expression $\|\cdot\|$ denotes the usual Euclidean norm, $\|u\|_\infty$ is the L_∞ norm of $u \in \mathcal{U}$ and for $t > 0$ and any measurable function $g : \mathbb{R} \to \mathbb{R}_0^+$ the expression $\operatorname{ess\,sup}_{\tau \in [0,t]} g(\tau)$ denotes the essential supremum of g on $[0, t]$.

Using these notations we can now formulate the concept of input–to-state dynamical stability.

Definition 1. *System (5) is called* input-to-state dynamically stable *(ISDS) provided there exists a function μ of class $\mathcal{KLD}$ and functions σ and γ of class $\mathcal{K}_\infty$ such that the inequality*

$$\|\varphi(t, x, u)\| \leq \max\{\mu(\sigma(\|x\|), t), \, \nu(u, t)\}.$$

holds for all $t \geq 0$, $x \in \mathbb{R}^n$ and all $u \in \mathcal{U}$, where ν is defined by

$$\nu(u, t) := \operatorname{ess\,sup}_{\tau \in [0,t]} \; \mu(\gamma(\|u(\tau)\|), t - \tau) \tag{6}$$

Here we call the function μ the decay rate, *the function σ the* overshoot gain *and the function γ the* robustness gain.

Since $\mu(\sigma(r), t)$ is of class $\mathcal{KL}$ and $\nu(u,t) \leq \gamma(\|u\|_\infty)$, ISDS implies ISS with $\beta(r,t) := \mu(\sigma(r), t)$ and robustness gain $\rho = \gamma$.

Conversely, a straightforward application of [19, Proposition 7] shows that any class $\mathcal{KL}$ function can be bounded from above by the composition of a class $\mathcal{KLD}$ and a class $\mathcal{K}_\infty$ function; see [5, Lemma B.1.4]. Hence the only real difference between ISS and ISDS is the decay property of the $\nu(u,t)$–term. The following theorem shows how one can pass from the ISS to the ISDS formulation. For the proof see [5, Proposition 3.4.4].

Theorem 1. *Assume that the system (5) is ISS for some β of class $\mathcal{KL}$ and ρ of class $\mathcal{K}_\infty$. Then for any class $\mathcal{K}_\infty$ function γ with $\gamma(r) > \rho(r)$ for all $r > 0$ there exists a class $\mathcal{KLD}$ function μ such that the system is ISDS with attraction rate μ, overshoot gain $\sigma(r) = \beta(r, 0)$ and robustness gain γ.*

For some results in this paper we will need the following assumption:

Assumption A. *The functions μ, σ and γ in Definition 1 are C^∞ on $\mathbb{R}^+ \times \mathbb{R}$ or $\mathbb{R}^+$, respectively, and the function μ solves the ordinary differential equation*

$$\frac{d}{dt}\mu(r,t) = -g(\mu(r,t))$$

for some locally Lipschitz continuous function $g : \mathbb{R}^+ \to \mathbb{R}^+$, all $r > 0$ and all $t \in \mathbb{R}$.

It was shown in [5, Appendix A] that for given nonsmooth rates and gains from Definition 1 one can find rates and gains arbitrarily close to the original ones, such that Assumption A holds and Definition 1 remains valid. Hence Assumption A is only a mild regularity condition.

It should be noted that the ISDS formulation and many of its properties can be generalized to arbitrary compact attracting sets, local attraction and systems with additional perturbation input; some results also carry over to systems with control input, leading to the definition of weak or controlled ISDS (wISDS/cISDS); see [5] for an extensive discussion and [6] for a shorter overview. Here, in order to make this short presentation concise, we stay with the basic ISDS formulation as given in Definition 1.

4 Lyapunov Function Characterization

One of the main tools for working with ISS systems is the ISS Lyapunov function whose existence is a necessary and sufficient condition for the ISS property; see [21]. In this section we provide two theorems on a Lyapunov function characterization of the ISDS property. We start with a version for discontinuous Lyapunov functions, which can exactly represent the rate and gains in the ISDS formulation. The proof of the following theorem is given in Section 7.

Theorem 2. *A system (5) is ISDS with rate μ of class $\mathcal{KLD}$ and gains σ and γ of class $\mathcal{K}_\infty$ if and only if there exists a (possibly discontinuous)* ISDS Lyapunov function, *i.e., a function $V : \mathbb{R}^n \to \mathbb{R}_0^+$ satisfying*

$$\|x\| \leq V(x) \leq \sigma(\|x\|) \tag{7}$$

and

$$V(\varphi(t, x, u)) \leq \max\{\mu(V(x), t),\ \nu(u, t)\} \tag{8}$$

for all $x \in \mathbb{R}^n$, $t \geq 0$ and all $u \in \mathcal{U}$, where ν is given by (6).

For many applications it might be desirable to have ISDS Lyapunov functions with some more regularity. The next theorem, which is also proved in Section 7, shows that if we slightly relax the sharp representation of the gains, then we can always find smooth (i.e., C^∞) Lyapunov functions, at least away from the origin.

Theorem 3. *A system (5) is ISDS with rate μ of class $\mathcal{KLD}$ and gains σ and γ of class $\mathcal{K}_\infty$ satisfying Assumption A if and only if for each $\varepsilon > 0$ there exists a continuous function $V : \mathbb{R}^n \to \mathbb{R}_0^+$ which is smooth on $\mathbb{R}^n \setminus \{0\}$ and satisfies*

$$\|x\|/(1+\varepsilon) \leq V(x) \leq \sigma(\|x\|) \tag{9}$$

and

$$\gamma(\|u\|) < V(x) \quad \Rightarrow \quad DV(x) \cdot f(x, u) \leq -(1-\varepsilon) g(V(x)) \tag{10}$$

for all $x \in \mathbb{R}^n \setminus \{0\}$ and all $u \in U$.

It should be noted that there exists an intermediate object between the discontinuous and the smooth ISDS Lyapunov function, namely a Lipschitz Lyapunov function which satisfies (10) in a suitable generalized sense using the theory of viscosity solutions; see [5] for details. While both smooth and Lipschitz Lyapunov functions characterize the optimal gains "in the limit", we conjecture that there are examples in which gains can be exactly characterized by Lipschitz but not by smooth ISDS Lyapunov functions, similar to what was shown recently for H_∞ Lyapunov functions in [15].

Theorem 3 gives rise to a constructive procedure of computing ISDS robustness gains from Lyapunov functions for the unperturbed system $f(x, 0)$. We illustrate this procedure by three examples.

Example 1. Consider a linear system $\dot{x} = f(x, u) = Ax + Bu$. If we assume this system is ISDS then the matrix A needs to be Hurwitz and we can find a quadratic Lyapunov function $W(x) = x^T P x$ for some positive definite matrix P satisfying $c_1\|x\|^2 \leq W(x) \leq c_2\|x\|^2$ and $DW(x)Ax \leq -c_3\|x\|^2$. Setting

$$V(x) = \sqrt{W(x)/c_1},$$

we obtain $\|x\| \leq V(x) \leq c_4\|x\|$, $DV(x)Ax \leq -c_5 V(x)$ and $\|DV(x)\| \leq c_4$ for

$$c_4 = \sqrt{c_2/c_1}, \quad c_5 = c_3/(2c_2).$$

Fixing some $\lambda \in (0,1)$ we set $\gamma(r) = c_4\|B\|r/(\lambda c_5)$. Then we obtain

$$\gamma(\|u\|) \le V(x) \quad \Rightarrow \quad DV(x) \cdot f(x,u) \le -(1-\lambda)c_5 V(x) =: -g(V(x)).$$

Hence V is an ISDS Lyapunov function in the sense of Theorem 3 (for each $\varepsilon > 0$) and we obtain ISDS with

$$\mu(r,t) = e^{-(1-\lambda)c_5 t} r, \quad \sigma(r) = c_4 r, \quad \gamma(r) = c_4\|B\|r/(\lambda c_5),$$

i.e., exponential convergence and linear overshoot and robustness gains.

This example nicely illustrates the (typical) tradeoff between the attraction rate μ and the robustness gain γ, which is represented here by the choice of λ: the smaller γ becomes, the slower the convergence that can be guaranteed. In the next two examples, showing ISDS estimates for two simple nonlinear systems, we set $\lambda = 3/4$.

Example 2. Consider the system $\dot{x} = f(x,u) = -x + u^3/2$ with $x \in \mathbb{R}$, $u \in \mathbb{R}$. Using the Lyapunov function $V(x) = |x|$ one obtains $DV(x)f(x,0) = -|x| = -V(x)$. We choose γ such that $\gamma(|u|) \le V(x) = |x|$ implies $|u^3/2| \le 3|x|/4$, i.e., $\gamma(r) = 2r^3/3$. Then we obtain

$$\gamma(\|u\|) \le V(x) \quad \Rightarrow \quad DV(x) \cdot f(x,u) \le -\frac{1}{4}V(x) =: -g(V(x)),$$

and consequently ISDS with $\mu(r,t) = e^{-t/4}r$, $\sigma(r) = r$ and $\gamma(r) = 2r^3/3$.

Example 3. Consider the system $\dot{x} = f(x,u) = -x^3 + u$ with $x \in \mathbb{R}$, $u \in \mathbb{R}$. Again using the Lyapunov function $V(x) = |x|$ one obtains $DV(x)f(x,0) = -|x|^3 = -V(x)^3$. Here we choose γ such that $\gamma(|u|) \le V(x) = |x|$ implies $|u| \le 3|x|^3/4$, i.e., $\gamma(r) = \sqrt[3]{4r/3}$. Then we obtain

$$\gamma(\|u\|) \le V(x) \quad \Rightarrow \quad DV(x) \cdot f(x,u) \le -\frac{1}{4}V(x)^3 =: -g(V(x)),$$

and consequently ISDS with $\mu(r,t) = \sqrt{2t + 4/r^2}/(t + 2/r^2)$ (the solution of $\dot{\mu} = -\mu^3/4$), $\sigma(r) = r$ and $\gamma(r) = \sqrt[3]{4r/3}$.

5 Applications in Numerical Analysis

In order to illustrate the way in which ISDS–like properties can be used in numerical analysis, we consider a problem from numerical dynamics. We briefly describe an algorithm for the computation of attractors developed by Dellnitz and Hohmann [2]; here we describe a version due to Junge [11].

Consider the differential equation (1) and its time–1 map $\Phi(x) := \varphi(1,x)$. Consider a rectangular domain $\Omega \subset \mathbb{R}^n$ and a partition $\mathcal{C}^0$ of Ω into N^0 rectangular cells $\mathcal{C}^0 = \{C_1^0, C_2^0, \ldots, C_{N^0}^0\}$.

Setting $k = 0$ we compute a collection of cells $\tilde{\mathcal{C}}^k \subset \mathcal{C}^k$ by defining

$$\tilde{\mathcal{C}}^k := \{C_i^k \in \mathcal{C}^k : \text{there exists } C \in \mathcal{C}^k \text{ with } \Phi(C) \cap C_i^k \neq \emptyset\}. \qquad (11)$$

For simplicity we assume here that $\Phi(C)$ can be computed, which will not be the case in general; cf. Remark 1 (ii), below. In the next step each cell contained in $\tilde{\mathcal{C}}^k$ is refined (e.g., by subdividing it into a number of finer rectangles) and the resulting collection of cells is denoted by $\mathcal{C}^{k+1}$. Now we set $k = k + 1$ and restart this procedure by going to step (11).

This generates a sequence of collections $\mathcal{C}^k$, $k = 0, 1, \ldots$, satisfying

$$\bigcup_i C_i^{k+1} \subseteq \bigcup_i C_i^k.$$

Now let $A \subset \Omega$ be an attractor, i.e., a minimal asymptotically stable set which attracts $\Omega \setminus \{A\}$. Then it is known that the convergence

$$d_H(C^k, A) \to 0$$

holds in the Hausdorff metric d_H for compact sets. However, estimates for the corresponding rate of convergence are difficult to obtain.

Such estimates can be derived from the ISDS property. Consider the perturbed system

$$\dot{x}(t) = f(x(t)) + u(t) \qquad (12)$$

with solution trajectories $\varphi(t, x, u)$ and assume that the attractor A has the ISDS property, i.e.,

$$\|\varphi(t, x, u)\|_A \leq \max\{\mu(\sigma(\|x\|_A), t), \; \nu(u, t)\},$$

where $\|x\|_A := \inf_{y \in A} \|x - y\|$. Let

$$\operatorname{diam}(\mathcal{C}^k) := \max_{i=1,\ldots,N^k} \max_{x,y \in C_i^k} \|x - y\|$$

be the maximal diameter of the cells in $\mathcal{C}^k$. Then we obtain the estimate

$$d_H(C^k, A) \leq \max\left\{\mu(\sigma(d_H(\Omega, A), k), \max_{j=0,\ldots,k-1} \mu(\gamma(\operatorname{diam}(\mathcal{C}^{k-j-1})), j)\right\}. \qquad (13)$$

For a proof of this estimate see [5, Theorem 6.3.3].

Remark 1. (i) In fact, for this estimate to hold we only need that the ISDS estimate is valid for $x \in \Omega$. It can be shown that any asymptotically stable set for the unperturbed system (1) for which Ω lies in its domain of attraction has this "local" ISDS property for the perturbed system (12) for suitable μ, σ and γ and suitable perturbation range U; see [5, Theorem 3.4.6]. Hence estimate (13) holds for all attractors without any additional asumptions for suitable functions μ, σ and γ.

(ii) One can incorporate numerical errors in the computation of the image $\Phi(C)$ in (11) in the analysis of the algorithm. We refer to [5, Section 6.3] for details.

6 Applications in Control Theory

As a first application, we derive an estimate on a nonlinear stability margin. In [21] it was shown that ISS implies the existence of a stability margin for a perturbed system. However, for ISS it is difficult to derive an estimate for this margin. In contrast to this, the ISDS property easily allows us to give an estimate based on the ISDS robustness gain.

Theorem 4. *Consider a system (5) which is ISDS with μ, σ and γ and $U = \mathbb{R}^m$ satisfying Assumption A. Consider a Lipschitz map $k : \mathbb{R}^n \to \mathbb{R}_0^+$ satisfying $k(x) \le \max\{\gamma^{-1}(\|x\|), k_0\}$ for some value $k_0 \ge 0$. Then for each $x \in \mathbb{R}^n$ and all $u \in \mathcal{U}$ with $\|u\|_\infty \le 1$ the trajectories $\varphi_k(t, x, u)$ of the system $\dot{x} = f_k(x, u) := f(x, k(x)u)$ satisfy*

$$\|\varphi_k(t, x, u)\| \le \max\{\mu(\sigma(\|x\|), t), \gamma(k_0)\}$$

for all $t \ge 0$.

The proof can be found in [7].

As a second application we consider the stability of coupled systems. The following theorem is a version of the generalized small gain theorem [10, Theorem 2.1] (in a simplified setting). As for Theorem 4, the qualitative result (i.e., asymptotic stability of the coupled system) can be proved using the ISS property. The advantage of ISDS lies in the estimates for the overshoot and the decay rates of the coupled system.

Theorem 5. *Consider two systems $\dot{x}_i = f(x_i, u_i)$, $i = 1, 2$, of type (5) where the f_i are Lipschitz in both x_i and u_i. Let $x_i \in \mathbb{R}^{n_i}$, $U_1 = \mathbb{R}^{n_2}$ and $U_2 = \mathbb{R}^{n_1}$. Assume that the systems are ISDS with rates μ_i and gains σ_i and γ_i and assume that the inequality $\gamma_1(\gamma_2(r)) \le r$ holds for all $r > 0$. Then the coupled system*

$$\dot{x}_1 = f_1(x_1, x_2), \quad \dot{x}_2 = f_2(x_2, x_1) \tag{14}$$

is globally asymptotically stable and the trajectories $(x_1(t), x_2(t))$ of (14) satisfy

$$\|x_i(t)\| \le \delta_i\Big(\max\{\sigma_i(\|x_i(0)\|), \gamma_i(\sigma_j(\|x_j(0)\|))\}, t\Big) \tag{15}$$

for $i = 1, 2$, $j = 3 - i$ and functions δ_i given by

$$\delta_i(r, t) := \sup\left\{ \theta_i^{t_1, s_1} \circ \ldots \circ \theta_i^{t_k, s_k}(r) \;\middle|\; k \ge 1,\, t_j, s_j \ge 0,\, \sum_{j=1}^{k} t_j + s_j = t \right\}$$

with $\theta_1^{t,s}(r) := \mu_1(\gamma_1(\mu_2(\gamma_1^{-1}(r), s)), t)$ and $\theta_2^{t,s}(r) := \mu_2(\gamma_2(\mu_1(\gamma_2^{-1}(r), s)), t)$. In particular, for all $t \ge 0$ we obtain from (15) the overshoot estimates

$$\|x_i(t)\| \le \max\{\sigma_i(\|x_i(0)\|), \gamma_i(\sigma_j(\|x_j(0)\|))\}.$$

Again, the proof can be found in [7]. Note that $\gamma_1(\gamma_2(r)) \le r$ for all $r > 0$ implies $\gamma_2(\gamma_1(r)) \le r$ for all $r > 0$.

Remark 2. A different characterization of the decay rates δ_i in Theorem 5 can be obtained if we assume that the gains γ_i and the class $\mathcal{KLD}$ functions μ_i satisfy Assumption A for functions g_i. In this case, differentiating the expressions in the definition of $\delta_i(r,t)$, $i = 1, 2$, with respect to t, one sees that the δ_i are bounded from above by the solutions of the one–dimensional differential equations $\dot r_i = \max\{-g_i(r_i), -\gamma_i'(\gamma_i^{-1}(r_i))g_j(\gamma_i^{-1}(r_i))\}$, $r_i(r,0) = r$, where γ_i' denotes the derivative of γ_i and $j = 3 - i$.

In the following example we illustrate the quantitative information one can obtain from Theorem 5 and Remark 2.

Example 4. Consider the two systems from Examples 2 and 3 with robustness gains $\gamma_1(r) = 2r^3/3$ and $\gamma_2(r) = \sqrt[3]{4r/3}$. Then the coupled system reads $\dot x_1(t) = -x_1(t) + x_2(t)^3/2$, $\dot x_2(t) = -x_2(t)^3 + x_1(t)$. One verifies that the gain condition of Theorem 5 is satisfied. Hence we can conclude asymptotic stability with overshoot estimates

$$\|x_1(t)\| \le \max\{\|x_1(0)\|, 2\|x_2(0)\|^3/3\},$$
$$\|x_2(t)\| \le \max\{\|x_2(0)\|, \sqrt[3]{4\|x_1(0)\|/3}\}.$$

Using the formula from Remark 2 we obtain

$$\dot r_1 = \max\{-c_1 r_1, -c_2 r_1^{5/3}\}, \quad \dot r_2 = \max\{-c_3 r_2^3, -c_4 r_2\}$$

for suitable constants $c_1, \ldots, c_4 > 0$. This shows that far away from the equilibrium exponential convergence can be expected, while in a neighborhood of 0 the rates of convergence in both components will slow down.

7 Proofs of the Main Results from Section 4

The following lemma will be crucial for all our proofs.

Lemma 1. *Consider a (possibly discontinuous) function $V : \mathbb{R}^n \to \mathbb{R}_0^+$. Then the following two statements are equivalent*

(i) $V(\varphi(t,x,u)) \le \max\{\mu(V(x),t), \nu(u,\tau)\}$ for all $t \ge 0$ and all $u \in \mathcal{U}$.
(ii) $V(\varphi(t,x,u)) \le \mu(a,t)$ for all times $t \ge 0$, all values $a \in \mathbb{R}$ with $a \ge V(x)$ and all $u \in \mathcal{U}$ satisfying $\gamma(\|u(\tau))\|) \le \mu(a,\tau)$ for almost all $\tau \in [0,t]$.

Proof: "(i) $\Rightarrow$ (ii)": The definition of ν immediately implies $\nu(u,t) \le \mu(a,t)$ for t, a and u satisfying the assumptions from (ii), hence (i) implies (ii).
"(ii) $\Rightarrow$ (i)": Consider an arbitrary $u \in \mathcal{U}$ and $t > 0$. We set $a = \max\{V(x), \mu(\nu(u,t), -t)\}$ which implies $\gamma(\|u(\tau))\|) \le \mu(a,\tau)$ for almost all $\tau \in [0,t]$. Now either $a = V(x)$ or $\mu(a,t) = \nu(u,t)$ holds. In the first case we obtain $V(\varphi(t,x,u)) \le \mu(a,t) = \mu(V(x),t)$ while in the second case we have $V(\varphi(t,x,u)) \le \mu(a,t) = \nu(u,t)$. Thus we can conclude (i).

Now we can turn to the **Proof of Theorem 2:**
"(i) ⇒ (ii)" We construct a function for which Lemma 1(ii) can be verified. We define

$$V(x) := \inf\{b \geq 0 \mid \|\varphi(t,x,u)\| \leq \max\{\mu(b,t),\, \nu(u,t)\} \text{ for all } u \in \mathcal{U},\, t \geq 0\}.$$

Clearly, the ISDS assumption implies $\|x\| \leq V(x) \leq \sigma(\|x\|)$. It remains to show Lemma 1(ii). To this end, fix $x \in \mathbb{R}^n$, $a \geq V(x)$, $t > 0$ and $u \in \mathcal{U}$ with $\gamma(\|u(\tau))\|) \leq \mu(a,\tau)$ for almost all $\tau \in [0,t]$. This implies

$$\nu(u(\cdot), t+s) \leq \max\{\mu(\mu(a,t),s),\, \nu(u(t+\cdot),s)\}$$

for each $s > 0$, so by the definition of V for any $b > a$ we obtain

$$\|\varphi(t{+}s,x,u)\| \leq \max\{\mu(b,t{+}s),\, \nu(u,t{+}s)\} \leq \max\{\mu(\mu(b,t),s),\, \nu(u(t{+}\cdot),s)\}$$

which implies $V(\varphi(t,x,u)) \leq \mu(a,t)$ and thus Lemma 1(ii).
"(ii) ⇒ (i)" This implication follows immediately using the assumed bounds on V.

Throughout the rest of this section we assume Assumption A. For the proof of Theorem 3 we need four preliminary lemmas.

Lemma 2. *Let μ be a class $\mathcal{KLD}$ function, let γ be a class $\mathcal{K}_\infty$ function and let $x \in \mathbb{R}^n$. If a continuous function $V : \mathbb{R}^n \to \mathbb{R}_0^+$, which is differentiable at some point $x \in \mathbb{R}^n$, satisfies the inequality*

$$V(\varphi(t,x,u)) \leq \max\{\mu(V(x),t),\, \nu(u,t)\}$$

for all $t \geq 0$, all $u \in \mathcal{U}$ and ν from (6), then it satisfies

$$\gamma(\|u\|) < V(x) \quad \Rightarrow \quad DV(x)\cdot f(x,u) \leq -g(V(x)). \tag{16}$$

for all $u \in U$.

Proof: Fix $u_0 \in U$ with $\gamma(\|u_0\|) < V(x)$ and consider the constant function $u(t) \equiv u_0$. By continuity, for all $\tau > 0$ small enough we obtain $V(\varphi(\tau,x,u)) \leq \mu(V(x),\tau)$, which implies

$$\begin{aligned} DV(x)\cdot f(x,u_0) &\leq \limsup_{\tau\to 0} \frac{V(\varphi(\tau,x,u)) - V(x)}{\tau} \\ &\leq \limsup_{\tau\to 0} \frac{\mu(V(x),\tau) - V(x)}{\tau} \quad = \quad -g(V(x)), \end{aligned}$$

and thus the claim.

We cannot in general conclude the result for $\gamma(\|u\|) = V(x)$ using continuity in u because U is an arbitrary set which might in particular be discrete. The following lemma shows that we can nevertheless obtain (16) for $\gamma(\|u\|) = V(x)$ if V is continuously differentiable. Furthermore, if V is smooth, then also the converse implication holds.

Lemma 3. *Let μ be a class $\mathcal{KLD}$ function satisfying Assumption A and let γ be a class $\mathcal{K}_\infty$ function. Then a continuous function $V : \mathbb{R}^n \to \mathbb{R}_0^+$ that is smooth on $\mathbb{R}^n \setminus \{0\}$ satisfies the inequality*

$$V(\varphi(t,x,u)) \le \max\{\mu(V(x),t),\ \nu(u,t)\} \tag{17}$$

for all $x \in \mathbb{R}^n$, $t \ge 0$ and all $u \in \mathcal{U}$, where ν is given by (6), if and only if it satisfies

$$\gamma(\|u\|) \le V(x) \quad \Rightarrow \quad DV(x)\cdot f(x,u) \le -g(V(x)) \tag{18}$$

for all $x \in \mathbb{R}^n \setminus \{0\}$ and all $u \in U$.

Proof: "(17) $\Rightarrow$ (18)": From (16) we already know the desired inequality for $\gamma(\|u\|) < V(x)$. Hence fix $u \in \mathcal{U}$ and $x \in \mathbb{R}^n \setminus \{0\}$ with $\gamma(\|u\|) = V(x)$. Since by (16) we know $DV(x) \ne 0$ the point x cannot be a local maximum. Hence there exists a sequence of points $x_i \to x$ with $V(x_i) > V(x) = \gamma(\|u\|)$. From (16) we obtain $DV(x_i)\cdot f(x_i,u) \le -g(V(x_i))$ for all $i \in \mathbb{N}$, which implies (18) by continuity.

"(18) $\Rightarrow$ (17)": Fix $x \in \mathbb{R}^n$ and $t > 0$. Integrating (18) we obtain

$$V(\varphi(t,x,u)) \le \mu(V(x),t) \tag{19}$$

for all $u \in \mathcal{U}$ with $\gamma(\|u(\tau)\|) \le \mu(V(x),t)$ f.a.a. $\tau \in [0,t]$, where μ solves $\dot\mu = -g(\mu)$, $\mu(r,0) = r$. We claim that (19) implies Lemma 1(ii).

In order to prove the assertion fix $x \in \mathbb{R}^n$, $a \ge V(x)$ and $t > 0$, let $u \in \mathcal{U}$ satisfy $\gamma(\|u(\tau))\|) \le \mu(a,\tau)$ for almost all $\tau \in [0,t]$ and assume $V(\varphi(t,x,u)) > \mu(a,t)$. Then there exists $\delta > 0$ such that $V(\varphi(t,x,u)) > \mu(a,t)+\delta$. Now pick an arbitrary $\varepsilon < \delta$ and choose $t^* > 0$ such that $V(\varphi(t^*,x,u)) = \mu(a,t^*)+\varepsilon$ and $V(\varphi(\tau,x,u)) > \mu(a,\tau)+\varepsilon$ for all $\tau \in [t^*,t]$. From the assumption on u we obtain $\gamma(\|u(\tau)\|) \le V(\varphi(\tau,x,u)) - \varepsilon$ for almost all $\tau \in [t^*,t]$. Using the continuity of $V(\varphi(\tau,x,u))$ in τ and the Lipschitz property of g we can now conclude the existence of times t_i, $i = 0,\dots,k$ such that $t_0 = t^*$, $t_k = t$ and $\mu(V(\varphi(t_i,x,u),t_{i+1}-t_i) \ge V(\varphi(t_i,x,u)) - \varepsilon$, which implies $\|u(\tau)\| \le \mu(V(\varphi(t_i,x,u))$ for almost all $\tau \in [t_i,t_{i+1}]$. Using (19) inductively and applying Gronwall's Lemma we obtain

$$V(\varphi(t,x,u)) \le \mu(V(\varphi(t^*,x,u)),t-t^*) \le \mu(\mu(a,t^*)+\varepsilon,t-t^*) \le \mu(a,t)+C\varepsilon$$

for some suitable $C > 0$ which contradicts $V(\varphi(t,x,u)) > \mu(a,t)+\delta$ as $\varepsilon \to 0$ and hence shows Lemma 1(ii) and thus the assertion.

The next lemma shows the existence of a Lipschitz ISDS Lyapunov function.

Lemma 4. *If a system (5) is ISDS with rate μ of class $\mathcal{KLD}$ satisfying Assumption A and gains σ and γ of class $\mathcal{K}_\infty$ then for each $\varepsilon > 0$ there exists a continuous function $V : \mathbb{R}^n \to \mathbb{R}_0^+$, which is Lipschitz on $\mathbb{R}^n \setminus \{0\}$ and satisfies*

$$\|x\|/(1+\varepsilon) \le V(x) \le \sigma(\|x\|) \tag{20}$$

for all $x \in \mathbb{R}^n$ *and*

$$\gamma(\|u\|) < V(x) \quad \Rightarrow \quad DV(x)\cdot f(x,u) \le -(1-\varepsilon)g(V(x)) \tag{21}$$

for almost all $x \in \mathbb{R}^n$ *and all* $u \in U$.

Proof: Fix some $\varepsilon > 0$ and set $\rho_\varepsilon(r) := \varepsilon(1-e^{-r}) + 1$. Then ρ_ε is strictly increasing for $r > 0$, $\rho_\varepsilon(0) = 1$ and $\rho_\varepsilon(r) \nearrow 1+\varepsilon$ as $r \to \infty$. Using this function we define

$$V(x) := \inf\left\{ b \ge 0 \;\middle|\; \begin{array}{l} \|\varphi(t,x,u)\| \le \\ \rho_\varepsilon(\mu(b,t)) \max\{\mu(b,(1-\varepsilon)t),\, \nu(u,t)\} \\ \text{for all } u \in \mathcal{U} \text{ and all } t \ge 0 \end{array} \right\}. \tag{22}$$

Similar to the proof of Theorem 2 one verifies (20) and (21).

We now show the Lipschitz property of V. In order to do this pick a compact set $N \subset \mathbb{R}^n$ not containing the origin. From the bounds on V we can conclude that there exists a compact interval $I = [c_1, c_2] \subset \mathbb{R}^+$ such that for $x \in N$ the infimum over $b \ge 0$ in (22) can be replaced by the infimum over $b \in I$. Now the ISDS property implies the existence of a constant $R > 0$ such that $\|\varphi(t,x,u)\| \le \max\{\mu(R,t),\, \nu(u,t)\}$ holds for all $x \in N$, all $u \in \mathcal{U}$ and all $t \ge 0$, which implies that we can restrict ourselves to those $u \in \mathcal{U}$ with $\|u\|_\infty \le R$. Furthermore, there exists $T > 0$ such that $\mu(R,t) < \mu(c_1,(1-\varepsilon)t)$ holds for all $t \ge T$, which implies that we only have to check the inequality for $\|\varphi(t,x,u)\|$ in (22) for $t \in [0,T]$. Thus the definition of V eventually reduces to

$$V(x) := \inf\left\{ b \in I \;\middle|\; \begin{array}{l} \|\varphi(t,x,u)\| \le \\ \rho_\varepsilon(\mu(b,t)) \max\{\mu(b,(1-\varepsilon)t),\, \nu(u,t)\} \\ \forall u \in \mathcal{U} \text{ with } \|u\|_\infty \le R \text{ and all } t \in [0,T] \end{array} \right\}. \tag{23}$$

Now we find constants $L_1 > 0$ and $C_1 > 0$ such that the inequalities

$$\begin{array}{l} \|\varphi(t,x_1,u) - \varphi(t,x_2,u)\| \le L_1\|x_1 - x_2\| \\ |\rho_\varepsilon(\mu(a_1,t)) - \rho_\varepsilon(\mu(a_2,t))| \ge C_1|a_1 - a_2| \end{array}$$

hold for all $u \in \mathcal{U}$ with $\|u\|_\infty \le R$, all $t \in [0,T]$, all $a_1, a_2 \in I$ and all $x_1, x_2 \in N$.

We set $L_N = L_1/(C_1\mu(c_1,T))$, pick $x_1, x_2 \in N$ and fix $\delta > 0$. From (23) we can conclude the existence of $b^* \in I$, $t^* \in [0,T]$ and $u^* \in \mathcal{U}$ with $\|u\|_\infty \le R$ such that $b^* \ge V(x_1) - \delta$ and $\|\varphi(t^*,x_1,u^*)\| > \rho_\varepsilon(\mu(b^*,t^*))\max\{\mu(b^*,(1-\varepsilon)t^*),\, \nu(u^*,t^*)\}$. Then $\|\varphi(t^*,x_2,u^*)\| \ge \rho_\varepsilon(\mu(b^{**},t^*))\max\{\mu(b^{**},(1-\varepsilon)t^*),\, \nu(u^*,t^*)\}$ holds for all $b^{**} < b^*$ with $|b^{**} - b^*| \ge L_N\|x_1 - x_2\|$, implying $V(x_2) \ge b^{**}$ and thus $V(x_1) - V(x_2) \le L_N\|x_1 - x_2\| + \delta$. Since $\delta > 0$ was arbitrary and this estimate is symmetric in x_1 and x_2 we obtain the desired Lipschitz estimate with constant L_N.

Finally, since by Rademacher's Theorem (see, e.g., [3, page 216]) a Lipschitz function is differentiable almost everywhere, inequality (21) follows from Lemma 2.

The following lemma gives a smoothing result for Lipschitz Lyapunov functions.

Lemma 5. *Consider a continuous function* $V : \mathbb{R}^n \to \mathbb{R}_0^+$ *that is Lipschitz on* $\mathbb{R}^n \setminus \{0\}$ *and satisfies*

$$\gamma(\|u\|) < V(x) \quad \Rightarrow \quad DV(x) \cdot f(x,u) \le -g(V(x))$$

for almost all $x \in \mathbb{R}^n$. *Then for any two continuous functions* $\alpha_1, \alpha_2 : \mathbb{R}^n \setminus \{0\} \to \mathbb{R}^+$ *there exists a continuous function* $\tilde{V} : \mathbb{R}^n \to \mathbb{R}_0^+$, *which is smooth on* $\mathbb{R}^n \setminus \{0\}$ *and satisfies*

$$\|V(x) - \tilde{V}(x)\| \le \alpha_1(x)$$

and

$$\gamma(\|u\|) \le V(x) \;\Rightarrow\; D\tilde{V}(x) \cdot f(x,u) \le -g(\tilde{V}(x)) + \alpha_2(x)$$

for all $x \in \mathbb{R}^n \setminus \{0\}$.

Proof: This follows from Theorem B.1 in [13], observing that the proof in [13] (which requires compact U) remains valid if for any compact subset $K \subset \mathbb{R}^n$ we can restrict ourselves to a compact subset of U, which is the case here since we only need to consider $\|u\| \le \gamma^{-1}(\max_{x\in K} V(x))$.

Finally, we can turn to the **Proof of Theorem 3:**
Assume ISDS, fix $\varepsilon > 0$ and let $\varepsilon_1 > 0$ be such that $1/(1+\varepsilon_1)^2 \ge (1-\varepsilon)$, $(1+\varepsilon_1)^2 \le (1+\varepsilon)$ and $(1-\varepsilon_1)^2 \ge (1-\varepsilon)$. Applying Lemma 4 with $\varepsilon = \varepsilon_1$ we can conclude the existence of a locally Lipschitz (away from 0) Lyapunov function V satisfying $\|x\|/(1+\varepsilon_1) \le V(x) \le \sigma(\|x\|)$ for all $x \in \mathbb{R}^n$ and $\gamma(\|u\|) < V(x)$ $\Rightarrow DV(x) \cdot f(x,u) \le -(1-\varepsilon_1)g(V(x))$ for almost all $x \in \mathbb{R}^n$. Applying Lemma 5 with $\alpha_1(x) = \min\{\gamma((1+\varepsilon)\gamma^{-1}(V(x))) - V(x), \varepsilon_1 V(x)\}$ and $\alpha_2(x) = \varepsilon_1 g(V(x))$ we obtain a smooth (away from 0) Lyapunov function $\tilde{V}$ satisfying the desired bounds and, since the choice of α_1 implies $\gamma((1+\varepsilon)\|u\|) \le \tilde{V}(x)$ $\Rightarrow \gamma(\|u\|) \le V(x)$, we obtain

$$\begin{aligned} &\gamma((1+\varepsilon)\|u\|) \le \tilde{V}(x) \\ &\Rightarrow \quad D\tilde{V}(x) \cdot f(x,u) \le -(1-\varepsilon_1)^2 g(\tilde{V}(x)) \le -(1-\varepsilon) g(\tilde{V}(x)) \end{aligned}$$

for all $x \in \mathbb{R}^n \setminus \{0\}$. Hence $\tilde{V}$ is the desired Lyapunov function.

Conversely, assume the existence of V for any $\varepsilon > 0$ and fix $t > 0$. By Lemma 3 we obtain $(1-\varepsilon)\|\varphi(t,x,u)\| \le \{\mu((1+\varepsilon)\sigma(\|x\|), (1-\varepsilon)t), \nu_\varepsilon(u,t)\}$ where $\nu_\varepsilon(u,t) := \operatorname{ess\,sup}_{\tau\in[0,t]} \mu(\gamma(\|(1+\varepsilon)u(\tau)\|), (1-\varepsilon)(t-\tau))$. Since all these expressions are continuous in ε we obtain the desired inequality.

References

1. Christofides P, Teel A (1996) Singular perturbations and input-to-state stability, IEEE Trans Autom Control 41:1645–1650

2. Dellnitz M, Hohmann A (1997) A subdivision algorithm for the computation of unstable manifolds and global attractors, Numer Math 75:293–317
3. Federer H (1969) Geometric Measure Theory. Springer–Verlag New York
4. Grüne L (1997) Input-to-state stability of exponentially stabilized semilinear control systems with inhomogenous perturbation, Syst Control Lett 38:27–35
5. Grüne L (2002) Asymptotic Behavior of Dynamical and Control Systems under Perturbation and Discretization (Lecture Notes in Mathematics Vol 1783). Springer–Verlag Heidelberg
6. Grüne L (2002) Gain preserving Lyapunov functions for perturbed and controlled systems. In: Proc 41st IEEE Conference Decision Control Las Vegas NV 707–712
7. Grüne L (2002) Input–to–state dynamical stability and its Lyapunov function characterization, IEEE Trans Autom Control 47:1499–1504
8. Grüne L, Sontag E, Wirth F (1999) Asymptotic stability equals exponential stability, and ISS equals finite energy gain—if you twist your eyes, Syst Control Lett 38:127–134
9. Isidori A, Global almost disturbance decoupling with stability for non minimum-phase single-input single-output nonlinear systems, Syst Control Lett 28:115–122
10. Jiang Z-P, Teel A, Praly L (1994) Small-gain theorem for ISS systems and applications, Math Control Signals Syst 7:95–120
11. Junge O (1999) Rigorous discretization of subdivision techniques. In: Fiedler B, Gröger K, Sprekels J (eds) EQUADIFF 99 (Proc International Congress held in Berlin, Germany). World Scientific Singapore 916–918
12. Krstić M, Deng H (1998) Stabilization of Nonlinear Uncertain Systems. Springer-Verlag London
13. Lin Y, Sontag E, Wang Y (1996) A smooth converse Lyapunov theorem for robust stability, SIAM J Control Optim 34:124–160
14. Praly L, Wang Y (1996) Stabilization in spite of matched unmodelled dynamics and an equivalent definition of input-to-state stability, Math Control Signals Systems 9:1–33
15. Rosier L, Sontag E (2000) Remarks regarding the gap between continuous, Lipschitz, and differentiable storage functions for dissipation inequalities, Syst Control Lett 41:237–249
16. Sepulchre R, Jankovic M, Kokotović P (1997) Constructive Nonlinear Control. Springer-Verlag Berlin
17. Sontag E (1989) Smooth stabilization implies coprime factorization, IEEE Trans Autom Control 34:435–443
18. Sontag E (1995) On the input-to-state stability property, Europ J Control 1:24–36
19. Sontag E (1998) Comments on integral variants of ISS, Syst Control Lett 34:93–100
20. Sontag E (2000) The ISS philosophy as a unifying framework for stability–like behavior. In: Isidori A, Lamnabhi-Lagarrigue F, Respondek W (eds) Nonlinear Control in the Year 2000 Volume 2 (Lecture Notes in Control and Information Sciences 259). Springer-Verlag London 443–468.
21. Sontag E, Wang Y (1995) On characterizations of the input-to-state stability property, Syst Control Lett 24:351–359

22. Tsinias J (1997) Input to state stability properties of nonlinear systems and applications to bounded feedback stabilization using saturation, ESAIM Control Optim Calc Var 2:57–85
23. Yoshizawa T (1966) Stability Theory by Lyapunov's Second Method. The Mathematical Society of Japan Tokyo

Explicit Formulae for Boundary Control of Parabolic PDEs

Andrey Smyshlyaev and Miroslav Krstic

Department of Mechanical and Aerospace Engineering, University of California at San Diego, La Jolla, CA 92093-0411. [asmyshly,krstic]@ucsd.edu

1 Introduction

Methods for boundary control of linear parabolic PDEs are well established (see, e.g., [1, 3, 7, 9, 10, 14, 15, 17, 18, 19]). However, even in the simplest case of constant coefficients the existing results are not explicit and require numerical solution, e.g., solving an operator Riccati equation in case of the LQR method.

In this paper, we present explicit (closed form) solutions to several application motivated parabolic problems, including the very important case of constant coefficients. The method we use to get those solutions is essentially the backstepping technique developed originally to handle nonlinear finite-dimensional systems [13]. Here we apply it to a linear infinite-dimensional plant. The approach exploits the structure of the parabolic PDE, namely, its 'tri-diagonal form'. As a result, the boundary stabilization problem is converted to a problem of solving a specific *linear hyperbolic* PDE. In several cases the solution to this PDE can be found explicitly; we show some of these cases.

We also give explicit solutions for the output feedback problem under boundary measurement. Both anti-collocated and collocated sensor/actuator pairs are considered. Duality of the observer design problem to the stabilization problem is shown, specifically meaning that the observer gains can be derived from the control gains. The closed form observers and controllers make it possible to design closed form frequency domain compensators.

We should note that the described method is not limited to plants for which an explicit solution can be found. It can be applied to a general class of parabolic PDEs. It can also be modified to obtain a controller that minimizes a reasonable cost functional that puts penalty on both state and control (so-called inverse optimal controller) giving stability margins and robustness. These extensions are beyond the scope of this paper.

M. de Queiroz et al. (Eds.): Optimal Control, Stabilization and Nonsmooth Analysis, LNCIS 301, pp. 231–249, 2004.

2 State-Feedback Designs

We will present the solutions to four distinct parabolic PDEs. For the first problem the procedure will be described in details and for the remaining cases most of the details will be omitted.

2.1 Unstable Heat Equation

Let us consider the following plant:

$$u_t(x,t) = \varepsilon u_{xx}(x,t) + \lambda_0 u(x,t)\,, \quad x \in (0,1)\,, \tag{1}$$

$$u(0,t) = 0\,, \tag{2}$$

where $\varepsilon > 0$ and λ_0 are arbitrary constants and the initial condition is $u(x,0) = u_0(x)$. This equation is controlled at $x = 1$ using $u(1,t)$ (Dirichlet actuation) or $u_x(1,t)$ (Neumann actuation) as a control input. The open-loop system (1)–(2) (with $u(1,t) = 0$ or $u_x(1,t) = 0$) is unstable with arbitrarily many unstable eigenvalues (for large λ_0/ε). The methods for the boundary control of (1)–(2) include pole placement [19], LQR [15], and finite-dimensional backstepping [2]. In [6], two stabilizing controllers (backstepping and pole placement) were constructed in a closed form, but only for $\lambda_0/\varepsilon < 3\pi^2/4$, i.e., in case of one unstable eigenvalue. However, the *explicit* (closed-form) boundary stabilization result in the case of *arbitrary* ε, λ_0 is not available in the literature even for this benchmark constant coefficient case.

We start with the coordinate transformation

$$w\,(x,t) = u\,(x,t) - \int_0^x k\,(x,y)\,u\,(y,t)\;dy \tag{3}$$

that transforms system (1)–(2) into the system

$$w_t(x,t) = \varepsilon w_{xx}(x,t) - cw(x,t)\,, \quad x \in (0,1)\,, \tag{4}$$

$$w(0,t) = 0\,, \tag{5}$$

$$w(1,t) = 0 \quad \text{or} \quad w_x(1,t) = 0, \tag{6}$$

which is exponentially stable for $c \geq 0$. Once we find the transformation (3) (namely $k(x,y)$), the boundary condition (6) gives the controller in the form

$$u(1,t) = \int_0^1 k_1(y)u(y,t)\,dy \tag{7}$$

for the Dirichlet actuation, $k_1(y) = k(1,y)$, and

$$u_x(1,t) = k_1(1)u(1,t) + \int_0^1 k_2(y)u(y,t)\,dy \tag{8}$$

for the Neumann actuation, $k_2(y) = k_x(1,y)$.

We are now going to derive $k(x,y)$. From (1), (3), and (4) we get

$$\begin{aligned} w_t(x,t) &= u_t(x,t) - \int_0^x k(x,y)(\varepsilon u_{yy}(y,t) + \lambda_0 u)\,dy \\ &= \varepsilon u_{xx}(x,t) + \lambda_0 u(x,t) - \varepsilon k(x,x)u_x(x,t) + \varepsilon k(x,0)u_x(0,t) \\ &\quad + \varepsilon k_y(x,x)u(x,t) - \int_0^x (\varepsilon k_{yy}(x,y) + \lambda_0 k(x,y))u(y,t)\,dy\,, \quad (9) \\ w_{xx}(x,t) &= u_{xx}(x,t) - k(x,x)u_x(x,t) - k_x(x,x)u(x,t) \\ &\quad - \frac{\mathrm{d}}{\mathrm{d}x}k(x,x) - \int_0^x k_{xx}(x,y)u(y,t)\,dy\,. \quad (10) \end{aligned}$$

Combining (9) and (10) gives

$$\begin{aligned} 0 &= w_t(x,t) - \varepsilon w_{xx}(x,t) + cw(x,t) \\ &= \left(2\varepsilon \frac{\mathrm{d}}{\mathrm{d}x}k(x,x) + \lambda_0 + c\right)u(x,t) + \varepsilon k(x,0)u_x(0,t) \\ &\quad + \int_0^x \{\varepsilon k_{xx}(x,y) - \varepsilon k_{yy}(x,y) - (\lambda_0 + c)k(x,y)\}\,u(y,t)\,dy \qquad (11) \end{aligned}$$

We can see now that the kernel $k(x,y)$ must satisfy the following hyperbolic PDE:

$$k_{xx}(x,y) - k_{yy}(x,y) = \lambda k(x,y)\,, \qquad (x,y) \in \mathcal{T}, \qquad (12)$$

$$k(x,0) = 0\,, \qquad (13)$$

$$k(x,x) = -\frac{\lambda x}{2}\,. \qquad (14)$$

where $\lambda = (\lambda_0 + c)/\varepsilon$ and $\mathcal{T} = \{x, y : 0 < y < x < 1\}$. To find the solution of this PDE, following Liu [16] and Colton [8], we use the transformation to an integral equation and the method of successive approximations. Introducing the change of variables

$$\xi = x + y\,, \quad \eta = x - y\,, \qquad (15)$$

and setting

$$G(\xi,\eta) = k(x,y) = k\left(\frac{\xi+\eta}{2}, \frac{\xi-\eta}{2}\right)\,, \qquad (16)$$

we get the PDE

$$G_{\xi\eta} = \frac{\lambda}{4}G(\xi,\eta)\,, \qquad (17)$$

$$G(\xi,\xi) = 0\,, \qquad (18)$$

$$G(\xi,0) = -\frac{\lambda}{4}\xi\,. \qquad (19)$$

Integrating (17) with respect to η from 0 to η and using (19) we obtain

$$G_\xi(\xi,\eta) = -\frac{\lambda}{4} + \frac{\lambda}{4}\int_0^\eta G(\xi,s)\,ds\,. \tag{20}$$

Integrating (20) with respect to ξ from η to ξ and using (18) gives the integral equation for $G(\xi,\eta)$

$$G(\xi,\eta) = -\frac{\lambda}{4}(\xi-\eta) + \frac{\lambda}{4}\int_\eta^\xi\int_0^\eta G(\tau,s)\,ds\,d\tau\,. \tag{21}$$

Now set

$$G_0(\xi,\eta) = -\frac{\lambda}{4}(\xi-\eta)\,, \qquad G_{n+1}(\xi,\eta) = \frac{\lambda}{4}\int_\eta^\xi\int_0^\eta G_n(\tau,s)\,ds\,d\tau\,. \tag{22}$$

We can find the general term G_n in closed form (which can be proved by induction):

$$G_n(\xi,\eta) = -\frac{(\xi-\eta)\,\xi^n\eta^n}{(n!)^2(n+1)}\left(\frac{\lambda}{4}\right)^{n+1}\,. \tag{23}$$

By the method of successive approximations, the sum of all the terms G_n gives the solution to (21):

$$G(\xi,\eta) = \sum_{n=0}^{\infty} G_n(\xi,\eta) = -\frac{\lambda(\xi-\eta)}{2}\frac{I_1(\sqrt{\lambda\xi\eta})}{\sqrt{\lambda\xi\eta}}\,, \tag{24}$$

where I_i is a modified Bessel function of order i. Writing (24) in terms of x, y gives the following solution for $k(x,y)$:

$$k(x,y) = -\lambda y\frac{I_1\left(\sqrt{\lambda(x^2-y^2)}\right)}{\sqrt{\lambda(x^2-y^2)}}\,. \tag{25}$$

Therefore, for Dirichlet actuation,

$$k_1(y) = k(1,y) = -\lambda y\frac{I_1\left(\sqrt{\lambda(1-y^2)}\right)}{\sqrt{\lambda(1-y^2)}}\,, \tag{26}$$

and for Neumann actuation,

$$k_2(y) = k_x(1,y) = -\lambda y\frac{I_2\left(\sqrt{\lambda(1-y^2)}\right)}{1-y^2}\,. \tag{27}$$

In Figure 1 the kernel $k_1(y)$ is plotted for several values of λ. We see that the maximum of the absolute value of the kernel moves to the left as λ grows.

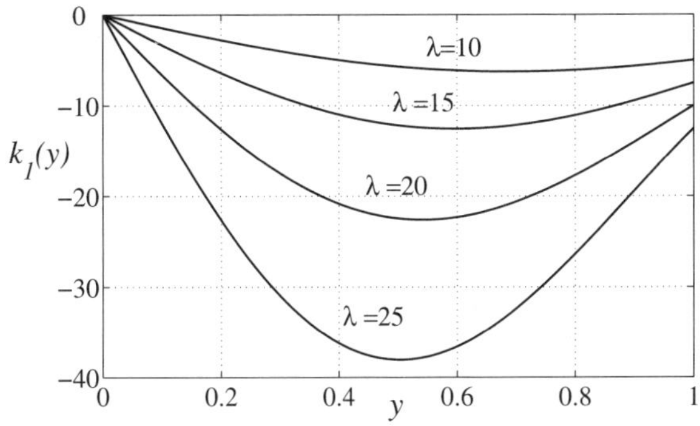

Fig. 1. Dependence of the gain kernel on the level of instability

We can actually calculate the area under the curves and estimate an amount of total gain effort required:

$$E = \int_0^1 |k_1(y)|\,dy = \lambda \int_0^1 y \frac{I_1\left(\sqrt{\lambda(1-y^2)}\right)}{\sqrt{\lambda(1-y^2)}}\,dy = \int_0^{\sqrt{\lambda}} I_1(z)\,dz = I_0\left(\sqrt{\lambda}\right) - 1\,. \tag{28}$$

Thus $E \sim \lambda$ for small λ and $E \sim e^{\sqrt{\lambda}}/(\sqrt{2\pi}\,\lambda^{1/4})$ for large λ.

For the case of a homogeneous Neumann boundary condition at $x = 0$ for the equation (1) it is easy to repeat all the steps we have done for the Dirichlet case and get the following closed form solution for the kernel:

$$k(x,y) = -\lambda x \frac{I_1(\sqrt{\lambda(x^2-y^2)})}{\sqrt{\lambda(x^2-y^2)}}\,. \tag{29}$$

Note that the leading factor here is x, versus y in (25). The maximum of the absolute value of the kernel is reached at $x = 0$. This makes sense because the control has to react the most aggressively to perturbations that are the farthest from it. The peak value is equal to

$$|k(1,0)| = \sqrt{\lambda}\, I_1(\sqrt{\lambda}) \sim e^{\sqrt{\lambda}}/\lambda^{1/4}$$

as $\lambda \to \infty$.

In order to prove stability we need to prove that the transformation (3) is invertible. Let us write the inverse transformation in the form

$$u(x,t) = w(x,t) + \int_0^x l(x,y)w(y,t)\,dy\,. \tag{30}$$

Substituting (30) into equations (4)–(6) and using (1)–(2) we obtain the following PDE governing $l(x,y)$:

$$l_{xx}(x,y) - l_{yy}(x,y) = -\lambda l(x,y)\,, \qquad (x,y) \in \mathcal{T}\,, \tag{31}$$

$$l(x,0) = 0\,, \tag{32}$$

$$l(x,x) = -\frac{\lambda x}{2}\,. \tag{33}$$

Noticing that in this case $l(x,y) = -k(x,y)$ when λ is replaced by $-\lambda$, we immediately obtain

$$l(x,y) = -\lambda y \frac{J_1(\sqrt{\lambda(x^2-y^2)})}{\sqrt{\lambda(x^2-y^2)}} \tag{34}$$

where J_1 is the usual (non-modified) Bessel function of the first order, which, incidentally, is a bounded function of $\lambda \geq 0$. The smoothness of the both $k(x,y)$ and $l(x,y)$ in x, y establishes the equivalence of norms of u and w in both L_2 and H_1. From the properties of the damped heat equation (4)–(6) exponential stability in both L_2 and H_1 follows.

Furthermore, the system (1)–(2), (7) is not only well posed but its solution is explicitly available. Let us show how it can be obtained. First we solve the damped heat equation (4)–(6):

$$w(x,t) = 2\sum_{n=1}^{\infty} e^{-(c+\pi^2 n^2)t} \sin(\pi n x) \int_0^1 w_0(\xi) \sin(\pi n \xi)\, d\xi. \tag{35}$$

The initial condition w_0 can be calculated explicitly from u_0 via (3), (25). Substituting the result into (30), (34), changing order of integration and calculating some integrals we obtain the explicit solution to closed loop system (1)–(2), (7):

$$u(x,t) = \sum_{n=1}^{\infty} e^{-(c+\pi^2 n^2)t} \phi_n(x) \int_0^1 \psi_n(\xi) u_0(\xi)\, d\xi\,, \tag{36}$$

where

$$\psi_n(x) = \sin(\pi n x) + \int_x^1 \lambda x \frac{I_1(\sqrt{\lambda(\xi^2-x^2)})}{\sqrt{\lambda(\xi^2-x^2)}} \sin(\pi n \xi)\, d\xi,$$

$$\phi_n(x) = \frac{2\pi n}{\sqrt{\lambda+\pi^2 n^2}} \sin(\sqrt{\lambda+\pi^2 n^2}\, x). \tag{37}$$

Several things can be noticed here. One can directly see from (36) that the backstepping controller has moved the eigenvalues from their open-loop (unstable) locations $\lambda_0 - \pi^2 n^2$ into locations of the damped heat equation $-(c+\pi^2 n^2)$. It is interesting to note that, although infinitely many eigenvalues cannot be arbitrarily assigned, our controller is able to assign all of them to the particular location $-(c+\pi^2 n^2)$. The eigenfunctions of the closed-loop system are assigned to $\phi_n(x)$. We can see from (37) that the controller has reduced the amplitude and increased the frequency of the open-loop eigenfunctions $2\sin(\pi n x)$.

We come to the following result:

Theorem 1. *For any initial condition $u_0(x) \in L_2(0,1)$ the system (1)–(2), (7) with the kernel $k_1(y)$ given by (26) has a unique classical solution $u(x,t) \in C^{2,1}((0,1)\times(0,\infty))$ given by (36)–(37) and is exponentially stable at the origin in the $L_2(0,1)$ and $H_1(0,1)$ norms. The same result holds in case of the Neumann type of actuation with $k_2(y)$ given by (27).*

For the rest of the paper we will not repeat the procedure of derivation of a PDE for the kernel and the arguments of previous paragraph proving stability and well posedness.

2.2 Heat Equation with Destabilizing Boundary Condition

We now consider a more complicated system

$$u_t(x,t) = \varepsilon u_{xx}(x,t) + \lambda_0 u(x,t)\,, \tag{38}$$
$$u_x(0,t) = qu(0,t)\,, \tag{39}$$
$$u(1,t) = \int_0^1 k(1,y)u(y)\,dy\,, \tag{40}$$

where the boundary condition on the uncontrolled end is mixed and can cause instability for $q<0$. We use the transformation (3) to map this system into the target system

$$w_t(x,t) = \varepsilon w_{xx}(x,t) - cw(x,t)\,, \tag{41}$$
$$w(0,t) = qw(0,t)\,, \tag{42}$$
$$w(1,t) = 0\,, \tag{43}$$

which is exponentially stable for $c > \max\{0, -\varepsilon q|q|\}$. It can be shown that the gain kernel should satisfy the following PDE:

$$k_{xx}(x,y) - k_{yy}(x,y) = \lambda k(x,y)\,, \tag{44}$$
$$k_y(x,0) = qk(x,0)\,, \tag{45}$$
$$k(x,x) = -\frac{\lambda x}{2}\,. \tag{46}$$

where $\lambda = (\lambda_0 + c)/\varepsilon$. We propose to search a solution in the following form

$$k(x,y) = -\lambda x \frac{I_1(\sqrt{\lambda(x^2-y^2)})}{\sqrt{\lambda(x^2-y^2)}} + \int_0^{x-y} I_0(\sqrt{\lambda(x+y)(x-y-\tau)})\sigma(\tau)\,d\tau. \tag{47}$$

Here the first term is a solution to (44)–(46) with $q=0$ which have been obtained in the previous subsection. The second term is just one of the solutions to (44), σ being an *arbitrary* function. We can see now that (47) is a solution to (44), (46) and we need only to choose $\sigma(\tau)$ so that (45) is satisfied. Substituting (47) into (45) we obtain the following integral equation for $\sigma(x)$:

$$\sigma(x) + \int_0^x \sigma(\tau)\left(\frac{\lambda}{2}\tau\frac{I_1(\sqrt{\lambda x(x-\tau)})}{\sqrt{\lambda x(x-\tau)}} + qI_0(\sqrt{\lambda x(x-\tau)})\right)d\tau = \sqrt{\lambda}qI_1(\sqrt{\lambda}x). \tag{48}$$

To solve this equation, we apply the Laplace transform with respect to x to both sides of (48) and get:

$$\sigma(s) + \int_0^\infty e^{-s\xi}\int_0^x \sigma(\tau)\left(\frac{\lambda}{2}\tau\frac{I_1(\sqrt{\lambda x(x-\tau)})}{\sqrt{\lambda x(x-\tau)}}\right.$$
$$\left. + qI_0(\sqrt{\lambda x(x-\tau)})\right) d\tau\, d\xi = q\frac{s-\sqrt{s^2-\lambda}}{\sqrt{s^2-\lambda}}. \tag{49}$$

After changing the order of integration, calculating the inner integral, and introducing $s' = (s+\sqrt{s^2-\lambda})/2$ we obtain:

$$\int_0^\infty \sigma(\tau)e^{-s'\tau}\, d\tau = q\frac{s-\sqrt{s^2-\lambda}}{q+\sqrt{s^2-\lambda}}, \tag{50}$$

Now using the relation $s = s' + \lambda/(4s')$ we get

$$\sigma(s') = \frac{2q\lambda}{(2s'+q)^2-(\lambda+q^2)}. \tag{51}$$

Taking inverse Laplace transform we get

$$\sigma(x) = \frac{q\lambda}{\sqrt{\lambda+q^2}}e^{-qx/2}\sinh\frac{\sqrt{\lambda+q^2}}{2}x. \tag{52}$$

So, the final solution to (44)–(46) is

$$k(x,y) = \frac{q\lambda}{\sqrt{\lambda+q^2}}\int_0^{x-y} e^{-q\tau/2}\sinh\left(\frac{\sqrt{\lambda+q^2}}{2}\tau\right)I_0(\sqrt{\lambda(x+y)(x-y-\tau)})\, d\tau$$
$$-\lambda x\frac{I_1(\sqrt{\lambda(x^2-y^2)})}{\sqrt{\lambda(x^2-y^2)}}. \tag{53}$$

2.3 'One-Peak' Family – Chemical Tubular Reactor

Consider the system

$$u_t(x,t) = \varepsilon u_{xx}(x,t) + \lambda_{\alpha\beta}(x)u(x,t), \quad x \in (0,1), \tag{54}$$
$$u(0,t) = 0, \tag{55}$$

where $\lambda_{\alpha\beta}(x)$ is given by

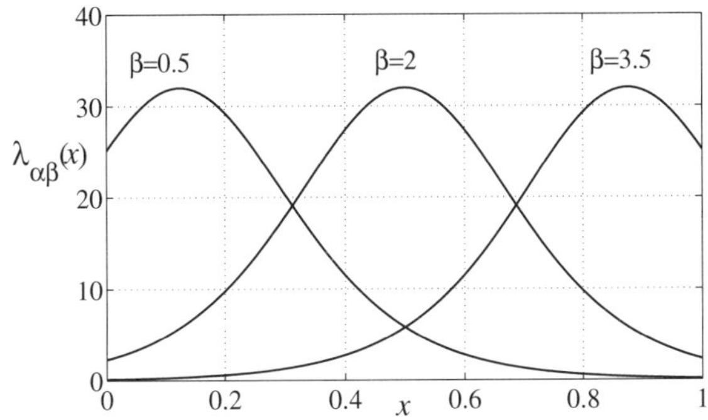

Fig. 2. 'One-peak' $\lambda_{\alpha\beta}(x)$ for $\alpha = 4$ and $\varepsilon = 1$.

$$\lambda_{\alpha\beta}(x) = \frac{2\varepsilon\alpha^2}{\cosh^2(\alpha x - \beta)} . \tag{56}$$

This function parametrizes a family of 'one-peak' functions. The maximum of $\lambda(x)$ is $2\varepsilon\alpha^2$ and is achieved at $x = \beta/\alpha$. The parameters α and β can be chosen to give the maximum an arbitrary value and location. Examples of $\lambda_{\alpha\beta}(x)$ for different values of α and β are shown in Figure 2. The 'sharpness' of the peak is not arbitrary and is given by $\lambda''_{\max} = -\lambda^2_{\max}/\varepsilon$. Despite the strange-looking expression for $\lambda_{\alpha\beta}(x)$, the system (54)–(55) can very well approximate the linearized model of a chemical tubular reactor (see [4] and references therein) which is open loop unstable. Our result on stabilization of (54)–(55) is given by the following theorem.

Theorem 2. *The controller*

$$u(1,t) = -\int_0^1 \alpha e^{\alpha \tanh \beta(1-y)} \left[\tanh\beta - \tanh(\beta - \alpha y)\right] u(y,t)\, dy \tag{57}$$

exponentially stabilizes the zero solution of the system (54)–(55).

Proof. Using the same procedure as in the previous subsection we map (54)–(55) into the system (4)–(6) using the transformation (3) Without repeating all the steps (9)–(11) we show the resulting PDE for the kernel $k(x,y)$:

$$k_{xx}(x,y) - k_{yy}(x,y) = \varepsilon^{-1}\lambda_{\alpha\beta}(y)k(x,y) , \tag{58}$$

$$k(x,0) = 0 , \tag{59}$$

$$k(x,x) = -\frac{1}{2\varepsilon}\int_0^x \lambda_{\alpha\beta}(\tau)\, d\tau . \tag{60}$$

Postulating $k(x,y) = X(x)Y(y)$, we have the following set of ODEs:

$$X''(x) = \mu X(x) , \tag{61}$$

$$Y''(y) = Y(y)(\mu + 2X(y)Y'(y) + 2X'(y)Y(y)) , \tag{62}$$

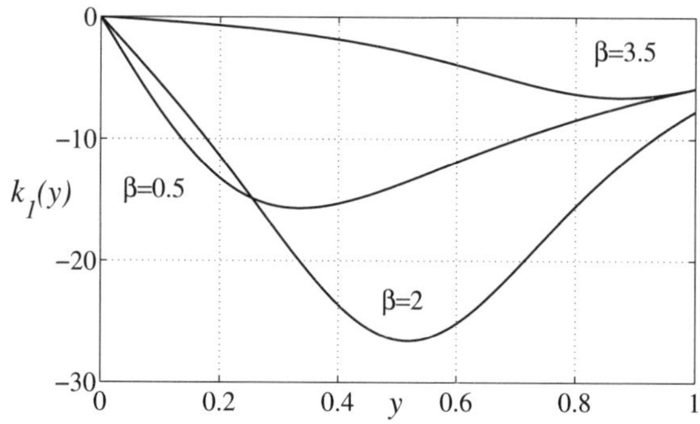

Fig. 3. Gain kernel $k_1(y)$ for (54)–(55) with $\alpha = 4$ and $\varepsilon = 1$.

with the additional conditions

$$Y(0) = 0, \quad (X(x)Y(x))' = -\lambda_{\alpha\beta}(x)/(2\varepsilon)\,, \tag{63}$$

where μ is an arbitrary parameter. Let us choose $X(x) = e^{\sqrt{\mu}x}$ and substitute it into (62). We get

$$Y''(y) - 2e^{\sqrt{\mu}y}Y'(y)Y(y) - 2\sqrt{\mu}\,e^{\sqrt{\mu}y}Y^2(y) - \mu\,Y(y) = 0\,. \tag{64}$$

Changing variables $Y = f(y)e^{-\sqrt{\mu}y}$ we arrive at the following ODE:

$$f''(y) - 2f'(y)f(y) - 2\sqrt{\mu}f'(y) = 0 \tag{65}$$

$$f(0) = 0\,, \tag{66}$$

$$f'(0) = \mu - \alpha^2\,. \tag{67}$$

with the additional condition

$$f'(x) = -\frac{\lambda_{\alpha\beta}(x)}{2\varepsilon}\,. \tag{68}$$

The solution to the problem (65)–(67) is

$$f(y) = -\alpha\left(\tanh(\alpha y - \beta) + \tanh\beta\right)\,, \tag{69}$$

where $\tanh\beta = \sqrt{\mu}/\alpha$. We can check that this solution satisfies (68) and gives the kernel

$$k(x,y) = -\alpha e^{\alpha\tanh\beta(x-y)}\left(\tanh\beta - \tanh(\beta - \alpha y)\right)\,. \tag{70}$$

Setting $x = 1$ in (70) concludes the proof. Q.E.D.

In Figure 3 the stabilizing kernels corresponding to $\lambda_{\alpha\beta}(x)$ from Figure 2 are shown. We can see that the controller effort depends very much on the location of the peak of $\lambda(x)$, which has an obvious explanation. When the peak is close to $x = 1$, the controller's influence is very high, when it is close to $x = 0$, the boundary condition helps to stabilize, so the worst case is the peak somewhere in the middle of the domain.

2.4 Solid Propellant Rocket Model

Consider the following system

$$u_t(x,t) = u_{xx}(x,t) + ge^{\gamma x}u(0,t), \quad x \in (0,1), \tag{71}$$
$$u_x(0,t) = 0. \tag{72}$$

Here g and γ are arbitrary constants. This equation represents a model of unstable burning in solid propellant rockets. (For more details see [5] and references therein.) This system is unstable (with $u(1) = 0$) for any $g > 2$, $\gamma \geq 0$.

The PDE for the gain kernel can be obtained using the same procedure as in Section 2.1 except that now the transformation (3) is used to convert the plant into system (4)–(6) with $w_x(0,t) = 0$ instead of $w(0,t) = 0$:

$$k_{xx}(x,y) - k_{yy}(x,y) = 0, \qquad (x,y) \in \mathcal{T}, \tag{73}$$
$$k_y(x,0) = ge^{\gamma x} - g\int_0^x k(x,y)e^{\gamma y}\,dy, \tag{74}$$
$$k(x,x) = 0. \tag{75}$$

Note the non-local character of the boundary condition (74). The structure of (73)–(75) suggests to search for the solution in the following form:

$$k(x,y) = C_1 e^{\gamma_1(x-y)} + C_2 e^{\gamma_2(x-y)}. \tag{76}$$

Substituting (76) into (73)–(75) we determine the constants C_1, C_2, γ_1, γ_2 and thus obtain the solution

$$k(x,y) = -\frac{g}{g_0}e^{\frac{\gamma}{2}(x-y)}\sinh\left(g_0(x-y)\right), \quad g_0 = \sqrt{g + \frac{\gamma^2}{4}}. \tag{77}$$

We arrive at the following result:

Theorem 3. *The controller*

$$u(1,t) = -\int_0^1 \frac{ge^{\frac{\gamma}{2}(1-y)}}{g_0}\sinh\left(g_0(1-y)\right)u(y,t)\,dy \tag{78}$$

exponentially stabilizes the zero solution of the system (71)–(72).

2.5 Combining Previous Results

The solutions presented in subsections 2.1– 2.4 can be combined to obtain the explicit result for the more complex systems. Consider the system

$$u_t(x,t) = \varepsilon u_{xx}(x,t) + (\lambda_{\alpha\beta}(x) + \lambda_0)\, u(x,t)\,, \tag{79}$$

$$u(0,t) = 0\,, \tag{80}$$

$$u(1,t) = \int_0^1 k_c(y)u(y)\,dy\,. \tag{81}$$

where $k_c(y)$ is sought to stabilize this system. Denote by $k^{\alpha\beta}(x,y)$ and $k^{\lambda}(x,y)$ the control gains for the equations (54)–(55) and (1)–(2), respectively. The transformation (3) with $k(x,y) = k^{\alpha\beta}(x,y)$ maps (79)–(81) into the system

$$w_t(x,t) = \varepsilon w_{xx}(x,t) + \lambda_0 w(x,t)\,, \tag{82}$$

$$w(0,t) = 0\,, \tag{83}$$

$$w(1,t) = \int_0^1 k_1^{\lambda}(y)w(y)\,dy\,. \tag{84}$$

which is stabilized by $k_1^{\lambda}(y)$. Thus we can get the expression for $k_c(y)$ in quadratures in terms of $k^{\alpha\beta}(x,y)$ and $k^{\lambda}(x,y)$:

$$k_c(y) = k^{\lambda}(1,y) + k^{\alpha\beta}(1,y) - \int_y^1 k^{\lambda}(1,\xi)k^{\alpha\beta}(\xi,y)\,d\xi \tag{85}$$

For example for $\beta = 0$ one can get the closed-form solution

$$k_c(y) = -\lambda y \frac{I_1\left(\sqrt{\lambda(1-y^2)}\right)}{\sqrt{\lambda(1-y^2)}} - \alpha \tanh(\alpha y) I_0\left(\sqrt{\lambda(1-y^2)}\right) \tag{86}$$

In the same fashion one can obtain the explicit stabilizing controllers for even more complicated plants. Here is the most general plant for which we have been able to do it:

$$u_t(x,t) = \varepsilon u_{xx}(x,t) + b u_x(x,t) + \lambda_0 u(x,t) + g e^{\gamma x} u(0,t)\,, \tag{87}$$

$$u_x(0,t) = q u(0,t)\,. \tag{88}$$

This is a system with 6 parameters, each of them can contribute to instability. It is remarkable that it can be stabilized by closed form control law.

3 Output-Feedback Designs

The stabilizing controllers developed thus far require complete measurements from the interior of the domain which are usually unavailable. So we look for the observers that estimate $u(x,t)$ inside the domain. We assume that sensing is available only at the boundary and thus consider two cases: anti-collocated, when the sensor is placed on the opposite boundary to the actuator (so $u(0,t)$ is measured, $u(1,t)$ is controlled), and collocated, when the sensor and the actuator are set at the same boundary (so $u(1,t)$ is measured, $u_x(1,t)$ is controlled).

3.1 Unstable Heat Equation

Consider the following equation

$$u_t(x,t) = \varepsilon u_{xx}(x,t) + \lambda_0 u(x,t)\,, \tag{89}$$
$$u_x(0,t) = 0\,, \tag{90}$$

with only $u(0)$ measured. We propose the following Luenberger-type observer motivated by a finite dimensional backstepping type observer of Krener and Wang [11]:

$$\hat{u}_t(x,t) = \varepsilon \hat{u}_{xx}(x,t) + \lambda_0 \hat{u}(x,t) + p_1(x)[u(0,t) - \hat{u}(0,t)]\,, \tag{91}$$
$$\hat{u}_x(0,t) = p_{10}[u(0,t) - \hat{u}(0,t)]\,, \tag{92}$$
$$\hat{u}(1,t) = \int_0^1 k_1(y)\hat{u}(y,t)\,dy\,, \tag{93}$$

and the controller

$$u(1,t) = \int_0^1 k_1(y)\hat{u}(y,t)\,dy\,. \tag{94}$$

Here $p_1(x)$ and p_{10} are output injection functions (p_{10} is a constant) to be designed. Note, that we introduce output injection not only in the equation (91) but also at the boundary where measurement is available. The observer error $\tilde{u}(x,t) = u(x,t) - \hat{u}(x,t)$ satisfies the following PDE:

$$\tilde{u}_t\,(x,t) = \varepsilon \tilde{u}_{xx}\,(x,t) + \lambda_0 \tilde{u}\,(x,t) - p_1(x)\tilde{u}(0,t)\,, \tag{95}$$
$$\tilde{u}_x\,(0,t) = -p_{10}\tilde{u}(0,t)\,, \tag{96}$$
$$\tilde{u}\,(1,t) = 0. \tag{97}$$

Observer gains $p_1(x)$ and p_{10} should now be chosen to stabilize the system (95)–(97). For linear finite dimensional systems the problem of finding the observer gains is dual to the problem of finding the control gains. This motivates us to try to solve the problem of stabilization of (95)–(97) by the same integral transformation approach as the (state feedback) boundary control problem. We look for a backstepping-like transformation

$$\tilde{u}\,(x,t) = \tilde{w}\,(x,t) - \int_0^x p\,(x,y)\,\tilde{w}\,(y,t)\;dy \tag{98}$$

that transforms system (95)–(97) into the exponentially stable system

$$\tilde{w}_t(x,t) = \varepsilon \tilde{w}_{xx}(x,t) - c\tilde{w}(x,t)\,, \quad x \in (0,1)\,, \tag{99}$$
$$\tilde{w}_x(0,t) = 0\,, \tag{100}$$
$$\tilde{w}(1,t) = 0. \tag{101}$$

By substituting (98) into (95)–(97) and using (99)–(101) it can be shown that the kernel $p(x,y)$ must satisfy the following hyperbolic PDE:

$$p_{yy}(x,y) - p_{xx}(x,y) = \lambda p(x,y)\,, \tag{102}$$

$$\frac{d}{dx}p(x,x) = \frac{\lambda}{2}, \tag{103}$$

$$p(1,y) = 0\,. \tag{104}$$

In addition, the following conditions must be satisfied:

$$p_1(x) = \varepsilon p_y(x,0), \qquad p_{10} = p(0,0). \tag{105}$$

Let us make a change of variables

$$\breve{x} = 1-y, \quad \breve{y} = 1-x, \quad \breve{p}(\breve{x},\breve{y}) = p(x,y). \tag{106}$$

It can be verified that in these new variables the problem (102)–(104) becomes exactly the same as (12)–(14) for $k(x,y)$ and we get

$$\breve{p}(\breve{x},\breve{y}) = -\lambda\breve{y}\frac{I_1(\sqrt{\lambda(\breve{x}^2-\breve{y}^2)})}{\sqrt{\lambda(\breve{x}^2-\breve{y}^2)}}\,. \tag{107}$$

Using (105) we obtain the following result:

Theorem 4. *The controller (94) with the observer (91)–(93) where $k_1(x)$ and $p_1(x)$ are given by*

$$k_1(y) = -\lambda\frac{I_1(\sqrt{\lambda(1-y^2)})}{\sqrt{\lambda(1-y^2)}}, \quad p_1(x) = \frac{\lambda(1-x)}{x(2-x)}I_2(\sqrt{\lambda x(2-x)})\,, \tag{108}$$

and $p_{10} = -\lambda/2$, exponentially stabilizes the zero solution of the system (89)–(90).

The above result can be easily extended for the Neumann type of actuation as well.

Since both the controller and the observer are explicit in our design we can derive again (following the same procedure as in section 2.1) the closed loop solution of (89)–(94). The presence of the second PDE (the observer) makes it more complicated:

$$u(x,t) = \sum_{n=0}^{\infty} e^{-(c+\mu_n^2)t}\phi_n(x)\left\{\int_0^1 \psi_n(\xi)u_0(\xi)\,d\xi \right.$$
$$\left. -(-\mu_n)^n\left(C_n t + \sum_{m\neq n}^{\infty} C_m \frac{1-e^{(\mu_n^2-\mu_m^2)t}}{\mu_n^2-\mu_m^2}\right)\right\}, \tag{109}$$

where $\mu_n = \pi(n+1/2)$,

$$C_n = \int_0^1 \lambda\frac{I_1(\sqrt{\lambda(1-\xi^2)})}{\sqrt{\lambda(1-\xi^2)}}\psi_n'(1-\xi)\,d\xi \int_0^1 \frac{\phi_n'(1-\xi)}{\lambda+\mu_n^2}(u_0(\xi)-\hat{u}_0(\xi))\,d\xi, \tag{110}$$

$$\psi_n(x) = \cos(\mu_n x) + \int_0^x \lambda \xi \frac{I_1(\sqrt{\lambda(x^2-\xi^2)})}{\sqrt{\lambda(x^2-\xi^2)}} \cos(\mu_n \xi)\, d\xi,$$
$$\phi_n(x) = 2\cos(\sqrt{\lambda+\mu_n^2}x). \tag{111}$$

3.2 Explicit Solution for a 'One-Peak' Family of $\lambda(x)$

We show now how the observer for the collocated case can be designed. Of course, for the problem to make sense we assume that the actuation is Neumann type. Consider the problem (54)–(56) with only $u(1,t)$ measured. We propose the observer

$$\hat{u}_t(x,t) = \varepsilon \hat{u}_{xx}(x,t) + \lambda_{\alpha\beta}(x)\,\hat{u}(x,t) + p_1(x)[u(1,t) - \hat{u}(1,t)], \tag{112}$$
$$\hat{u}(0,t) = 0, \tag{113}$$
$$\hat{u}_x(1,t) = -p_{10}[u(1,t) - \hat{u}(1,t)] + k_1(1)\hat{u}(1,t) + \int_0^1 k_2(y)\hat{u}(y,t)\, dy, \tag{114}$$

with the controller

$$u_x(1,t) = k_1(1)\hat{u}(1,t) + \int_0^1 k_2(y)\hat{u}(y,t)\, dy. \tag{115}$$

Control gains k_1 and k_2 have been determined in Section 2.3.

The observer error $\tilde{u}(x)$ satisfies the equation

$$\tilde{u}_t(x,t) = \varepsilon \tilde{u}_{xx}(x,t) + \lambda_{\alpha\beta}(x)\tilde{u}(x,t) - p_1(x)\tilde{u}(1,t) \tag{116}$$
$$\tilde{u}(0,t) = 0, \tag{117}$$
$$\tilde{u}_x(1,t) = p_{10}\tilde{u}(1,t). \tag{118}$$

We are looking for the transformation

$$\tilde{u}(x,t) = \tilde{w}(x,t) - \int_x^1 p(x,y)\,\tilde{w}(y,t)\, dy \tag{119}$$

that transforms (116)–(118) into the exponentially stable target system

$$\tilde{w}_t(x,t) = \varepsilon \tilde{w}_{xx}(x,t), \quad x \in (0,1), \tag{120}$$
$$\tilde{w}(0,t) = 0, \tag{121}$$
$$\tilde{w}_x(1,t) = 0. \tag{122}$$

Note, that the transformation (119) is in upper-triangular form. By substituting (119) into (116)–(118) and using (120)–(122) it can be shown that the kernel $p(x,y)$ must satisfy the following hyperbolic PDE:

$$\varepsilon p_{yy}(x,y) - \varepsilon p_{xx}(x,y) = \lambda_{\alpha\beta}(x)p(x,y), \tag{123}$$
$$p_x(0,y) = qp(0,y), \tag{124}$$
$$p(x,x) = -\frac{1}{2\varepsilon}\int_0^x (\lambda(\xi) + c)\, d\xi. \tag{125}$$

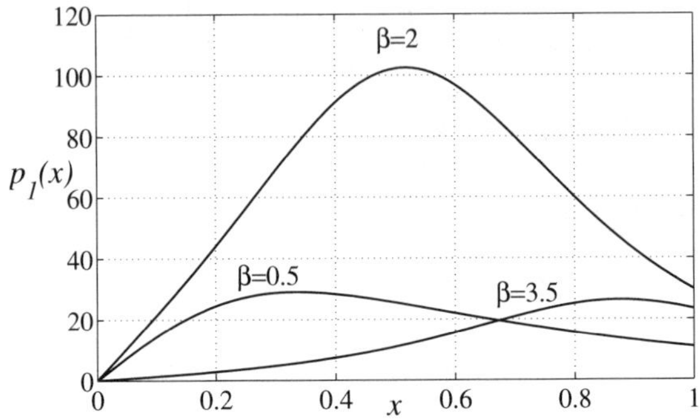

Fig. 4. Observer gain for $\alpha = 4$.

In addition, the following conditions must be satisfied:

$$p_1(x) = -\varepsilon p_y(x,1), \qquad p_{10} = p(1,1). \tag{126}$$

Once the solution $p(x,y)$ to the problem (123)–(125) is found, the observer gains can be obtained from (126). It can be verified that the solution of this PDE is $p(x,y) = k(x,y)$ where $k(x,y)$ is given by (70). We immediately get $p_1(x) = \varepsilon k_2(x)$, $p_{10} = k_1(1)$ and the following result:

Theorem 5. *The controller (115) with the observer (112)–(114) where $k_2(y)$ and $p_1(x)$ are given by*

$$k_2(y) = p_1(x) = \varepsilon\alpha^2 \tanh(\beta) e^{(1-x)\alpha\tanh\beta} \left(\tanh\beta - \tanh(\beta - \alpha x)\right), \tag{127}$$

stabilizes the zero solution of the system (54)–(55).

In Figure 4 the observer gains corresponding to $\lambda_{\alpha\beta}(x)$ from Figure 2 are shown.

3.3 Combining Results

The solutions from Sections 3.1 and 3.2 can be combined to obtain an explicit solution to (79)–(81). Denote by $p^{\alpha\beta}(x,y)$ and $p^{\lambda}(x,y)$ the observer gains for (54)–(55) and (1)–(2), respectively. In the same fashion as it was done in Section 2.5 we can obtain the observer gain for the equation (79)-(81):

$$p_1(x) = p_1^{\lambda}(x) + p_1^{\alpha\beta}(x) + \varepsilon p_{10}^{\lambda} p^{\alpha\beta}(x,1) - \int_x^1 p^{\alpha\beta}(x,\xi) p_1^{\lambda}(\xi)\, d\xi\,, \tag{128}$$

$$p_{10} = p_{10}^{\lambda} + p_{10}^{\alpha\beta}\,. \tag{129}$$

For example for $\beta = 0$ one can get the closed-form solution

$$p_1(x) = \frac{\varepsilon\lambda}{1-y^2} I_2(\sqrt{\lambda(1-x^2)}) + \varepsilon\lambda\alpha\tanh(\alpha x) \frac{I_1(\sqrt{\lambda(1-x^2)})}{\sqrt{\lambda(1-x^2)}}\,. \tag{130}$$

4 Explicit Compensator Design (Frequency Domain)

The solutions obtained in previous sections can be used to get the explicit compensator transfer functions (treating $u(0,t)$ or $u(1,t)$ as an input and $u(1,t)$ or $u_x(1,t)$ as an output). We will illustrate this point with the system from Section 2.4 with $\gamma = 0$ and $u(0,t)$ measured:

$$u_t(x,t) = u_{xx}(x,t) + gu(0,t)\,, \tag{131}$$
$$u_x(0,t) = 0\,. \tag{132}$$

Consider the following observer:

$$\hat{u}_t(x,t) = \hat{u}_{xx}(x,t) + gu(0,t)\,, \tag{133}$$
$$\hat{u}_x(0,t) = 0\,, \tag{134}$$
$$\hat{u}(1,t) = -\sqrt{g}\int_0^1 \sinh(\sqrt{g}(1-y))\hat{u}(y,t)\,dy\,. \tag{135}$$

The controller (and the output for the observer) is (see (78)):

$$u(1,t) = -\sqrt{g}\int_0^1 \sinh(\sqrt{g}(1-y))\hat{u}(y,t)\,dy\,. \tag{136}$$

We want to find a transfer function from the input $u(0,t)$ to the output $u(1,t)$, i.e., $u(1,s) = -C(s)u(0,s)$. Taking the Laplace transform of (133)–(135), setting the initial condition to zero, $\hat{u}(x,0) = 0$, we have (for simplicity of notation we denote by $\hat{u}(x,s)$ and $u(0,s)$ the Laplace transforms of $\hat{u}(x,t)$ and $u(0,t)$, respectively):

$$s\hat{u}(x,s) = \hat{u}_{xx}(x,s) + gu(0,s)\,, \tag{137}$$
$$\hat{u}_x(0,s) = 0\,, \tag{138}$$
$$\hat{u}(1,s) = -\sqrt{g}\int_0^1 \sinh(\sqrt{g}(1-y))\hat{u}(y,s)\,dy\,. \tag{139}$$

The equation (137) with boundary conditions (138)–(139) is an ODE with respect to x (we regard s as a parameter). The solution of (137) satisfying (138) is:

$$\hat{u}(x,s) = \hat{u}(0,s)\cosh(\sqrt{s}x) + \frac{g}{s}(1-\cosh(\sqrt{s}x))u(0,s) \tag{140}$$

Using boundary condition (139) we obtain $\hat{u}(0,s)$:

$$\hat{u}(0,s) = \frac{\cosh(\sqrt{s}) - \cosh(\sqrt{g})}{s\cosh(\sqrt{s}) - g\cosh(\sqrt{g})}\,gu(0,s) \tag{141}$$

Substituting now (141) into (140) with $x = 1$ we obtain the following result:

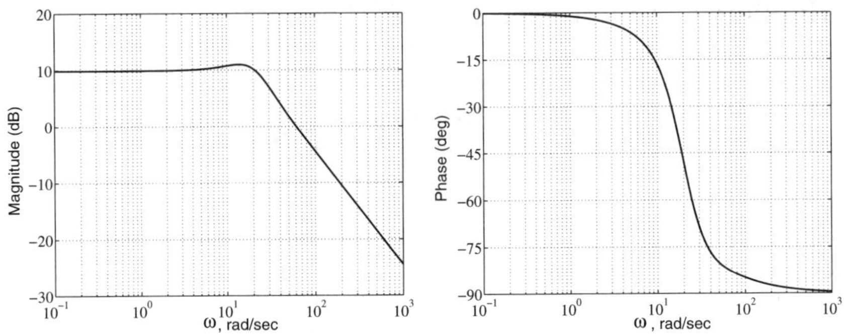

Fig. 5. Bode plot of $C(j\omega)$ for $g = 8$.

Theorem 6. *The transfer function of the system (133)–(136) with $u(0,t)$ as an input and $u(1,t)$ as an output is*

$$C(s) = \frac{g}{s}\left(-1 + \frac{(s-g)\cosh(\sqrt{s})\cosh(\sqrt{g})}{s\cosh(\sqrt{s}) - g\cosh(\sqrt{g})}\right) \tag{142}$$

The validation of application of the above procedure for linear parabolic PDEs (which proves that $C(s)$ is indeed a transfer function) can be found in [9, Chapter 4]. Note that $s = 0$ is not the pole:

$$C(0) = \frac{g}{2} + \frac{1}{\cosh(\sqrt{g})} - 1 \tag{143}$$

The transfer function (142) has infinitely many poles, all of them are real and negative. The Bode plots of $C(s)$ for $g = 8$ are presented in Figure 5. It is evident from the Bode plots that $C(s)$ can be approximated by a second order, relative degree one transfer function. For example, a rough estimate would be

$$C(s) \approx 60\frac{s+17}{s^2+25s+320}\,. \tag{144}$$

The relative degree one nature of the compensator is the result of employing a full order (rather than a reduced order) observer.

References

1. Amann H (1989) Feedback Stabilization of Linear and Semilinear Parabolic Systems. In: Clément P, Invernizzi S, Mitidieri E, Vrabie I (eds) Semigroup Theory and Applications Lecture Notes in Pure and Applied Mathematics 116. Marcel Dekker New York
2. Balogh A, Krstic M (2002) Infinite dimensional backstepping-style feedback transformations for a heat equation with an arbitrary level of instability, European Journal of Control 8:165-175

3. Bensoussan A, Da Prato G, Delfour M, Mitter S (1993) Representation and control of infinite–dimensional systems Vol II. Birkhauser Boston
4. Boskovic D, Krstic M (2002) Backstepping control of chemical tubular reactors, Computers and Chemical Engineering 26:1077-1085
5. Boskovic D, Krstic M (2003) Stabilization of a solid propellant rocket instability by state feedback, Int J of Robust and Nonlinear Control 13:483-95
6. Boskovic D, Krstic M, Liu W (2001) Boundary control of an unstable heat equation via measurement of domain-averaged temperature, IEEE Transactions on Automatic Control 46:2022-2028
7. Christofides P (2001) Nonlinear and Robust Control of Partial Differential Equation Systems Methods and Applications to Transport-Reaction Processes. Birkhäuser Boston
8. Colton D (1977) The solution of initial-boundary value problems for parabolic equations by the method of integral operators, Journal of Differential Equations 26:181-190
9. Curtain R, Zwart H (1995) An Introduction to Infinite Dimensional Linear Systems Theory. Springer-Verlag New York
10. Fattorini H (1968) Boundary control systems, SIAM J Control Optim 6:349–385
11. Krener A, Kang W (2003) Locally convergent nonlinear observers, SIAM J Control Optim 42:155-177
12. Krstic M, Deng H (1998) Stabilization of Nonlinear Uncertain Systems. Springer-Verlag New York
13. Krstic M, Kanellakopoulos I, Kokotovic P (1995) Nonlinear and Adaptive Control Design. Wiley New York
14. Lasiecka I, Triggiani R (1983) Stabilization and structural assignment of Dirichlet boundary feedback parabolic equations, SIAM J Control Optim 21:766–803
15. Lasiecka I, Triggiani R (2000) Control Theory for Partial Differential Equations Continuous and Approximation Theories. Cambridge Univ Press Cambridge UK
16. Liu W (2003) Boundary feedback stabilization of an unstable heat equation, SIAM J Control Optim 42:1033-1043
17. Nambu T (1984) On the stabilization of diffusion equations: boundary observation and feedback, Journal of Differential Equations 52:204–233
18. Russel D (1978) Controllability and stabilizability theory for linear partial differential equations recent progress and open questions, SIAM Rev 20:639-739
19. Triggiani R (1980) Boundary feedback stabilization of parabolic equations, Applied Mathematics and Optimization 6:201-220

A Continuous Control Mechanism for Uncertain Nonlinear Systems

Bin Xian[1], Marcio S. de Queiroz[2], and Darren M. Dawson[1]

[1] Dept. Elect. and Comp. Engg., Clemson University, Clemson, SC 29634-0915. [xbin, ddawson]@ces.clemson.edu

[2] Dept. Mech. Engg., Louisiana State University, Baton Rouge, LA 70803-6413. dequeiroz@me.lsu.edu

1 Introduction

The control of uncertain nonlinear systems is a topic that continues to challenge control theoreticians. This topic is also of practical importance since many real-world systems exhibit nonlinear dynamic behavior. To make matters even more difficult, in many practical cases, the mathematical model is poorly known or uncertain. To cope with the uncertainty issue, an abundance of design tools have emerged during the last two decades that facilitate the systematic construction of controllers for various classes of uncertain nonlinear systems. As one might expect, the choice of a specific control design method is strongly influenced by the type of uncertainty associated with the system model. To illustrate this point, consider the following first-order, scalar system:

$$\dot{w} = -w + \varphi(t) - \hat{\varphi}(t) \tag{1}$$

where $\hat{\varphi}(t)$ represents a control term injected to compensate for the unknown additive disturbance $\varphi(t)$. The main issue here is: *How does one design $\hat{\varphi}(t)$ such that $w(t)$ is regulated to zero?* If $\varphi(t)$ can be linearly parameterized (i.e., $\varphi(t) = Y(t)\theta$ where the nonlinearity $Y(t)$ is known, and θ is a constant unknown parameter), adaptive control [10, 17] is often considered to be the method of choice. If $\varphi(t)$ can be upper bounded by a norm-based inequality, sliding mode control (SMC) [20, 21] (i.e., discontinuous control strategies) can be applied. If $\varphi(t)$ is periodic and the period is known, learning control schemes can be designed [14]. The approach of [14] is also applicable to the case where $\varphi(t)$ can be represented as an integral of a predefined kernel function multiplied by a unknown influence function. While this brief discussion is not intended to be a comprehensive review of all the various methods of control, we also note that there has been much progress in the areas of adaptive control for time-varying systems [5], robust control [15] (i.e., the continuous counterpart of SMC), control of exosystems [19] (i.e., $\varphi(t)$ is the solution of a

M. de Queiroz et al. (Eds.): Optimal Control, Stabilization and Nonsmooth Analysis, LNCIS 301, pp. 251–264, 2004.

linear time-invariant system with unknown coefficients), adaptive control for classes of nonlinearly parameterized systems [1, 4, 13], and neural network-based control [12].

In addition to the problem of modeling uncertainty, the obstacles associated with dealing with high-order, multivariable nonlinear systems represent additional control design challenges. Since this work addresses the control of a class of uncertain, multi-input/multi-output (MIMO) nonlinear systems, a brief review of related work is also warranted. For example, recently in [22], a continuous, robust adaptive controller was formulated for a class of MIMO nonlinear systems that guarantees the semi-global convergence of the tracking error to a small residual set. In [9], a general procedure was presented for designing robust switching adaptive controllers for a broad class of multi-input nonlinear systems which includes feedback linearizable systems, parametric-pure feedback systems, and systems with a linear-in-the-parameters control Lyapunov function. A more restrictive class of multi-input systems than [9] was considered in [23] in the design of an adaptive robust controller. Adaptive controllers were proposed for a class of feedback linearizable systems in [3, 6, 18].

In this work, we present a full-state feedback tracking controller for a class of uncertain MIMO nonlinear systems. Specifically, we examine the class of high-order MIMO systems

$$\frac{d^n x}{dt^n} = H(x, \dot{x}, ..., \frac{d^{n-1}x}{dt^{n-1}}, \phi(t), t) + G(x, \dot{x}, ..., \frac{d^{n-1}x}{dt^{n-1}}, \phi(t), t)u. \tag{2}$$

where $x(t)$ is the output vector, $u(t)$ is the control input vector, $\phi(t)$ denotes an unknown, time-varying parameter or disturbance vector, $H(\cdot)$ is an uncertain nonlinear vector function, and $G(\cdot)$ is an uncertain nonlinear matrix function. The proposed controller produces semi-global asymptotic tracking under the assumption that $H(\cdot)$ and $G(\cdot)$ are second-order differentiable, and $G(\cdot)$ is a positive-definite symmetric matrix. To achieve this result, we utilize a new *continuous* control mechanism which is inspired by a simple example provided in [16].[3] The control approach compensates for the uncertainty associated with $H(\cdot)$ and $G(\cdot)$ with limited assumptions with regard to structure of the nonlinearities.

To briefly explain the control mechanism, we return to the system given by (1). If the proposed control approach is applied to (1), the continuous mechanism that compensates for $\varphi(t)$ takes the form

$$\hat{\varphi}(t) = \int_0^t (\delta_o w(\tau) + \delta_1 \mathrm{sgn}(w(\tau)))\, d\tau \tag{3}$$

where δ_o, δ_1 are positive control gains, and sgn$(\cdot)$ is the standard signum function. For the system given by (1), the control in (3) can be shown to

[3] In [16], this control mechanism, which was referred to as dynamic robust control, was shown only for the simple system of (1).

achieve global asymptotic regulation of $w(t)$ provided $d^i\varphi(t)/dt^i$, $i = 0, 1, 2$ are bounded, and the control gains δ_o, δ_1 are selected large enough relative to the bounds associated with $d^i\varphi(t)/dt^i$, $i = 0, 1, 2$. The proposed solution when applied to (1) also leads to the interesting fact that the control law identifies the unknown disturbance in the sense that $\varphi(t) - \hat{\varphi}(t) \to 0$ as $t \to \infty$. A careful examination of (3) indicates that the control mechanism is reminiscent of the so-called second-order SMC technique (see [2, 11] and the references therein). This technique, which is used to avoid the chattering phenomenon in standard SMC, consists of increasing the order of the original system such that the derivative of the control input appears. This signal is then designed to be discontinuous, which results in the actual control input being continuous. With this said, it is important that we highlight the contributions of this work vis-à-vis the second-order SMC literature. First, unlike previous second-order SMC results (e.g., [2]), we provide a constructive Lyapunov-based design and analysis approach to the problem. Second, the proposed solution is not limited to first-order, multi-input systems (i.e., when $n = 1$ in (2)) as in [2]. Finally, we would like to note that we are unaware of any continuous control strategy that achieves asymptotic tracking for the examined class of systems under the modeling assumptions used in this work.

2 Problem Statement

We consider a class of high-order MIMO nonlinear systems having the general form

$$M(x, \dot{x}, ..., x^{(n-1)})x^{(n)} + f(x, \dot{x}, ..., x^{(n-1)}) = u \tag{4}$$

where $(\cdot)^{(i)}(t)$ denotes the ith derivative with respect to time, $x^{(i)}(t) \in \mathbb{R}^m$, $i = 0, ..., n-1$ are the system states, $u(t) \in \mathbb{R}^m$ is the control input, and $M(\cdot) \in \mathbb{R}^{m \times m}$ and $f(\cdot) \in \mathbb{R}^m$ are uncertain nonlinear $\mathcal{C}^2$ functions.[4] We assume $M(\cdot)$ is symmetric and positive-definite, and bounded by

$$\underline{m}\,\|\xi\|^2 \leq \xi^T M(\cdot)\xi \leq \overline{m}(\cdot)\,\|\xi\|^2 \qquad \forall \xi \in \mathbb{R}^m \tag{5}$$

where $\underline{m}$ denotes a positive constant, and $\overline{m}(\cdot)$ denotes a positive nondecreasing function.

Let $x_d(t) \in \mathbb{R}^m$ be a $\mathcal{C}^{n+2}$ reference trajectory such that

$$x_d^{(i)}(t) \in \mathcal{L}_\infty \quad \text{for} \quad i = 0, 1, ..., n+2, \tag{6}$$

and let the tracking error $e_1(t) \in \mathbb{R}^m$ be defined as follows:

$$e_1 := x_d - x. \tag{7}$$

[4] The symbol "·" is used from now on to indicate the function's dependency on $x, \dot{x}, ..., x^{(n-1)}$.

Our control objective is to obtain asymptotic tracking with a continuous control law with full-state feedback (i.e., $x^{(i)}$, $i = 0, ..., n-1$ are assumed measurable) and only using (6) and norm-based, inequality bounds on the functions

$$M^{(i)}(x_d, \dot{x}_d, ..., x_d^{(n-1)}), \quad f^{(i)}(x_d, \dot{x}_d, ..., x_d^{(n-1)}),$$

$i = 0, 1, 2$.

Remark 1. For simplicity of presentation, we have assumed $M(\cdot)$ and $f(\cdot)$ do not depend explicitly on time or on unknown time-varying parameters. However, it should be emphasized that the proposed control approach can compensate for these phenomena provided the time-varying effects satisfy second-order differentiability conditions. That is, the functions $M(\cdot)$ and $f(\cdot)$ could be easily replaced by $M(\cdot, \theta(t), t)$ and $f(\cdot, \theta(t), t)$ where $\theta(t)$ denotes unknown, time-varying parameter vectors and other time-varying disturbances that may appear nonlinearly in the model; hence, the system given by (4) represents the same class of systems introduced in (2).

3 Control Law

Before presenting the control law we introduce the following auxiliary error signals $e_i(t) \in \mathbb{R}^m$, $i = 2, 3, ..., n$:

$$\begin{aligned} e_2 &:= \dot{e}_1 + e_1 \\ e_3 &:= \dot{e}_2 + e_2 + e_1 \\ e_4 &:= \dot{e}_3 + e_3 + e_2 \\ &\vdots \\ e_n &:= \dot{e}_{n-1} + e_{n-1} + e_{n-2} \end{aligned} \tag{8}$$

where $e_1(t)$ was defined in (7). An expression can be derived for $e_i(t)$, $i = 3, 4, ..., n$ in terms of $e_1(t)$ and its derivatives as follows (see Appendix B):

$$e_i = \sum_{j=0}^{i-1} a_{ij} e_1^{(j)} \tag{9}$$

where the constant coefficients a_{ij} are defined by

$$\begin{aligned} &a_{i0} = B(i) = \tfrac{1}{\sqrt{5}} \left[\left(\tfrac{1+\sqrt{5}}{2} \right)^i - \left(\tfrac{1-\sqrt{5}}{2} \right)^i \right], \quad i = 2, 3, ..., n \\ &a_{ij} = \sum_{k=1}^{i-1} B(i-k-j+1) a_{k+j-1, j-1}, \quad i = 3, 4, ..., n \\ &\text{and} \quad j = 1, 2, ..., i-2; \\ &a_{i,i-1} = 1, \quad i = 1, 2, ..., n. \end{aligned} \tag{10}$$

Note that $e_n(t)$ in (9) is measurable since it is a function of the system states and the reference trajectory.

Based on the subsequent stability analysis, we propose the following control law to achieve the stated control objective:

$$u(t) = (K_s + I_m)e_n(t) - (K_s + I_m)e_n(t_0) + \int_{t_0}^{t} \left[(K_s + I_m)\Lambda e_n(\tau) + \Gamma \text{sgn}(e_n(\tau))\right] d\tau \tag{11}$$

where $K_s, \Lambda, \Gamma \in \mathbb{R}^{m\times m}$ are positive-definite, diagonal control gain matrices, $I_m \in \mathbb{R}^{m\times m}$ represents the identity matrix, and $\text{sgn}(\cdot)$ is defined as follows:

$$\text{sgn}(\xi) := \left[sgn(\xi_1)\ sgn(\xi_2)\ ...\ sgn(\xi_m) \right]^T \qquad \forall \xi = \left[\xi_1\ \xi_2\ ...\ \xi_m \right]^T. \tag{12}$$

Note that the second term in (11) is used to ensure that $u(t_0) = 0$.

4 Error System Development

We begin by defining the filtered tracking error $r(t) \in \mathbb{R}^m$ as

$$r = \dot{e}_n + \Lambda e_n \tag{13}$$

where Λ was defined in (11). After differentiating (13), multiplying both sides of the resulting equation by $M(\cdot)$, and then substituting from the second derivative of (9) with $i = n$, we have

$$M(\cdot)\dot{r} = M(\cdot)\sum_{j=0}^{n-1} a_{nj}e_1^{(j+2)} + M(\cdot)\Lambda\dot{e}_n. \tag{14}$$

After applying (7) and the last equation of (10) to (14), we obtain

$$M(\cdot)\dot{r} = M(\cdot)\left(x_d^{(n+1)} + \sum_{j=0}^{n-2} a_{nj}e_1^{(j+2)} + \Lambda\dot{e}_n \right) + \dot{M}(\cdot)x^{(n)} + \dot{f}(\cdot) - \dot{u} \tag{15}$$

where the time derivative of (4) was used. We now arrange (15) into the following form:

$$M(\cdot)\dot{r} = -\frac{1}{2}\dot{M}(\cdot)r - e_n - \dot{u} + N(\cdot, t) \tag{16}$$

where the auxiliary function $N(\cdot, t) \in \mathbb{R}^m$ is defined as

$$N(\cdot, t) := M(\cdot)\left(x_d^{(n+1)} + \sum_{j=0}^{n-2} a_{nj}e_1^{(j+2)} + \Lambda\dot{e}_n \right) + \dot{M}(\cdot)\left(x^{(n)} + \frac{1}{2}r \right) + e_n + \dot{f}(\cdot). \tag{17}$$

After taking the time derivative of (11), we obtain

$$\dot{u} = (K_s + I_m)r + \Gamma \mathrm{sgn}(e_n) \tag{18}$$

upon use of (13). After substituting (18) into (16), we obtain the following closed-loop system

$$M(\cdot)\dot{r} = -\frac{1}{2}\dot{M}(\cdot)r - e_n - (K_s + I_m)r - \Gamma \mathrm{sgn}(e_n) + N(\cdot, t). \tag{19}$$

Remark 2. The structure of the closed-loop error system of (19) may look familiar to the control designer who is acquainted with SMC; however, there are some significant differences that are important to highlight. Specifically, note that the variable structure-like term $\Gamma \mathrm{sgn}(e_n)$ in (19) would need to be $\Gamma \mathrm{sgn}(r)$ to facilitate the stability analysis for a standard SMC. However, it is not clear how the control can be designed to obtain a $\Gamma \mathrm{sgn}(r)$-like term in (19). That is, note that the variable $r(t)$ is unmeasurable since $x^{(n)}(t)$ is not usually considered measurable for an nth-order system similar to (4) (i.e., the direct use of $x^{(n)}(t)$ in a feedback control law can cause the problem to be ill-posed). For this reason, the control given in (11) has been designed to only depend on measurement of the state variables. With this explanation as a background, the primary questions for this new type of control design that must answered are as follows: i) Can the control term $\Gamma \mathrm{sgn}(e_n)$ compensate for the unknown, unmeasurable nonlinearity $N(\cdot, t)$ through proper selection of the control gain Γ?; and ii) Can asymptotic tracking be proven via the construction of an appropriate stability analysis argument? These questions are answered in the affirmative in the next section.

Before analyzing the stability of the closed-loop system, we perform the following manipulation on (19). Let $N_d(t) := N(x_d, \dot{x}_d, ..., x_d^{(n)}, t)$; hence, from (17), we have that

$$\begin{aligned} N_d(t) = {} & M(x_d, \dot{x}_d, ..., x_d^{(n-1)})x_d^{(n+1)} + \dot{M}(x_d, \dot{x}_d, ..., x_d^{(n-1)})x_d^{(n)} \\ & + \dot{f}(x_d, \dot{x}_d, ..., x_d^{(n-1)}). \end{aligned} \tag{20}$$

Note that $N_d(t), \dot{N}_d(t) \in \mathcal{L}_\infty$ due to (6) and the $\mathcal{C}^2$ condition on $M(\cdot)$ and $f(\cdot)$. Now, after adding and subtracting $N_d(t)$ to the right-hand side of (19), we obtain

$$M(\cdot)\dot{r} = -\frac{1}{2}\dot{M}(\cdot)r - e_n - (K_s + I_m)r - \Gamma \mathrm{sgn}(e_n) + \tilde{N}(\cdot, t) + N_d \tag{21}$$

where

$$\tilde{N} := N - N_d. \tag{22}$$

Remark 3. Since $N(\cdot, t)$ defined in (17) is continuously differentiable, we can show $\tilde{N}(\cdot, t)$ can be upper bounded as follows:

$$\left\|\tilde{N}\right\| \leq \rho(\|z\|)\,\|z\| \tag{23}$$

where $z(t) \in \mathbb{R}^{m(n+1)}$ is defined as

$$z := [\,e_1^T\; e_2^T\; ...\; e_n^T\; r^T\,]^T, \tag{24}$$

and the bounding function $\rho(\|z\|)$ is nondecreasing in $\|z\|$.

5 Stability Analysis

Before presenting the main result of this section, we state the following two lemmas which will be invoked later:

Lemma 1. *Let the auxiliary function $L(t) \in \mathbb{R}$ be defined as follows:*

$$L := r^T\left(N_d - \Gamma\operatorname{sgn}(e_n)\right). \tag{25}$$

If the control gain Γ is selected to satisfy the following sufficient condition

$$\Gamma_i > \|N_{di}(t)\|_\infty + \frac{1}{\Lambda_i}\left\|\dot{N}_{di}(t)\right\|_\infty, \tag{26}$$

then

$$\int_{t_0}^{t} L(\tau)d\tau \leq \zeta_b \tag{27}$$

where the positive constant ζ_b is defined as

$$\zeta_b := \sum_{i=1}^{m}\Gamma_i\,|e_{ni}(t_0)| - e_n^T(t_0)N_d(t_0) \tag{28}$$

and the subscript $i = 1, 2, .., m$ denotes the ith element of the vector or diagonal matrix.

Proof. See Appendix A.

Lemma 2. *Let the region $\mathcal{D}$ in the state space be defined as follows:*

$$\mathcal{D} := \{\xi \in \mathbb{R}^n \mid \|\xi\| < \varepsilon\},$$

where $\varepsilon \in \mathbb{R}$ is some positive constant and let $V(\xi, t)$: $\mathcal{D} \times \mathbb{R}_+ \rightarrow \mathbb{R}_+$ be a continuously differentiable function such that

$$\begin{aligned} W_1(\xi) \leq V(\xi, t) \leq W_2(\xi) \\ \dot{V}(\xi, t) \leq -W(\xi) \end{aligned} \tag{29}$$

$\forall t \geq t_0$ and $\forall \xi \in \mathcal{D}$ where $W_1(\xi), W_2(\xi) \in \mathbb{R}$ are continuous positive definite functions, $W(\xi) \in \mathbb{R}$ is a uniformly continuous positive semi-definite function, and t_0 denotes the initial time. Provided (29) is satisfied and $\xi(t_0) \in \mathcal{S}$, we have

$$W(\xi(t)) \to 0 \quad \text{as} \quad t \to \infty \tag{30}$$

where the region in the state space denoted by $\mathcal{S}$ is defined as follows

$$\mathcal{S} := \{\xi \in \mathcal{D} \mid W_2(\xi) \leq \delta\} \quad \text{where} \quad \delta < \min_{\|\xi\|=\varepsilon} W_1(\xi) \tag{31}$$

with δ being some positive constant.

Proof. See proof of Theorem 4.4 in [7].

We now state the stability result for the proposed controller.

Theorem 1 *The control law of (11) ensures that all system signals are bounded under closed-loop operation and*

$$e_1^{(i)}(t) \to 0 \quad \text{as} \quad t \to \infty, \quad i = 0, 1, ..., n, \tag{32}$$

provided the control gain Γ is adjusted according to (26) and the control gain K_s is selected sufficiently large.[5]

Proof. Let the auxiliary function $P(t) \in \mathbb{R}$ be defined as follows:

$$P(t) := \zeta_b - \int_{t_0}^{t} L(\tau)d\tau \tag{33}$$

where ζ_b and $L(t)$ were defined in Lemma 1. It is easy to see that the use of Lemma 1 ensures $P(t) \geq 0$. We now define the function $V(y, t)$: $\mathbb{R}^{m(n+1)} \times \mathbb{R}_+ \times \mathbb{R}_+ \to \mathbb{R}_+$ as follows:

$$V := \frac{1}{2}\sum_{j=1}^{n} e_j^T e_j + \frac{1}{2} r^T M(\cdot) r + P \tag{34}$$

where $y(t)$ is defined as

$$y := [z^T \quad \sqrt{P}]^T \tag{35}$$

and $z(t)$ was defined in (24). Note that (34) can be bounded as

$$\lambda_1 \|y\|^2 \leq V \leq \lambda_2(\cdot) \|y\|^2 \tag{36}$$

where $\lambda_1 := \frac{1}{2}\min\{1, \underline{m}\}$ and $\lambda_2(\cdot) := \max\left\{\frac{1}{2}\overline{m}(\cdot), 1\right\}$.

After taking the time derivative of (34) and substituting from (8), (13), and (21), we obtain

[5] A sufficient condition on K_s is explicitly given in the subsequent proof.

$$\dot{V} = -\sum_{j=1}^{n-1} e_j^T e_j - e_n^T \Lambda e_n + e_{n-1}^T e_n - r^T r + r^T \tilde{N} - r^T K_s r \tag{37}$$

upon use of (25). By using the fact that $e_{n-1}^T e_n \leq \frac{1}{2}(\|e_{n-1}\|^2 + \|e_n\|^2)$, we can obtain an upper bound on (37) as follows:

$$\begin{aligned} \dot{V} &\leq -\lambda_3 \|z\|^2 + \|r\| \rho(\|z\|) \|z\| - \lambda_{\min}(K_s) \|r\|^2 \\ &\leq -\left(\lambda_3 - \frac{\rho^2(\|z\|)}{4\lambda_{\min}(K_s)}\right) \|z\|^2 \end{aligned} \tag{38}$$

where (23) was used, $\lambda_{\min}(\cdot)$ represents the minimum eigenvalue of the matrix,

$$\lambda_{\min}(\Lambda) > \frac{1}{2}, \tag{39}$$

and $\lambda_3 := \min\left\{\frac{1}{2}, \lambda_{\min}(\Lambda) - \frac{1}{2}\right\}$. From (38), we can state that

$$\dot{V} \leq -\gamma \|z\|^2 \quad \text{for} \quad \begin{array}{l} \lambda_{\min}(K_s) > \frac{1}{4\lambda_3}\rho^2(\|z\|) \\ (\text{or } \|z\| < \rho^{-1}(2\sqrt{\lambda_3 \lambda_{\min}(K_s)}) \end{array} \tag{40}$$

where γ is some positive constant.

We now apply Lemma 2 to (36) and (40). Specifically, we see from (36) that lower and upper bounds for (34) are given as follows:

$$W_1(y) = \lambda_1 \|y\|^2 \qquad W_2(y) = \lambda_2(x) \|y\|^2 \tag{41}$$

while the upper bound for the derivative of (34) is given by

$$W(y) = \gamma \|z\|^2 . \tag{42}$$

Following Lemma 2, we utilize the right-most inequality in (40) to define the region $\mathcal{D}$ as

$$\mathcal{D} := \left\{ y \in \mathbb{R}^{m(n+1)} \times \mathbb{R}_+ \mid \|y\| < \rho^{-1}(2\sqrt{\lambda_3 \lambda_{\min}(K_s)}) \right\}. \tag{43}$$

From (34) and (40), we know $V(y(t),t) \in \mathcal{L}_\infty$; hence, $e_i(t), r(t) \in \mathcal{L}_\infty$, $i = 1, 2, ..., n$. From (13), we then know $\dot{e}_n(t) \in \mathcal{L}_\infty$. Using (6) and (7), we can show that $x^{(i)}(t) \in \mathcal{L}_\infty$, $i = 0, 1, ..., n$. We now know that $M(\cdot), f(\cdot) \in \mathcal{L}_\infty$. Finally, we can use (4) to show that $u(t) \in \mathcal{L}_\infty$.

Using the above boundedness statements, it is clear from (21), (24), and (42) that $\dot{W}(y(t)) \in \mathcal{L}_\infty$, which is a sufficient condition for $W(y(t))$ being uniformly continuous. If we define the region $\mathcal{S}$ by

$$\mathcal{S} := \left\{ y \in \mathcal{D} \mid W_2(y) < \lambda_1 \left(\rho^{-1}(2\sqrt{\lambda_3 \lambda_{\min}(K_s)})\right)^2 \right\}, \tag{44}$$

then we can now invoke Lemma 2 to state that $\|z(t)\|^2 \to 0$ as $t \to \infty$ $\forall y(t_0) \in \mathcal{S}$. From (24), we then know that $e_i(t), r(t) \to 0$ as $t \to \infty$ $\forall y(t_0) \in \mathcal{S}$

and $i = 1, 2, ..., n$. From (13), we know $\dot{e}_n(t) \to 0$ as $t \to \infty$. Now, we can recursively use (9) to show

$$e_1^{(i)}(t) \to 0 \text{ as } t \to \infty,$$

$i = 1, 2, ..., n-1$. Finally, from the time derivative of (9) with $i = n$, we can show $e_1^{(n)}(t) \to 0$ as $t \to \infty$.

Note that the region of attraction in (44) can be made arbitrarily large to include any initial conditions by increasing the control gain K_s (i.e., a semi-global stability result). Specifically, we can use the right-most equality in (41) and the definition given by (44) to calculate the region of attraction as follows

$$\|y(t_0)\| < \sqrt{\frac{\lambda_1}{\lambda_2(t_0)}}\rho^{-1}(2\sqrt{\lambda_3\lambda_{\min}(K_s)}), \tag{45}$$

which can be rearranged as

$$\lambda_{\min}(K_s) > \frac{1}{4\lambda_3}\rho^2\left(\sqrt{\frac{\lambda_2(t_0)}{\lambda_1}}\|y(t_0)\|\right) \tag{46}$$

where the following shorthand notation has been used $\lambda_2(t_0) = \lambda_2(x(t_0), \dot{x}(t_0), ..., x^{(n-1)}(t_0))$. By using (13) and (35), we can write an explicit expression for $\|y(t_0)\|$ as follows:

$$\|y(t_0)\| = \sqrt{\sum_{i=1}^{n} e_i^T(t_0)e_i(t_0) + (\dot{e}_n(t_0) + \Lambda e(t_0))^T(\dot{e}_n(t_0) + \Lambda e_n(t_0)) + P(t_0)}$$

where from (4), (7), the fact that $u(t_0) = 0$, and the time derivative of (9) with $i = n$

$$\dot{e}_n(t_0) = \sum_{j=0}^{n-2} a_{nj}e_1^{(j+1)}(t_0) + \left(x_d^{(n)}(t_0) + M^{-1}(t_0)f(t_0)\right). \tag{47}$$

Remark 4. An important extension of the proposed control strategy that merits a brief discussion is for the case where the functional form of $M(\cdot, \theta, t)$ and $f(\cdot, \theta, t)$ in (4) is known, but the constant parameter vector θ is not known precisely. In this case, a certainty equivalence-type term can be added to the control law of (11) as shown below

$$\begin{aligned} u(t) = &(K_s + I_m)e_n(t) - (K_s + I_m)e_n(t_0) \\ &+ \int_{t_0}^{t}\left[\hat{N}_d(\tau) + (K_s + I_m)\Lambda e_n(\tau) + \Gamma\text{sgn}(e_n(\tau))\right]d\tau \end{aligned} \tag{48}$$

where $\hat{N}_d(t) := N(x_d, \dot{x}_d, ..., x_d^{(n)}, \hat{\theta}, t)$, $N(\cdot)$ was defined in (17), and $\hat{\theta}$ contains best-guess estimates of the unknown parameter vector. Although the addition of the term $\hat{N}_d(t)$ is not necessary to prove asymptotic tracking, it is expected that the $\hat{N}_d(t)$ term will improve the tracking performance and reduce the magnitude of the control gain Γ (i.e., the right-hand side of (26) would be in all likelihood smaller because $N_{di}(t)$ would now be replaced by $N_{di}(t) - \hat{N}_{di}(t)$).

References

1. Annaswanny A, Skantze F, Loh A (1998) Adaptive control of continuous-time systems with convex/concave parametrizations, Automatica 34:33-49
2. Bartolini G, Ferrara A, Usai E, Utkin V (2000) On multi-input chattering-free second-order sliding mode control, IEEE Trans Autom Control 45:1711-1717
3. Campion G, Bastin G (1990) Indirect adaptive state feedback control of linearly parameterized nonlinear systems, Int J Adaptive Control and Signal Processing 4:345-358
4. Fradkov A, Ortega R, Bastin G (2001) On semi-adaptive control of convexly parametrized systems, Int J Adaptive Control and Signal Processing 15:415-426
5. Ioannou P, Sun J (1996) Stable and Robust Adaptive Control. Prentice Hall Englewood Cliffs NJ
6. Kanellakopoulos I, Kokotovic P, Marino R (1991) An extended direct scheme for robust adaptive nonlinear control, Automatica 27:247-255
7. Khalil H (2002) Nonlinear Systems. Prentice Hall Englewood Cliffs NJ
8. Koshy T (2001) Fibonacci and Lucas Numbers with Applications. John Wiley and Sons New York
9. Kosmatopoulos E, Ioannou P(2002) Robust switching adaptive control of multi-input nonlinear systems, IEEE Trans Autom Control 47:610-624
10. Krstic M, Kanellakopoulos I, Kokotovic P (1995) Nonlinear and Adaptive Control Design. John Wiley and Sons New York
11. Levant A (1993) A sliding order and sliding accuracy in sliding mode control, Int J Control 58:1247-1263
12. Lewis F, Jagannathan S, Yesildirak A (1999) Neural Network Control of Robot Manipulators and Nonlinear Systems. Taylor and Francis New York
13. Lin W, Qian C (2002) Adaptive control of nonlinearly parameterized systems: A nonsmooth feedback framework, IEEE Trans Autom Control 47:757-774
14. Messner W, Horowitz R, Kao W, Boals M (1991) A new adaptive learning rule, IEEE Trans Automat Control 36:188-197
15. Qu Z (1998) Robust Control of Nonlinear Uncertain Systems. John Wiley and Sons New York
16. Qu Z, Xu J (2002) Model-based learning controls and their comparisons using Lyapunov direct method, Asian J Control 4:99-110
17. Sastry S, Bodson M (1989) Adaptive Control Stability Convergence and Robustness. Prentice Hall Englewood Cliffs NJ
18. Sastry S, Isidori A (1989) Adaptive Control of Linearizable Systems, IEEE Trans Autom Control 34:405-412
19. Serrani A, Isidori A, Marconi L (2001) Semiglobal nonlinear output regulation with adaptive internal model, IEEE Trans Autom Control 46:1178-1194
20. Slotine J, Li W (1991) Applied Nonlinear Control. Prentice Hall Englewood Cliffs NJ

21. Utkin V (1992) Sliding Modes in Control and Optimization. Springer-Verlag New York
22. Xu H, Ioannou P (2003) Robust adaptive control for a class of MIMO nonlinear systems with guaranteed error bounds, IEEE Trans Autom Control 48:728-742
23. Yao B, Tomizuka M (1996) Adaptive robust control of a class of multivariable nonlinear systems. In: Proc IFAC World Congress San Francisco CA 335-340

A Proof of Lemma 1

After substituting (13) into (25) and then integrating in time, we obtain

$$\int_{t_0}^{t} L(\tau)d\tau = \int_{t_0}^{t} e_n^T(\tau)\Lambda\left(N_d(\tau) - \Gamma\operatorname{sgn}(e_n(\tau))\right)d\tau + \int_{t_0}^{t} \frac{de_n^T(\tau)}{d\tau}N_d(\tau)d\tau \\ - \int_{t_0}^{t} \frac{de_n^T(\tau)}{d\tau}\Gamma\operatorname{sgn}(e_n(\tau))d\tau. \tag{49}$$

After integrating the second integral on the right-hand side of (49) by parts, we have

$$\int_{t_0}^{t} L(\tau)d\tau = \int_{t_0}^{t} e_n^T(\tau)\Lambda\left(N_d(\tau) - \Gamma\operatorname{sgn}(e_n(\tau))\right)d\tau + e_n^T(\tau)N_d(\tau)\Big|_{t_0}^{t} \\ - \int_{t_0}^{t} e_n^T(\tau)\frac{dN_d(\tau)}{d\tau}d\tau - \sum_{i=1}^{n}\Gamma_i\left|e_{ni}(\tau)\right|\Bigg|_{t_0}^{t} \\ = \int_{t_0}^{t} e_n^T(\tau)\Lambda\left(N_d(\tau) - \Lambda^{-1}\frac{dN_d(\tau)}{d\tau} - \Gamma\operatorname{sgn}(e_n(\tau))\right)d\tau \\ + e_n^T(t)N_d(t) - e_n^T(t_0)N_d(t_0) - \sum_{i=1}^{n}\Gamma_i\left|e_{ni}(t)\right| \\ + \sum_{i=1}^{n}\Gamma_i\left|e_{ni}(t_0)\right|. \tag{50}$$

We now upper bound the right-hand side of (50) as follows:

$$\int_{t_0}^{t} L(\tau)d\tau \leq \int_{t_0}^{t}\sum_{i=1}^{n}\left|e_{ni}(\tau)\right|\Lambda_i\left(\left|N_{di}(\tau)\right| + \frac{1}{\Lambda_i}\left|\frac{dN_{di}(\tau)}{d\tau}\right| - \Gamma_i\right)d\tau \\ + \sum_{i=1}^{n}\left|e_{ni}(t)\right|\left(\left|N_{di}(t)\right| - \Gamma_i\right) \\ + \sum_{i=1}^{n}\Gamma_i\left|e_{ni}(t_0)\right| - e_n^T(t_0)N_d(t_0). \tag{51}$$

From (51), it is easy to see that if Γ is chosen according to (26), then (27) holds.

B Calculation of Auxiliary Error Signals

Note that the second equation in (8) can be rewritten as

$$e_3 = \dot{e}_2 + e_2 + e_1 = (\ddot{e}_1 + \dot{e}_1) + (\dot{e}_1 + e_1) + e_1 = \ddot{e}_1 + 2\dot{e}_1 + 2e_1 \tag{52}$$

upon application of the first equation in (8) and its derivative. Similar derivations can be applied to the other error signal definitions in (8) such that, in general,

$$e_i = \sum_{j=0}^{i-1} a_{ij} e_1^{(j)}, \quad i = 1, 2, ..., n. \tag{53}$$

In order to determine a general expression for the coefficients a_{ij}, we first apply (53) to the definition of $e_i(t)$ in (8) to obtain

$$e_i = \sum_{j=0}^{(i-1)-1} a_{i-1,j} e_1^{(j+1)} + \sum_{j=0}^{(i-1)-1} a_{i-1,j} e_1^{(j)} + \sum_{j=0}^{(i-2)-1} a_{i-2,j} e_1^{(j)}. \tag{54}$$

After equating the coefficients in (53) and (54), we have

$$\begin{aligned} a_{i0} &= a_{i-1,0} + a_{i-2,0}, \quad i = 3, 4, ..., n \\ a_{ij} &= a_{i-1,j-1} + a_{i-1,j} + a_{i-2,j}, \quad i = 4, 5, ..., n \text{ and } j = 1, 2, ..., i-2 \\ a_{i,i-1} &= 1, \quad i = 1, 2, ..., n \end{aligned} \tag{55}$$

where $a_{20} = 1$ from the first equation in (8) and $a_{31} = 2$ from (52).

For the Fibonacci numbers generated via the following series [8]

$$B(1) = B(2) = 1 \qquad B(k) = B(k-1) + B(k-2), \quad k = 3, 4, ..., \tag{56}$$

the general formula is given by

$$B(k) = \frac{1}{\sqrt{5}} \left[\left(\frac{1+\sqrt{5}}{2} \right)^k - \left(\frac{1-\sqrt{5}}{2} \right)^k \right], \quad k = 1, 2, ... \tag{57}$$

Using (55) through (57), it is easy to see that

$$a_{i0} = \frac{1}{\sqrt{5}} \left[\left(\frac{1+\sqrt{5}}{2} \right)^i - \left(\frac{1-\sqrt{5}}{2} \right)^i \right], \quad i = 2, 3, ..., n. \tag{58}$$

Furthermore, we can deduce the expression for a_{ij}, $i = 3, 4, ..., n$ and $j = 1, 2, ..., i-2$ as follows:

$$a_{ij} = \sum_{k=1}^{i-1} B(i-k-j+1) a_{k+j-1,j-1}. \tag{59}$$

Part III

Nonsmooth Analysis and Applications

Lyapunov Functions and Feedback in Nonlinear Control

Francis Clarke

Professeur à l'Institut Universitaire de France. Institut Desargues, Université de Lyon 1, 69622 Villeurbanne, France. clarke@igd.univ-lyon1.fr

Summary. The method of Lyapunov functions plays a central role in the study of the controllability and stabilizability of control systems. For nonlinear systems, it turns out to be essential to consider nonsmooth Lyapunov functions, even if the underlying control dynamics are themselves smooth. We synthesize in this article a number of recent developments bearing upon the regularity properties of Lyapunov functions. A novel feature of our approach is that the guidability and stability issues are decoupled. For each of these issues, we identify various regularity classes of Lyapunov functions and the system properties to which they correspond. We show how such regularity properties are relevant to the construction of stabilizing feedbacks. Such feedbacks, which must be discontinuous in general, are implemented in the sample-and-hold sense. We discuss the equivalence between open-loop controllability, feedback stabilizability, and the existence of Lyapunov functions with appropriate regularity properties. The extent of the equivalence confirms the cogency of the new approach summarized here.

1 Introduction

We consider a system governed by the standard control dynamics

$$\dot{x}(t) = f(x(t), u(t)) \text{ a.e., } u(t) \in \mathcal{U} \text{ a.e.}$$

or equivalently (under mild conditions) by the differential inclusion

$$\dot{x}(t) \in F(x(t)) \text{ a.e.}$$

The issue under consideration is that of guiding the state x to the origin. (The use of more general target sets presents no difficulties in the results presented here.)

A century ago, for the uncontrolled case in which the multifunction F is given by a (smooth) single-valued function (that is, $F(x) = \{f(x)\}$), Lyapunov introduced a criterion for the *stability* of the system, a property whereby all the trajectories $x(t)$ of the system tend to the origin (in a certain sense

M. de Queiroz et al. (Eds.): Optimal Control, Stabilization and Nonsmooth Analysis, LNCIS 301, pp. 267–282, 2004.

which we gloss over for now). This criterion involves the existence of a certain function V, now known as a Lyapunov function. Later, in the classical works of Massera, Barbashin and Krasovskii, and Kurzweil, this sufficient condition for stability was also shown to be necessary (under various sets of hypotheses).

In extending the technique of Lyapunov functions to control systems, a number of new issues arise. To begin with, we can distinguish two cases: we may require that *all* trajectories go to the origin (strong stability) or that (for a suitable choice of the control function) *some* trajectory goes to zero (weak stability, or controllability). In the latter case, unlike the former, it turns out that characterizing stability in terms of smooth Lyapunov functions is not possible; thus elements of *nonsmooth analysis* become essential. Finally, the issue of stabilizing *feedback design* must be considered, for this is one of the main reasons to introduce control Lyapunov functions. Here again regularity intervenes: in general, such feedbacks must be discontinuous, so that a method of implementing them must be devised, and new issues such as robustness addressed.

While these issues have been considered for decades, they have only recently been resolved in a unified and (we believe) satisfactory way. Several new tools have contributed to the analysis, notably: proximal analysis and attendant Hamilton-Jacobi characterizations of monotonicity properties of trajectories, semiconcavity, and sample-and-hold implementation of discontinuous feedbacks. The point of view in which the issues of guidability and stability are decoupled is also very recent. Our purpose here is to sketch the complete picture of these related developments for the first time, thereby synthesizing a guide for their comprehension. The principal results being summarized here appear in the half-dozen joint articles of Clarke, Ledyaev, Rifford and Stern cited in the references, and in the several works by Rifford; the article [8] of Clarke, Ledyaev, Sontag and Subbotin is also called upon. The necessary background in nonsmooth analysis is provided by the monograph of Clarke, Ledyaev, Stern and Wolenski [10].

Of course there is an extensive literature on the issues discussed here, with contributions by Ancona, Artstein, Bressan, Brockett, Coron, Kellett, Kokotovic, Praly, Rosier, Ryan, Sontag, Sussmann, Teel, and many others; these are discussed and cited in the introductions of the articles mentioned above. General references for Lyapunov functions in control include [2] and [14].

2 Strong Stability

We shall say that the control system $\dot{x}(t) \in F(x(t))$ a.e. is *strongly asymptotically stable* if every trajectory $x(t)$ is defined for all $t \geq 0$ and satisfies $\lim_{t\to+\infty} x(t) = 0$, and if in addition the origin has the familiar local property known as 'Lyapunov stability'. The following result, which unifies and extends several classical theorems dealing with the uncontrolled case, is due to Clarke, Ledyaev and Stern [9]:

Theorem 1. *Let F have compact convex nonempty values and closed graph. Then the system is strongly asymptotically stable if and only if there exists a pair of C^∞ functions*

$$V : \mathbb{R}^n \to \mathbb{R}, \quad W : \mathbb{R}^n \backslash \{0\} \to \mathbb{R}$$

satisfying the following conditions:

1. Positive Definiteness:

$$V(x) > 0 \text{ and } W(x) > 0 \quad \forall x \neq 0, \text{ and } V(0) \geq 0.$$

2. Properness: *The sublevel sets $\{x : V(x) \leq c\}$ are bounded* $\forall c$.
3. Strong Infinitesimal Decrease:

$$\max_{v \in F(x)} \langle \nabla V(x), v \rangle \;\leq\; -W(x) \quad x \neq 0.$$

We refer to the function (V, W) as a *strong Lyapunov function* for the system. Note that in this result, whose somewhat technical proof we shall not revisit here, the system multifunction F itself need not even be continuous, yet strong stability is equivalent to the existence of a smooth Lyapunov function: this is a surprising aspect of these results. As we shall see, this is in sharp contrast to the case of weak stability, where stronger hypotheses on the underlying system are required. In fact, in addition to the hypotheses of Theorem 1, we shall suppose henceforth that F is locally Lipschitz with linear growth. Even so, Lyapunov functions will need to be nondifferentiable in the controllability context.

Finally, we remark that in the positive definiteness condition, the inequality $V(0) \geq 0$ is superfluous when V is continuous (which will not be the case later); also, it could be replaced by the more traditional condition $V(0) = 0$ in the present context.

3 Guidability and Controllability

The Case for Less Regular Lyapunov Functions

Strong stability is most often of interest when F arises from a perturbation of an ordinary (uncontrolled) differential equation. In most control settings, it is weak (open loop) stability that is of interest: the possibility of guiding *some* trajectory to 0 in a suitable fashion. It is possible to distinguish two distinct aspects of the question: on the one hand, the possibility of guiding the state from any prescribed initial condition to 0 (or to an arbitrary neighborhood of 0), and on the other hand, that of keeping the state close to 0 when the initial condition is already near 0. In a departure from the usual route, we choose to decouple these two issues, introducing the term 'guidability' for the first. We believe that in so doing, a new level of clarity emerges in connection with Lyapunov theory.

A point α is *asymptotically guidable* to the origin if there is a trajectory x satisfying $x(0) = \alpha$ and $\lim_{t\to\infty} x(t) = 0$. When every point has this property, and when additionally the origin has the familiar local stability property known as *Lyapunov stability*, it is said in the literature to be *GAC*: (open loop) *globally asymptotically controllable* (to 0). A well-known *sufficient* condition for this property is the existence of a smooth (C^1, say) pair (V, W) of functions satisfying the positive definiteness and properness conditions of Theorem 1, together with *weak infinitesimal decrease:*

$$\min_{v\in F(x)} \langle \nabla V(x), v\rangle \leq -W(x) \quad x \neq 0.$$

Note the presence of a minimum in this expression rather than a maximum.

It is a fact, however, that as demonstrated by simple examples (see [6] or [23]), the existence of a smooth function V with the above properties fails to be a necessary condition for global asymptotic controllability; that is, the familiar converse Lyapunov theorems of Massera, Barbashin and Krasovskii, and Kurzweil do not extend to this weak controllability setting, at least not in smooth terms.

It is natural therefore to seek to weaken the smoothness requirement on V so as to obtain a necessary (and still sufficient) condition for a system to be GAC. This necessitates the use of some construct of nonsmooth analysis to replace the gradient of V that appears in the infinitesimal decrease condition. In this connection we use the *proximal subgradient* $\partial_P V(x)$, which requires only that the (extended-valued) function V be lower semicontinuous. In proximal terms, the weak infinitesimal decrease condition becomes

$$\sup_{\zeta\in\partial_P V(x)} \min_{v\in F(x)} \langle \zeta, v\rangle \leq -W(x) \quad x \neq 0.$$

Note that this last condition is trivially satisfied when x is such that $\partial_P V(x)$ is empty, in particular when $V(x) = +\infty$. (The supremum over the empty set is $-\infty$.) Henceforth, a *general Lyapunov pair* (V, W) refers to extended-valued lower semicontinuous functions $V : \mathbb{R}^n \to \mathbb{R} \cup \{+\infty\}$ and $W : \mathbb{R}^n \backslash \{0\} \to \mathbb{R} \cup \{+\infty\}$ satisfying the positive definiteness and properness conditions of Theorem 1, together with proximal weak infinitesimal decrease.

The following is proved in [10]:

Theorem 2. *Let (V, W) be a general Lyapunov pair for the system. Then any $\alpha \in \operatorname{dom} V$ is asymptotically guidable to* 0.

We proceed to make some comments on the proof. To show that any initial condition can be steered towards zero (in the presence of a Lyapunov function), one can invoke the infinitesimal decrease condition to deduce that the function $V(x)+y$ is *weakly decreasing* for the augmented dynamics $F(x) \times \{W(x)\}$ (see pp. 213-214 of [10] for details); this implies the existence of a trajectory x such that the function

$$t \mapsto V(x(t)) + \int_0^t W(x(\tau))\, d\tau$$

is nonincreasing, which in turn implies that $x(t) \to 0$. We remark that viability theory can also be used in this type of argument; see for example [1].

It follows from the theorem that the existence of a lower semicontinuous Lyapunov pair (V, W) with V everywhere finite-valued implies the global asymptotic guidability to 0 of the system. This does not imply Lyapunov stability at the origin, however, so it cannot characterize global asymptotic controllability. An early and seminal result due to Sontag [22] considers *continuous* functions V, with the infinitesimal decrease condition expressed in terms of Dini derivates. Here is a version of it in proximal subdifferential terms:

Theorem 3. *The system is GAC if and only if there exists a* continuous *Lyapunov pair* (V, W).

For the sufficiency, the requisite guidability evidently follows from the previous theorem. The continuity of V provides the required local stability: roughly speaking, once $V(x(t))$ is small, its value cannot take an upward jump, so $x(t)$ remains near 0.

The proof of the converse theorem (that a continuous Lyapunov function must exist when the system is globally asymptotically controllable) is more challenging. One route is as follows: In [7] it was shown that certain locally Lipschitz *value functions* give rise to *practical* Lyapunov functions (that is, assuring stable controllability to arbitrary neighborhoods of 0, as in Theorem 4 below). Building upon this, Rifford [18, 19] was able to combine a countable family of such functions in order to construct a global *locally Lipschitz* Lyapunov function. This answered a long-standing open question in the subject. Rifford also went on to show the existence of a *semiconcave* Lyapunov function, a property whose relevance to feedback construction will be seen in the following sections.

Finally, we remark that the equivalence of the Dini derivate and of the proximal subdifferential forms of the infinitesimal decrease condition is a consequence of Subbotin's Theorem (see [10]).

Practical guidability.

The system is said to be (open-loop) *globally practically guidable* (to the origin) if for each initial condition α and for every $\varepsilon > 0$ there exists a trajectory x and a time T (both depending on α and ε) such that $|x(T)| \leq \varepsilon$. We wish to characterize this property in Lyapunov terms. For this purpose we need an extension of the Lyapunov function concept.

ε-Lyapunov functions.

An *ε-Lyapunov pair* for the system refers to lower semicontinuous functions $V : \mathbb{R}^n \to \mathbb{R} \cup \{+\infty\}$ and $W : \mathbb{R}^n \backslash \overline{B}(0,\varepsilon) \to \mathbb{R} \cup \{+\infty\}$ satisfying the usual properties of a Lyapunov pair, but with the role of the origin replaced by the closed ball $\overline{B}(0,\varepsilon)$:

1. *Positive Definiteness:*

$$V(x) > 0 \text{ and } W(x) > 0 \quad \forall x \notin \overline{B}(0,\varepsilon), \text{ and } V \geq 0 \text{ on } \overline{B}(0,\varepsilon).$$

2. *Properness:* The sublevel sets $\{x : V(x) \leq c\}$ are bounded $\forall c$.
3. *Weak Infinitesimal Decrease:*

$$\min_{v \in F(x)} \langle \nabla V(x), v \rangle \leq -W(x) \quad x \notin \overline{B}(0,\varepsilon).$$

The results in [7] imply:

Theorem 4. *The system is globally practically guidable to the origin if and only if there exists a locally Lipschitz ε-Lyapunov function for each $\varepsilon > 0$.*

We do not know whether global asymptotic guidability can be characterized in analogous terms, or whether practical guidability can be characterized by means of a *single* Lyapunov function. However, it is possible to do so for *finite-time* guidability (see Section 6 below).

4 Feedback

The Case for More Regular Lyapunov Functions

The need to consider discontinuous feedback in nonlinear control is now well established, together with the attendant need to define an appropriate solution concept for a differential equation in which the dynamics fail to be continuous in the state. The best-known solution concept in this regard is that of Filippov. For the stabilization issue, and using the standard formulation

$$\dot{x}(t) = f(x(t), u(t)) \text{ a.e., } u(t) \in \mathcal{U} \text{ a.e.}$$

rather than the differential inclusion, the issue becomes that of finding a feedback contol function $k(x)$ (having values in $\mathcal{U}$) such that the ensuing differential equation

$$\dot{x} = g(x), \quad \text{where} \quad g(x) := f(x, k(x))$$

has the required stability. The *central question* in the subject has long been: If the system is open loop globally asymptotically controllable (to the origin), is there a feedback k such that the resulting g exhibits global asymptotic stability (of the origin)? It has long been known that continuous feedback laws cannot suffice for this to be the case; it also turns out that admitting discontinuous feedbacks interpreted in the Filippov sense is also inadequate.

The question was settled by Clarke, Ledyaev, Sontag and Subbotin [8], who used the proximal aiming method (see also [11]) to show that the answer is positive if the (discontinuous) feedbacks are implemented in the *closed-loop system sampling* sense (also referred to as *sample-and-hold*). We proceed now to describe the sample-and-hold implementation of a feedback.

Let $\pi = \{t_i\}_{i \geq 0}$ be a partition of $[0, \infty)$, by which we mean a countable, strictly increasing sequence t_i with $t_0 = 0$ such that $t_i \to +\infty$ as $i \to \infty$. The *diameter* of π, denoted $\operatorname{diam}(\pi)$, is defined as $sup_{i \geq 0}(t_{i+1} - t_i)$. Given an initial condition x_0, the π-trajectory $x(\cdot)$ corresponding to π and an arbitrary feedback law $k : \mathbb{R}^n \to \mathcal{U}$ is defined in a step-by-step fashion as follows. Between t_0 and t_1, x is a classical solution of the differential equation

$$\dot{x}(t) = f(x(t), k(x_0)), \quad x(0) = x_0, \quad t_0 \leq t \leq t_1.$$

(Of course in general we do not have uniqueness of the solution, nor is there necessarily even one solution, although nonexistence can be ruled out when blow-up of the solution in finite time cannot occur, as is the case in the stabilization problem.) We then set $x_1 := x(t_1)$ and restart the system at $t = t_1$ with control value $k(x_1)$:

$$\dot{x}(t) = f(x(t), k(x_1)), \quad x(t_1) = x_1, \quad t_1 \leq t \leq t_2,$$

and so on in this fashion. The trajectory x that results from this procedure is an actual state trajectory corresponding to a piecewise constant open-loop control; thus it is a physically meaningful one. When results are couched in terms of π-trajectories, the issue of defining a solution concept for discontinuous differential equations is effectively sidestepped. Making the diameter of the partition smaller corresponds to increasing the sampling rate in the implementation.

We remark that the use of possibly discontinuous feedback has arisen in other contexts. In linear time-optimal control, one can find discontinuous feedback syntheses as far back as the classical book of Pontryagin et al [17]; in these cases the feedback is invariably piecewise constant relative to certain partitions of state space, and solutions either follow the switching surfaces or cross them transversally, so the issue of defining the solution in other than a classical sense does not arise. Somewhat related to this is the approach that defines a multivalued feedback law [4]. In stochastic control, discontinuous feedbacks are the norm, with the solution understood in terms of stochastic differential equations. In a similar vein, in the control of certain linear partial differential equations, discontinuous feedbacks can be interpreted in a distributional sense. These cases are all unrelated to the one under discussion. We remark too that the use of discontinuous pursuit strategies in differential games [15] is well-known, together with examples to show that, in general, it is not possible to achieve the result of a discontinuous optimal strategy to within any tolerance by means of a continuous stategy (so there can be a positive unbridgeable gap between the performance of continuous and discontinuous feedbacks).

We can use the π-trajectory formulation to implement feedbacks for either guidability or stabilization (see [12]); we limit attention here to the latter issue.

It is natural to say that a feedback $k(x)$ (continuous or not) *stabilizes* the system in the sample-and-hold sense provided that for every initial value x_0, for all $\epsilon > 0$, there exists $\delta > 0$ and $T > 0$ such that if the diameter of the partition π is less than δ, then the corresponding π-trajectory x beginning at x_0 satisfies

$$\|x(t)\| \leq \epsilon \quad \forall t \geq T.$$

The following theorem is proven in [8]:

Theorem 5. *The system is open loop globally asymptotically controllable if and only if there exists a (possibly discontinuous) feedback $k : \mathbb{R}^n \to \mathcal{U}$ which stabilizes it in the sample-and-hold sense.*

The proof of the theorem actually yields precise estimates regarding how small the step size diam (π) must be for a prescribed stabilization tolerance to ensue, and of the resulting stabilization time, in terms of the given data. These estimates are uniform on bounded sets of initial conditions, and are a consequence of the method of proximal aiming. The latter, which can be viewed as a geometric version of the Lyapunov technique, appears to be difficult to implement in practice, however. One of our principal goals is to show how stabilizing feedbacks can be defined much more conveniently if one has at hand a sufficiently regular Lyapunov function.

The Smooth Case

We begin with the case in which a C^1 smooth Lyapunov function exists, and show how the natural 'pointwise feedback' described below stabilizes the system (in the sample-and-hold sense). For $x \neq 0$, we define $k(x)$ to be any element $u \in \mathcal{U}$ satisfying

$$\langle \nabla V(x), f(x,u) \rangle \leq -W(x).$$

Note that at least one such u does exist, in light of the infinitesimal decrease condition. We mention two more definitions that work: take u to be the element minimizing the inner product above over $\mathcal{U}$, or take any $u \in \mathcal{U}$ satisfying $\langle \nabla V(x), f(x,u) \rangle \leq -W(x)/2$.

Theorem 6. *The pointwise feedback k described above stabilizes the system in the sense of closed-loop system sampling.*

We proceed to sketch the elementary proof of this theorem, which we deem to be a basic result in the theory of control Lyapunov functions.

We begin with a remark: for any $R > 0$, there exists $\delta_R > 0$ such that for all $\alpha \in B(0,R)$ and for all $u \in \mathcal{U}$, any solution x of $\dot{x} = f(x,u),\ x(0) = \alpha$ satisfies

$$|x(t)| \leq R+1 \quad \forall t \in [0, \delta_R]$$

(this is a simple consequence of the linear growth hypothesis and Gronwall's Lemma).

Now let positive numbers r and ε be given; we show that for any $\alpha \in B(0,r)$ there is a trajectory x beginning at α that enters the ball $B(0,\varepsilon)$ in finite time.

Let $R \geq r$ be chosen so that

$$V(x) \leq \max_{B(0,r)} V \Longrightarrow x \in B(0,R).$$

For simplicity, let us assume that ∇V is locally Lipschitz (as otherwise, the argument is carried out with a modulus of continuity). We proceed to choose $K > 0$ such that for every $u \in \mathcal{U}$, the function

$$x \mapsto \langle \nabla V(x), f(x,u) \rangle$$

is Lipschitz on $B(0, R+1)$ with constant K, together with positive numbers M and m satisfying

$$|f(x,u)| \leq M \quad \forall x \in B(0,R+1),\ \forall u \in \mathcal{U},$$

and

$$W(x) \geq m \quad \forall x \in B(0,R+1) \backslash B(0,\varepsilon).$$

Now let $\pi = \{t_i\}_{0 \leq i \leq N}$ be a partition (taken to be uniform for simplicity) of step size $\delta \leq \delta_R$, of an interval $[0,T]$, where $t_0 = 0, t_N = T, T = N\delta$. We apply the pointwise feedback k relative to this partition, and with initial condition $x(0) = x_0 := \alpha$. We proceed to compare the values of V at the first two nodes:

$$V(x(t_1)) - V(x(t_0)) = \langle \nabla V(x(t^*)), \dot{x}(t^*) \rangle (t_1 - t_0)$$

(by the mean value theorem, for some $t^* \in (0,\delta)$)

$$\begin{aligned} &= \langle \nabla V(x(t^*)), f(x(t^*), k(t_0)) \rangle (t_1 - t_0) \\ &\leq \langle \nabla V(x(t_0)), f(x(t_0), k(t_0)) \rangle (t_1 - t_0) \\ &\quad + K\, |x(t^*) - x(t_0)|\, (t_1 - t_0) \end{aligned}$$

(by the Lipschitz condition)

$$\leq -W(x(t_0))\delta + KM\delta^2$$

(by the way k is defined)

$$\le -m\delta + KM\delta^2.$$

Note that these estimates apply because $x(t_1)$ and $x(t^*)$ remain in $B(0, R+1)$, and, in the case of the last step, provided that x_0 does *not* lie in the ball $B(0,\varepsilon)$. Inspection of the final term above shows that if δ is taken less than $m/(2KM)$, then the value of V between the two nodes has decreased by at least $m\delta/2$. It follows from the definition of R that $x(t_1) \in B(0,R)$. Consequently, the same argument as above can be applied to the next partition subinterval, and so on. Iteration then yields:

$$V(x(N\delta)) - V(x_0) \le -mN/2.$$

This will contradict the nonnegativity of V when N exceeds $2V(x_0)/m$, so it follows that the argument must fail at some point, which it can only do when a node $x(t_i)$ lies in $B(0,\varepsilon)$.

This proves that any sample-and-hold trajectory generated by the feedback k enters $B(0,\varepsilon)$ in a time that is bounded above in a way that depends only upon ε and $|\alpha|$ (and V), provided only that the step size is sufficiently small, as measured in a way that depends only on $|\alpha|$.

That k stabilizes the system in the sense of closed-loop system sampling now follows.

Remark.

Rifford [20] has shown that the existence of a smooth Lyapunov pair is equivalent to the existence of a locally Lipschitz one satisfying weak decrease in the sense of generalized gradients (rather than proximal subgradients), which in turn is equivalent to the existence of a stabilizing feedback in the Filippov (rather than sample-and-hold) sense.

5 Semiconcavity

The 'Right' Regularity for Lyapunov Functions

We have seen that a smooth Lyapunov function generates a stabilizing feedback in a very simple and natural way. But since a smooth Lyapunov function does not necessarily exist, we still require a way to handle the general case. It turns out that the two issues can be reconciled through the notion of *semiconcavity.* This is a certain regularity property (not implying smoothness) which can always be guaranteed to hold for some Lyapunov function (if the system is globally asymptotically controllable, of course), and which permits a natural extension of the pointwise definition of a stabilizing feedback.

A function $\phi : \mathbb{R}^n \to \mathbb{R}$ is said to be (globally) semiconcave provided that for every ball $B(0,r)$ there exists $\gamma = \gamma(r) \ge 0$ such that the function $x \mapsto \phi(x) - \gamma|x|^2$ is (finite and) concave on $B(0,r)$. (Hence ϕ is locally the sum of a concave function and a quadratic one.) Observe that any function of class C^2 is semiconcave; also, any semiconcave function is locally Lipschitz, since both concave functions and smooth functions have that property. (There is a local definition of semiconcavity that we omit for present purposes.)

Semiconcavity is an important regularity property in partial differential equations; see for example [5]. The fact that the semiconcavity of a Lyapunov function V turns out to be useful in stabilization is a new observation, and may be counterintuitive: V often has an interpretation in terms of energy, and it may seem more appropriate to seek a *convex* Lyapunov function V. We proceed now to explain why semiconcavity is a highly desirable property, and why a convex V would be of less interest (unless it were smooth, but then it would be semiconcave too).

Recall the ideal case discussed above, in which (for a smooth V) we select a function $k(x)$ such that

$$\langle \nabla V(x), f(x, k(x)) \rangle \leq -W(x).$$

How might this appealing idea be adapted to the case in which V is nonsmooth? We cannot use the proximal subdifferential $\partial_P V(x)$ directly, since it may be empty for 'many' x. We are led to consider the *limiting subdifferential* $\partial_L V(x)$, which, when V is continuous, is defined by applying a natural limiting operation to $\partial_P V$:

$$\partial_L V(x) := \left\{ \zeta = \lim_{i \to \infty} \zeta_i : \zeta_i \in \partial_P V(x_i), \lim_{i \to \infty} x_i = x \right\}.$$

It follows readily that, when V is locally Lipschitz, $\partial_L V(x)$ is nonempty for all x. By passing to the limit, the weak infinitesimal decrease condition for proximal subgradients implies the following:

$$\min_{u \in U} \langle f(x,u), \zeta \rangle \leq -W(x) \; \forall \zeta \in \partial_L V(x), \forall x \neq 0.$$

Accordingly, let us consider the following idea: for each $x \neq 0$, choose some element $\zeta \in \partial_L V(x)$, then choose $k(x) \in U$ such that

$$\langle f(x, k(x)), \zeta \rangle \leq -W(x).$$

Does this lead to a stabilizing feedback, when (of course) the discontinuous differential equation is interpreted in the sample-and-hold sense? When V is smooth, the answer is 'yes', as we have seen. But when V is merely locally Lipschitz, a certain 'dithering' phenomenon may arise to prevent k from being stabilizing. However, if V is semiconcave (on $\mathbb{R}^n \backslash \{0\}$), this does not occur, and stabilization is guaranteed. This accounts in part for the desirability of a semiconcave Lyapunov function, and the importance of knowing one always exists.

The proof that the pointwise feedback defined above is stabilizing hinges upon the following fact in nonsmooth analysis:

Lemma.

Suppose that $V(x) = g(x) + \gamma |x|^2$, where g is a concave function. Then for any $\zeta \in \partial_L V(x)$, we have

$$V(y) - V(x) \leq \langle \zeta, y - x \rangle + \gamma |y - x|^2 \ \forall\, y.$$

The proof of Theorem 6 can be mimicked when V is semiconcave rather than smooth, by invoking the 'decrease property' described in the lemma at a certain point. The essential step remains the comparison of the values of V at successive nodes; for the first two, for example, we have

$$V(x(t_1)) - V(x(t_0)) \leq \langle \zeta, x(t_1) - x(t_0) \rangle (t_1 - t_0) + \gamma |x(t_1) - x(t_0)|^2$$

(where $\zeta \in \partial_L V(x(t_0)$, by the lemma)

$$= \langle \zeta, f(x(t^*), k(t_0)) \rangle (t_1 - t_0) + \gamma |x(t_1) - x(t_0)|^2$$

(for some $t^* \in (t_0, t_1)$, by the mean value theorem)

$$\leq \langle \zeta, f(x(t_0), k(t_0)) \rangle (t_1 - t_0)$$
$$+ K_V K_f |x(t^*) - x(t_0)| (t_1 - t_0) + \gamma M^2 \delta^2$$

(where K_V and K_f are suitable Lipschitz constants for V and f)

$$\leq -W(x(t_0))\delta + K_V K_f M(1 + \gamma M)\delta^2,$$

by the way k is defined. Then, as before, a decrease in the value of V can be guaranteed by taking δ sufficiently small, and the proof proceeds as before. (The detailed argument must take account of the fact that V is only semiconcave away from the origin, and that a parameter γ as used above is available only on bounded subsets of $\mathbb{R}^n \backslash \{0\}$.)

6 Finite-Time Guidability

So far we have been concerned with possibly asymptotic approach to the origin. There is interest in being able to assert that the origin can be reached in finite time. If such is the case from any initial condition, then we say that the system is *globally guidable* in finite time (to 0). There is a well-studied local version of this property that bears the name *small-time local controllability* (STLC for short). A number of verifiable criteria exist which imply that the system has property STLC, which is stronger than Lyapunov stability; see [3].

Theorem 7. *The system is globally guidable in finite time if and only if there exists a general Lyapunov pair (V, W) with V finite-valued and $W \equiv 1$. If the system has the property STLC, then it is globally guidable in finite time iff there exists a Lyapunov pair (V, W) with V continuous and $W \equiv 1$.*

The proof of the theorem revolves around the much-studied minimal time function $T(\cdot)$. If the system is globally guidable in finite time, then $(T, 1)$ is the required Lyapunov pair: positive definiteness and properness are easily checked, and weak infinitesimal decrease follows from the (now well-known) fact that T satisfies the proximal Hamilton-Jacobi equation

$$h(x, \partial_P T(x)) + 1 = 0, \quad x \neq 0.$$

This is equivalent to the assertion that T is a viscosity solution of a related equation; see [10].

The sufficiency in the first part of the theorem follows much as in the proof of Theorem 2: we deduce the existence of a trajectory x for which $V(x(t)) + t$ is nonincreasing as long as $x(t) \neq 0$; this implies that $x(\tau)$ equals 0 for some $\tau \leq V(x(0))$. As for the second part of the theorem, it follows from the fact that, in the presence of STLC, the minimal time function is continuous.

7 An Equivalence Theorem

The following result combines and summarizes many of the ones given above concerning the regularity of Lyapunov functions and the presence of certain system properties.

Theorem 8. *The following are equivalent:*

1. *The system is open-loop globally asymptotically controllable.*
2. *There exists a continuous Lyapunov pair (V, W).*
3. *There exists a locally Lipschitz Lyapunov pair (V, W) with V semiconcave on $\mathbb{R}^n \backslash \{0\}$.*
4. *There exists a globally stabilizing sample-and-hold feedback.*

If, a priori, the system has Lyapunov stability at 0, then the following item may be added to the list:

5. *There exists for each positive ε a locally Lipschitz ε-Lyapunov function.*

If, a priori, the system has the property STLC, the following further item may be added to the list:

6. *There exists a continuous Lyapunov pair (V, W) with $W \equiv 1$.*

In this last case, the system is globally guidable in finite time.

8 Some Related Issues

Robustness.

It may be thought in view of the above that there is no advantage in having a smooth Lyapunov function, except the greater ease of dealing with derivatives rather than subdifferentials. In any case, stabilizing feedbacks will be discontinuous; and they can be conveniently defined in a pointwise fashion if the Lyapunov function is semiconcave. In fact, however, there is a robustness consequence to the existence of a smooth Lyapunov function.

The robustness of which we speak here is with respect to possible error e in state measurement when the feedback law is implemented: we are at x, but measure the state as $x + e$, and therefore apply the control $k(x + e)$ instead of the correct value $k(x)$. When k is continuous, then for e small enough this error will have only a small effect: the state may not approach the origin, but will remain in a neighborhood of it, a neighborhood that shrinks to the origin as e goes to zero; that is, we get practical stabilization. This feature of continuous feedback laws is highly desirable,and in some sense essential, since some imprecision seems inevitable in practice. One might worry that a discontinuous feedback law might not have this robustness property, since an arbitrarily small but nonzero e could cause $k(x)$ and $k(x + e)$ to differ significantly.

It is a fact that the (generally discontinuous) feedback laws constructed above do possess a *relative robustness* property: if, in the sample-and-hold implementation, the measurement error is at most of the same order of magnitude as the partition diameter, then practical stabilization is obtained. To put this another way, the step size may have to be big enough relative to the potential errors (to avoid dithering, for example). At the same time, the step size must be sufficiently small for stabilization to take place, so there is here a conflict that may or may not be reconcilable. It appears to us to be a great virtue of the sample-and-hold method that it allows, apparently for the first time, a precise error analysis of this type.

There is another, stronger type of robustness (called absolute robustness), in which the presence of small errors preserves practical stabilization independently of the step size. Ledyaev and Sontag [16] have shown that there exists an absolutely robust stabilizing feedback if and only if there exists a smooth Lyapunov pair. This, then, is an advantage that such systems have. Recall that the nonholonomic integrator, though stabilizable, does not admit a smooth Lyapunov function and hence fails to admit an absolutely robust stabilizing feedback.

State constraints.

There are situations in which the state x is naturally constrained to lie in a given closed set S, so that in steering the state to the origin, we must respect the condition $x(t) \in S$. The same questions arise as in the unconstrained case: is the possibility of doing this in the open-loop sense characterized by some kind of Lyapunov function, and would such a function lead to the definition of a stabilizing feedback that respects the state constraint? The more challenging case is that in which the origin lies on the boundary of S, but the case in which 0 lies in the interior of S is also of interest, since it localizes around the origin the global and constraint-free situation that has been the focus of this article.

An important consideration in dealing with state constraints is to identify a class of sets S for which meaningful results can be obtained. Recently Clarke and Stern [13, 12], for what appears to have been the first time, have extended many of the Lyapunov and stabilization methods discussed above to the case of state constraints specified by a set S which is *wedged* (see [10]). This rather large class of sets includes smooth manifolds with boundaries and convex bodies (as well as their closed complements). A set is wedged (or epi-Lipschitz) when its (Clarke) tangent cone at each point has nonempty interior, which is equivalent to the condition that locally (and after a change of coordinates), it is the epigraph of a Lipschitz function.

A further hypothesis is made regarding the consistency of the state constraint with the dynamics of the system: for every nonzero vector ζ in the (Clarke) normal cone to a point $x \in \operatorname{bdry} S$, there exists $u \in \mathcal{U}$ such that $\langle f(x,u), \zeta \rangle < 0$. Thus an 'inward-pointing' velocity vector is always available.

Under these conditions, and in terms of suitably defined extensions to the state-constrained case of the underlying definitions, one can prove an equivalence between open-loop controllability, closed-loop stabilization, and the existence of more or less regular (and in particular semiconcave) Lyapunov functions.

Regular and essentially stabilizing feedbacks.

In view of the fact that a GAC system need not admit a continuous stabilizing feedback, the question arises of the extent to which the discontinuities can be minimized. Ludovic Rifford has exploited the existence of a semiconcave Lyapunov function, together with both proximal and generalized gradient calculus, to show that when the system is affine in the control, there exists a stabilizing feedback whose discontinuities form a set of measure zero. Moreover, the discontinuity set is *repulsive* for the trajectories generated by the feedback: the trajectories lie in that set at most initially. This means that in applying the feedback, the solutions can be understood in the usual Carathéodory sense; robustness ensues as well. In the case of planar systems, Rifford has gone on to settle an open problem of Bressan by classifying the types of discontinuity that must occur in stabilizing feedbacks.

More recently, Rifford [21] has introduced the concept of *stratified* semiconcave Lyapunov functions, and has shown that every GAC system must admit one. Building upon this, he proves that there then exists a *smooth* feedback which *almost* stabilizes the system (that is, from almost all initial values). This highly interesting result is presented in Rifford's article in the present collection.

References

1. Aubin J (1991) Viability theory. Birkhäuser Boston

2. Bacciotti A, Rosier L (2001) Lyapunov functions and stability in control theory. Springer-Verlag London
3. Bardi M, Capuzzo-Dolcetta I (1997) Optimal control and viscosity solutions of Hamilton-Jacobi-Bellman equations. Birkhäuser Boston
4. Berkovitz L (1989) Optimal feedback controls, SIAM J Control Optim 27:991–1006
5. Cannarsa P, Sinestrari C (2003) Semiconcave functions Hamilton-Jacobi equations and optimal control. Birkhäuser Boston to appear
6. Clarke F (2001) Nonsmooth analysis in control theory a survey, European J Control 7:63–78
7. Clarke F, Ledyaev Y, Rifford L, Stern R (2000) Feedback stabilization and Lyapunov functions, SIAM J Control Optim 39:25–48
8. Clarke F, Ledyaev Y, Sontag E, Subbotin A (1997) Asymptotic controllability implies feedback stabilization, IEEE Trans Aut Control 42:1394–1407
9. Clarke F, Ledyaev Y, Stern R (1998) Asymptotic stability and smooth Lyapunov functions, J Differential Equations 149:69–114
10. Clarke F, Ledyaev Y, Stern R, Wolenski P (1998) Nonsmooth analysis and control theory. Springer-Verlag New York
11. Clarke F, Ledyaev Y, Subbotin A (1997) The synthesis of universal feedback pursuit strategies in differential games, SIAM J Control Optim 35:552–561
12. Clarke F, Stern R (2003) CLF and feedback characterizations of state constrained controllability and stabilization, preprint
13. Clarke F, Stern R (2003) State constrained feedback stabilization, SIAM J Control Optim 42:422-441
14. Freeman R, Kokotović P (1996) Robust nonlinear control design state space and Lyapunov techniques. Birkhäuser Boston
15. Krasovskii N, Subbotin A (1988) Game-theoretical control problems. Springer-Verlag New York
16. Ledyaev Y, Sontag E (1999) A Lyapunov characterization of robust stabilization, Nonlinear Analysis 37:813–840
17. Pontryagin L, Boltyanskii R, Gamkrelidze R, Mischenko E (1962) The mathematical theory of optimal processes. Wiley-Interscience New York
18. Rifford L (2000) Existence of Lipschitz and semiconcave control-Lyapunov functions, SIAM J Control Optim 39:1043–1064
19. Rifford L (2000) Problèmes de stabilisation en théorie du contrôle. PhD Thesis Université Claude Bernard Lyon I
20. Rifford L (2001) On the existence of nonsmooth control-Lyapunov functions in the sense of generalized gradients, ESAIM Control Optim Calc Var 6:593–611
21. Rifford L (2003) Stratified semiconcave control-Lyapunov functions and the stabilization problem, preprint
22. Sontag E (1983) A Lyapunov-like characterization of asymptotic controllability, SIAM J Control Optim 21:462–471
23. Sontag E (1999) Stability and stabilization discontinuities and the effect of disturbances. In: Clarke F, Stern R (eds) Nonlinear analysis differential equations and control. Kluwer Acad Publ Dordrecht

Techniques for Nonsmooth Analysis on Smooth Manifolds I: Local Problems

Yu. S. Ledyaev[1] and Qiji J. Zhu[2]

[1] Department of Mathematics, Western Michigan University, Kalamazoo, MI 49008 and Steklov Institute of Mathematics, Moscow 117966, Russia. ledyaev@wmich.edu

[2] Department of Mathematics, Western Michigan University, Kalamazoo, MI 49008. zhu@wmich.edu

1 Introduction

Nonsmooth analysis, differential analysis for functions without differentiability, has witnessed a rapid growth in the past several decades stimulated by intrinsic nonsmooth phenomena in control theory, optimization, mathematical economics and many other fields. In the past several years many problems in control theory, matrix analysis and geometry naturally led to an increasing interest in nondifferentiable functions on smooth manifolds. Since a smooth manifold only locally resembles a Euclidean space and, in general, lacks of a linear structure, new techniques are needed to adequately address these problems. A number of results and techniques for dealing with such problems have emerged recently [8, 16, 18, 32]. The purpose of this paper is to report some useful techniques that we developed in the past several years for studying nonsmooth functions on smooth manifolds.

Many competing definitions of generalized derivatives have been developed for nonsmooth functions on Banach spaces [2, 9, 10, 21, 22, 24, 25, 29, 30, 31, 33, 34]. We choose to focus on a generalization of the Fréchet subdifferential and its related limiting subdifferentials. These objects naturally fit the variational technique that we use extensively (see [27] for a comprehensive introduction to the variational techniques in finite dimensional spaces). Using these generalized derivative concepts in dealing various problems involving nondifferentiable functions on smooth manifolds, we find three techniques particularly helpful. They are (i) using a chain rule to handle local problems, (ii) finding forms of essential results in nonsmooth analysis on Banach spaces that do not rely on the linear structure of such spaces so that they can be developed on a smooth manifold and (iii) using flows. We divide our paper in two parts. In this first part, we discuss local problems such as deriving necessary optimality conditions for local minimization problems and calculus rules for subdifferentials. For these problems, it is often efficient to use the

M. de Queiroz et al. (Eds.): Optimal Control, Stabilization and Nonsmooth Analysis, LNCIS 301, pp. 283–297, 2004.

following scheme. First use a local coordinate system of the smooth manifold to project the problem into a Euclidean space. Then use results in nonsmooth analysis that have been established in such a Euclidean space to derive results for this corresponding problem in the Euclidean space. Finally, use the local coordinate mapping and its induced mapping between the tangent or cotangent spaces to lift the results back to the original problem on the manifold. Results related to methods (ii) and (iii) will be discussed in the second part of the paper [17].

We will introduce notation and necessary preliminaries about smooth manifolds in the next section. In Section 3 we introduce subdifferentials for lower semicontinuous functions on smooth manifolds and the related concepts of normal cones to closed sets. In Sections 4 we discuss how to use a chain rule to handle nonsmooth problems on smooth manifolds that are local in nature by a chain rule.

2 Preliminaries and Notation

In this section we recall some pertinent concepts and results related to a smooth manifold. Our main references are [5, 20, 28].

Let M be an N-dimensional C^∞ complex manifold (paracompact Hausdorff space) with a C^∞ atlas $\{(U_a, \psi_a)\}_{a\in A}$. For each a, the set of N components $(x_a^1, ..., x_a^N)$ of ψ_a is called a local coordinate system on (U_a, ψ_a). A function $g : M \to \mathbb{R}$ is C^r at $m \in M$ if $m \in U_a$ and $g \circ \psi_a^{-1}$ is a C^r function in a neighborhood of $\psi_a(m)$. Here r is a nonnegative integer or ∞. As usual C^0 represents the collection of continuous functions. It is well known that this definition is independent on the coordinate systems. If g is C^∞ at all $m \in M$, then we say g is C^∞ on M. The collection of all C^∞ (resp., C^r) functions on M is denoted by $C^\infty(M)$ (resp., $C^r(M)$). A map $v : C^\infty(M) \to \mathbb{R}$ is called a *tangent vector* of M at m provided that, for any $f, g \in C^\infty(M)$, (1) $v(\lambda f + \mu g) = \lambda v(f) + \mu v(g)$ for all $\lambda, \mu \in \mathbb{R}$ and (2) $v(f \cdot g) = v(f)g(m) + f(m)v(g)$. The collection of all the tangent vectors of M at m form an (N-dimensional) vector space and is denoted by $T_m(M)$. The union $\bigcup_{m\in M}(m, T_m(M))$ forms a new space called the *tangent bundle* to M, and denoted by $T(M)$. The dual space of $T_m(M)$ is called the *cotangent space* of M at m, denoted by $T_m^*(M)$. The *cotangent bundle* to M then is $T^*(M) := \bigcup_{m\in M}(m, T_m^*(M))$. We will use π (resp., π^*) to denote the canonical projection on $T(M)$ (resp., $T^*(M)$) defined by $\pi(m, T_m(M)) = m$ (resp., $\pi^*(m, T_m^*(M)) = m$). A mapping $X : M \to T(M)$ is called a *vector field* provided that $\pi(X(m)) = m$. A vector field X is C^∞ (resp., continuous) at $m \in M$ provided so is $X(g)$ for any $g \in C^\infty$. If a vector field X is C^∞ (resp., continuous) for all $m \in M$ we say it is C^∞ (resp., continuous) on M. The collection of all C^∞ vector fields on M is denoted by $V^\infty(M)$.

In particular, if (U, ψ) is a local coordinate neighborhood with $m \in U$ and $(x^1, ..., x^N)$ is the corresponding local coordinate system on (U, ψ) then

$(\partial/\partial x^n)_m$, $n = 1, ..., N$ defined by

$$(\partial/\partial x^n)_m g = \frac{\partial g \circ \psi^{-1}}{\partial x^n}(\psi(m))$$

is a basis of $T_m(M)$. Let g be a C^1 function at m. The differential of g at m, $dg(m)$, is an element of $T^*_m(M)$, and is defined by

$$dg(m)(v) = v(g) \quad \forall v \in T_m(M).$$

Let M_1 and M_2 be two C^∞ manifolds. Consider a map $\phi : M_1 \to M_2$. Then for every function $g \in C^\infty(M_2)$, ϕ induces a function $\phi^* g$ on M_1 defined by $\phi^* g = g \circ \phi$. A map $\phi : M_1 \to M_2$ is called C^r at $m \in M_1$ (on $S \subset M_1$) provided that so is $\phi^* g$ for any $g \in C^\infty(M_2)$. Let $\phi : M_1 \to M_2$ be a C^1 map and let $m \in M_1$ be a fixed element. Define, for $v \in T_m(M_1)$ and $g \in C^\infty(M_2)$, $((\phi_*)_m v)(g) = v(\phi^* g)$. Then $(\phi_*)_m : T_m(M_1) \to T_{\phi(m)}(M_2)$ is a linear map. The dual map of $(\phi_*)_m$ is denoted by ϕ^*_m. It is a map from $T^*_{\phi(m)}(M_2) \to T^*_m(M_1)$ and has the property that, for any $g \in C^1(M_2)$, $\phi^*_m dg(\phi(m)) = d(\phi^* g)(m)$.

Let $v^*_i \in T^*_{m_i}(M)$, $i = 1, 2, ...$ be a sequence of cotangent vectors of M and let $v^* \in T^*_m(M)$. We say v^*_i *converges to* v^*, denoted by $\lim v^*_i = v^*$, provided that $m_i \to m$ and, for any $X \in V^\infty(M)$, $\langle v^*_i, X(m_i)\rangle \to \langle v^*, X(m)\rangle$. Let (U, ψ) be a local coordinate neighborhood with $m \in U$. Since $m_i \to m$ we may assume without loss of generality that $m_i \in U$ for all i. Then $\lim v^*_i = v^*$ if and only if $\langle v^*_i, (\partial/\partial x^n)_{m_i}\rangle \to \langle v^*, (\partial/\partial x^n)_m\rangle$ for $n = 1, ..., N$. Another equivalent description is $(\psi^{-1})^*_{\psi(m_i)} v^*_i \to (\psi^{-1})^*_{\psi(m)} v^*$ (in the dual of $\mathbb{R}^N$).

3 Subdifferentials of Nonsmooth Functions on Manifolds

Now we turn to nonsmooth functions on a manifold and their subdifferentials. We denote the extended real line by $\bar{\mathbb{R}} := \mathbb{R} \cup \{+\infty\}$. For an extended-valued function $f : M \to \bar{\mathbb{R}}$ the domain of f is defined by $\mathrm{dom}(f) := \{m \in M : f(m) < \infty\}$.

Definition 1. *Let $f : M \to \bar{\mathbb{R}}$ be an extended-valued lower semicontinuous function. We define the* Fréchet-subdifferential *of f at $m \in \mathrm{dom}(f)$ by*

$$\partial_F f(m) := \{dg(m) : g \in C^1(M) \text{ and } f - g \textit{ attains a local minimum at } m\}.$$

We define the (limiting) subdifferential *and* singular subdifferential *of f at $m \in M$ by*

$$\partial_L f(m) := \{\lim v^*_i : v^*_i \in \partial_F f(m_i), (m_i, f(m_i)) \to (m, f(m))\}$$

and

$$\partial^\infty f(m) := \{\lim t_i v^*_i : v^*_i \in \partial_F f(m_i), (m_i, f(m_i)) \to (m, f(m)) \text{ and } t_i \to 0^+\},$$

respectively.

Remark 1. Note that the Fréchet-subdifferential of f depends only on the local behavior of the function f. Moreover, every local C^1 function can be extended to a C^1 function on M (see e.g. [20, Lemmas 1 and 2]). Therefore, the support function g in the definition of the Fréchet subdifferential need only be C^1 in a neighborhood of m.

The elements of a Fréchet subdifferential are called *Fréchet subgradients.* The Fréchet superdifferential $\partial^F f(m)$ coincides with the set $-\partial_F(-f)(m)$ and its elements are called *supergradients.* An alternative definition of a supergradient $v^* \in \partial^F f(m)$ is $v^* = dg$ for some C^1 function g such that $f - g$ attains a local maximum at m on M.

If $M = \mathbb{R}^n$, then a function g in the definition of Fréchet subgradient can be chosen to be a quadratic one. In this case the definition becomes a definition of a proximal subgradient which has a natural geometric interpretation in terms of normal vectors to an epigraph of function f. (It is useful to recall that in the case of a smooth function f the vector $(f'(x), -1)$ is a normal vector to its epigraph.) This geometric interpretation of subgradients also explains their relationship to a concept of generalized solutions of Hamilton-Jacobi equations in view of normal vector characterization of invariance properties of solutions of differential inclusions. For details on proximal calculus and related results, we refer to the textbook [13]. Corresponding proximal subdifferentials can be defined on a manifold under additional smoothness assumptions on the manifold (cf. [8]). However, discussions in [1, 36] show that the two definitions do not much differ in terms of applications. The other subdifferentials defined in Definition 1 coincide with the usual limiting and singular subdifferentials [22, 24]. Note that $\text{co}\,\{\partial_L f(x) + \partial^\infty f(x)\}$ coincides with the original Clarke's generalized gradient (cf. [10]).

Returning to the manifold case we note that it follows directly from the definition that $\partial_F f(m) \subset \partial_L f(m)$ and $0 \in \partial^\infty f(m)$. Note $\partial_F f(m)$ may be empty. However, if f attains a local minimum at m then $0 \in \partial_F f(m) \subset \partial_L f(m)$. These are the usual properties to be expected for a subdifferential.

The geometric concept of normal cones to a closed set can be naturally established as usual by using the subdifferential for the corresponding indicator function.

Definition 2. *Let S be a closed subset of M with $s \in S$. We define the* Fréchet-normal cone of S at s *by*

$$N_F(s; S) := \partial_F \delta_S(s).$$

We define the (limiting) normal cone of S at s *by*

$$N_L(s; S) := \partial_L \delta_S(s).$$

Here δ_S is the indicator function of S defined by $\delta_S(s) = 0$ if $s \in S$ and $\delta_S(s) = \infty$ if $s \notin S$.

Remark 2. (a) It is easy to verify that $N_F(s;S)$ is a cone and $N_L(s;S) := \partial_L \delta_S(s) = \partial^\infty \delta_S(s)$.

(b) It follows from the definition that $\{0\} \in N_F(s;S) \subset N_L(s;S)$ for any $s \in S$ and $N_F(s;S) = N_L(s;S) = \{0\}$ for any $s \in \text{int } S$. Thus, we will mainly be interested in nonzero normal cones. They are necessarily normal cones for the boundary points of S, denoted by $\text{bdy}(S)$.

The normal cone of a submanifold is dual to its tangent cone.

Proposition 1. [16] (Normal Cone of a Submanifold). *Let S be a C^1 submanifold of M. Then, for any $s \in S$,*

$$N_F(s;S) = T_s(S)^\perp := \{v^* \in T_s^*(M) : \langle v^*, v\rangle = 0, \forall v \in T_s(S)\}.$$

4 Local Problems

A useful technique in dealing problems that are local by nature is by using known results in a Euclidean space along the following lines. First, convert the problem into one in a Euclidean space via a local coordinate system. Then apply corresponding results for subdifferentials in a Euclidean space to the problem. Finally, lift the conclusion back onto the manifold.

4.1 A Chain Rule

The following simple chain rule is crucial for implementing the scheme alluded to above:

Theorem 1. *Let M and N be smooth manifolds, let $g : N \to M$ be a C^1 mapping and let $f : M \to \bar{\mathbb{R}}$ be a lower semicontinuous function. Suppose that $m = g(n)$. Then*

$$g_m^* \partial_F f(m) \subset \partial_F (f \circ g)(n), \tag{1}$$

$$g_m^* \partial_L f(m) \subset \partial_L (f \circ g)(n), \tag{2}$$

and

$$g_m^* \partial^\infty f(m) \subset \partial^\infty (f \circ g)(n). \tag{3}$$

Moreover, if g is a C^1 diffeomorphism then both sides of (1), (2) and (3) are equal.

Proof. Since (2) and (3) follow directly from (1) by taking limits, we prove (1). Let $y^* \in \partial_F f(m)$. Then there exists a C^1 function h such that $dh(m) = y^*$ and $f - h$ attains a local minimum at m. It follows that $f \circ g - h \circ g$ attains a local minimum at n. Observing that $h \circ g$ is a C^1 function on N, we have

$$\partial_F(f \circ g)(n) \ni d(h \circ g)(n) = g_m^* dh(m) = g_m^* y^*.$$

Thus,

$$g_m^* \partial_F f(m) \subset \partial_F(f \circ g)(n).$$

When g is a diffeomorphism, applying (1) to $f \circ g$ and g^{-1} yields the opposite inclusion. Q.E.D.

Applying Theorem 1 to $g = \psi^{-1}$ for a local coordinate mapping ψ yields the following corollary:

Corollary 1. *Let $f : M \to \bar{\mathbb{R}}$ be a lower semicontinuous function. Suppose that (U, ψ) is a local coordinate neighborhood and $m \in U$. Then*

$$\partial_F f(m) = \psi_m^* \partial_F(f \circ \psi^{-1})(\psi(m)),$$

$$\partial_L f(m) = \psi_m^* \partial_L(f \circ \psi^{-1})(\psi(m)),$$

and

$$\partial^\infty f(m) = \psi_m^* \partial^\infty(f \circ \psi^{-1})(\psi(m)).$$

4.2 Fuzzy Sum Rules

The fuzzy sum rule is one of the equivalent forms of several fundamental principles for subdifferentials (cf. [36]) which plays an important role in many applications of the subdifferentials. Its prototype appeared in [14]. We illustrate how to use the technique described in this section to deduce a version of the fuzzy sum rule on a smooth manifold.

Theorem 2. (Fuzzy Sum Rule) *Let $f_1, ..., f_L$ be lower semicontinuous functions on manifold M and let $\xi \in \partial_F(f_1 + ... + f_L)(\bar{m})$. Then, for any $\varepsilon > 0$, any neighborhood V of $\bar{m}$ and any $v \in V^\infty(M)$, there exist $m_l \in V$, $l = 1, 2, ..., L$ and $\xi_l \in \partial_F f_l(m_l)$ such that $|f_l(m_l) - f_l(\bar{m})| < \varepsilon$ and*

$$\left| \langle \xi, v \rangle(\bar{m}) - \sum_{l=1}^{L} \langle \xi_l, v \rangle(m_l) \right| < \varepsilon.$$

Proof. Without loss of generality we may assume that $(f_1 + ... + f_L)$ attains a local minimum at $\bar{m}$. Let (U, ψ) be a local coordinate system of M with $\bar{m} \in U$ and $\psi = (x^1, ..., x^N)$. Without loss of generality we may assume that $U \subset V$. It is obvious that the function $f_1 \circ \psi^{-1} + ... + f_L \circ \psi^{-1}$ attains a local minimum at $\bar{x} = \psi(\bar{m})$. Let $v \in V^\infty(M)$. Then $v(x^n) \circ \psi^{-1}, n = 1, 2, ..., N$ are C^∞ functions on $\mathbb{R}^n$. In particular, they are Lipschitz on $\psi(U)$, say with a uniform rank K. Set $\varepsilon' = \varepsilon / \{2(LK + \sup_{x \in \psi(U)} \|(v(x^1) \circ \psi^{-1}, ..., v(x^N) \circ \psi^{-1})(x)\|)\}$.

Applying the fuzzy sum rule [4, Theorem 2.6] on $\mathbb{R}^N$ we have that there exist $(x_l, f_l \circ \psi^{-1}(x_l)) \in (\bar{x}, f_l \circ \psi^{-1}(\bar{x})) + \varepsilon' B_{\mathbb{R}^{N+1}}$ satisfying

$$\|x_l^*\| \mathrm{diam}(x_1, x_2, ..., x_L) < \varepsilon'$$

and

$$\left|\sum_{l=1}^{L} x_l^*\right| < \varepsilon'.$$

Suppose

$$x_l^* = (a_l^1, a_l^2, ..., a_l^N).$$

Define

$$\xi_l := \sum_{n=1}^{N} a_l^n dx^n.$$

Then

$$\xi_l = \psi_{x_l}^* x_l^* \in \partial_F f_l(m_l), \text{ where } m_l = \psi^{-1}(x_l) \in V.$$

We then have

$$\sum_{l=1}^{L}\langle \xi_l, v\rangle = \sum_{l=1}^{L}\sum_{n=1}^{N} a_l^n \langle dx^n, v\rangle_{m_l}$$
$$= \sum_{l=1}^{L}\sum_{n=1}^{N} a_l^n v(x^n)_{m_l} = \sum_{l=1}^{L}\sum_{n=1}^{N} a_l^n v(x^n)\circ\psi^{-1}(x_l)$$
$$= \left\langle \sum_{l=1}^{L} x_l^*, (v(x^1)\circ\psi^{-1}, ..., v(x^N)\circ\psi^{-1})\right\rangle (x_1)$$
$$+ \sum_{l=1}^{L}\langle x_l^*, (v(x^1)\circ\psi^{-1}(x_l) - v(x^1)\circ\psi^{-1}(x_1), ..., v(x^N)\circ\psi^{-1}(x_l)$$
$$- v(x^N)\circ\psi^{-1}(x_1))\rangle.$$

Thus,

$$\left|\sum_{l=1}^{L}\langle \xi_l, v\rangle\right|$$
$$\leq \left\|\sum_{l=1}^{L} x_l^*\right\| \cdot \sup_{x\in\psi(U)} \|(v(x^1)\circ\psi^{-1}, ..., v(x^N)\circ\psi^{-1})(x)\|$$
$$+ \sum_{l=1}^{L} \|x_l^*\| K \operatorname{diam}(x_1, ..., x_L)$$
$$< \varepsilon.$$

Q.E.D.

Remark 3. In a Banach space this fuzzy sum rule can be refined to include additional information on the 'size' of the subderivatives involved (see [3]). This feature is lost here due to the lack of a metric in $T(M)$. However, a similar estimate can be established on a Riemann manifold.

Using the same technique we can also derive the following sum rule for limiting subdifferentials that generalizes corresponding results in Euclidean spaces (see [23, 4]):

Theorem 3. (Sum Rule) *Let $f_1, ..., f_L : M \to \bar{\mathbb{R}}$ be lower semicontinuous functions. Then, for any $m \in M$, either*

(A1) $$\partial_L \left(\sum_{l=1}^{L} f_l \right)(m) \subset \sum_{l=1}^{L} \partial_L f_l(m)$$

and

$$\partial^\infty \left(\sum_{l=1}^{L} f_l \right)(m) \subset \sum_{l=1}^{L} \partial^\infty f_l(m);$$

or there exist $v_l^\infty \in \partial^\infty f_l(m)$, $l = 1, ..., L$ not all zero such that

(A2) $$0 = \sum_{l=1}^{L} v_l^\infty.$$

The technique illustrated here applies to most 'local' results in nonsmooth analysis. Examples include other calculus rules such as the chain rule, product rule, quotient rule and subdifferential of the max function; the density of the domain of Fréchet subdifferential to a lower semicontinuous function; necessary optimization conditions for constrained minimization problems; a subdifferential representation of the superdifferential; etc... Details can be found in [16].

4.3 Mean Value Inequalities

Some problems that do not appear to be local can be reduced to local problems, so that the above techniques still apply. As an illustration, we deduce a smooth manifold version of Zagrodny's mean value inequality in [35].

Theorem 4. (Mean Value Inequality) *Let $f : M \to \bar{\mathbb{R}}$ be a lower semicontinuous function, let $v \in V^\infty(M)$ and let c be a curve corresponding to this vector field, i.e., $\dot{c}(t) = v(c(t)), t \in [0, 1]$, with $f(c(0)) < \infty$. Then, for any*

$$r < f(c(1)) - f(c(0)),$$

and any open neighborhood U of $c([0, 1])$, there exist $\bar{m} \in U$ and $\xi \in \partial_F f(\bar{m})$ such that $f(\bar{m}) < \min\{f(c(0)), f(c(1)) + \max\{0, r\}\}$ and

$$r < \langle \xi, v(\bar{m}) \rangle.$$

Proof. We consider first the simple case when (U, ψ) is a local coordinate system with local coordinates $(x^1, ..., x^N)$. For any $r < f(c(1)) - f(c(0))$, choose $\bar{r} \in (r, f(c(1)) - f(c(0)))$ and choose $\bar{\alpha} > 0$ such that $\alpha \in (0, \bar{\alpha})$ and

$$\frac{1}{\alpha^2}|\psi(m) - \psi(c(t))|^2 < f(c(0)) - \inf_M f + |\bar{r}|$$

for some $t \in [0,1]$ implies that $m \in U$. Consider the function

$$\varphi_\alpha(t) := f_\alpha(c(t)) - \bar{r}t, \alpha \in (0, \bar{\alpha}),$$

where

$$f_\alpha(m) := \min_{m' \in U}\{f(m') + \frac{1}{\alpha^2}|\psi(m) - \psi(m')|^2\}.$$

Then

$$\varphi_\alpha(1) - \varphi_\alpha(0) = f_\alpha(c(1)) - f_\alpha(c(0)) - \bar{r}.$$

Since $f_\alpha \to f$ as $\alpha \to 0$, when α is sufficiently small, we have

$$\varphi_\alpha(0) < \varphi_\alpha(1).$$

Assume that φ_α attains a minimum at $t_\alpha \in [0,1)$ over $[0,1]$. Consider the right Dini derivative

$$\begin{aligned}(\varphi_\alpha)'_+(t_\alpha) &= \liminf_{\lambda \to 0^+} \frac{\varphi_\alpha(t_\alpha + \lambda) - \varphi_\alpha(t_\alpha)}{\lambda} \\ &= \liminf_{\lambda \to 0^+} \Big[\frac{f_\alpha(c(t_\alpha + \lambda)) - f_\alpha(c(t_\alpha))}{\lambda} - \bar{r}\Big] \\ &\leq \liminf_{\lambda \to 0^+} \frac{1}{\lambda}\Big[(f(m_\alpha) + \frac{1}{\alpha^2}|\psi(c(t_\alpha + \lambda)) - \psi(m_\alpha)|^2) \\ &\quad - (f(m_\alpha) + \frac{1}{\alpha^2}|\psi(c(t_\alpha)) - \psi(m_\alpha)|^2)\Big] - \bar{r}\end{aligned}$$

where m_α is a point of minimum for the function

$$m \to f(m) + \frac{1}{\alpha^2}|\psi(c(t_\alpha)) - \psi(m)|^2.$$

Observing that

$$\begin{aligned}\frac{1}{\alpha^2}|\psi(c(t_\alpha)) - \psi(m_\alpha)|^2 &= \varphi_\alpha(t_\alpha) - f(m_\alpha) + \bar{r}t_\alpha \\ &\leq \varphi_\alpha(0) - f(m_\alpha) + \bar{r}t_\alpha \leq f(c(0)) - \inf_M f + |\bar{r}|.\end{aligned}$$

we can conclude that $m_\alpha \in U$. Then we have

$$\begin{aligned}(\varphi_\alpha)'_+(t_\alpha) &\leq \liminf_{\lambda \to 0^+} \frac{1}{\lambda\alpha^2}\Big[|\psi(c(t_\alpha + \lambda)) - \psi(m_\alpha)|^2 - |\psi(c(t_\alpha)) - \psi(m_\alpha)|^2\Big] \\ &\quad - \bar{r} \\ &= \frac{2}{\alpha^2}\sum_{n=1}^{N}\langle dx^n(c(t_\alpha)), \dot{c}(t_\alpha)\rangle(x^n(c(t_\alpha)) - x^n(m_\alpha)) - \bar{r}.\end{aligned}$$

Since $(\varphi_\alpha)'_+(t_\alpha) \geq 0$ we have

$$\frac{2}{\alpha^2}\sum_{n=1}^{N}\langle dx^n(c(t_\alpha)), \dot{c}(t_\alpha)\rangle(x^n(c(t_\alpha)) - x^n(m_\alpha)) \geq \bar{r}.$$

By the definition of subdifferentials we have

$$\xi_\alpha := \frac{2}{\alpha^2}\sum_{n=1}^{N}(x^n(c(t_\alpha)) - x^n(m_\alpha))dx^n(m_\alpha) \in \partial_F f(m_\alpha).$$

Then we have the following inequality:

$$\begin{aligned}\bar{r} &\leq \langle \xi_\alpha, v(m_\alpha)\rangle + \frac{2}{\alpha^2}\sum_{n=1}^{N}(x^n(c(t_\alpha)) - x^n(m_\alpha))[\langle dx^n(c(t_\alpha)), v(c(t_\alpha))\rangle \\ &- \langle dx^n(m_\alpha), v(m_\alpha)\rangle] \\ &= \langle \xi_\alpha, v(m_\alpha)\rangle + \frac{2}{\alpha^2}\sum_{n=1}^{N}(x^n(c(t_\alpha)) - x^n(m_\alpha))[v(x^n)(c(t_\alpha)) - v(x^n)(m_\alpha)] \\ &= \langle \xi_\alpha, v(m_\alpha)\rangle + \frac{2}{\alpha^2}\sum_{n=1}^{N}(x^n(c(t_\alpha)) - x^n(m_\alpha))[(v(x^n)\circ\psi^{-1})(\psi(c(t_\alpha))) \\ &- (v(x^n)\circ\psi^{-1})(\psi(m_\alpha))]\end{aligned}$$

Since $v(x^n)\circ\psi^{-1}$ is C^1 and, therefore, locally Lipschitz around $\varphi(c(t_\alpha))$ we can conclude that there exists a constant K such that

$$|(v(x^n)\circ\psi^{-1})(\psi(c(t_\alpha))) - (v(x^n)\circ\psi^{-1})(\psi(m_\alpha))| \leq K|\psi(c(t_\alpha)) - \psi(m_\alpha)|.$$

Thus we have

$$\bar{r} \leq \langle \xi_\alpha, v(m_\alpha)\rangle + \frac{2KN}{\alpha^2}|\psi(c(t_\alpha)) - \psi(m_\alpha)|^2. \tag{4}$$

Next we show that

$$\lim_{\alpha\to 0^+}\frac{1}{\alpha^2}|\psi(c(t_\alpha)) - \psi(m_\alpha)|^2 = 0. \tag{5}$$

By the definition of φ_α it is increasing when $\alpha \to 0^+$ and so is $\varphi_\alpha(t_\alpha)$. Moreover $\varphi_\alpha(t_\alpha) \leq \varphi_\alpha(0) \leq f(c(0))$ and, therefore, $\lim_{\alpha\to 0^+}\varphi_\alpha(t_\alpha)$ exists. By the definition of t_α and m_α we have

$$\begin{aligned}\varphi_{2\alpha}(t_{2\alpha}) \leq \varphi_{2\alpha}(t_\alpha) &\leq f(m_\alpha) + \frac{1}{4\alpha^2}|\psi(c(t_\alpha)) - \psi(m_\alpha)|^2 - \bar{r}t_\alpha \\ &= \varphi_\alpha(t_\alpha) - \frac{3}{4\alpha^2}|\psi(c(t_\alpha)) - \psi(m_\alpha)|^2,\end{aligned}$$

that is to say

$$\frac{1}{\alpha^2}|\psi(c(t_\alpha)) - \psi(m_\alpha)|^2 \le \frac{4}{3}(\varphi_\alpha(t_\alpha) - \varphi_{2\alpha}(t_{2\alpha}))).$$

Taking limits as $\alpha \to 0^+$ yields (5). Now taking α small enough so that $m_\alpha \in U$ and

$$\bar{r} - \frac{2KN}{\alpha^2}|\psi(c(t_\alpha)) - \psi(m_\alpha)|^2 > r$$

we have, by (4), that

$$r < \langle \xi_\alpha, v(m_\alpha)\rangle.$$

Moreover,

$$\begin{aligned} f(m_\alpha) &\le f(m_\alpha) + \frac{1}{\alpha^2}|\psi(c(t_\alpha)) - \psi(m_\alpha)|^2 \\ &= \varphi_\alpha(t_\alpha) \le \min\{\varphi_\alpha(0), \varphi_\alpha(1)\} \\ &\le \min\{f(c(0)), f(c(1)) + \max\{0, r\}\}. \end{aligned}$$

It remains to take $\bar{m} = m_\alpha$ and $\xi = \xi_\alpha$.

For the general case, since there are a finite number of coordinate neighborhoods covering the compact set $c([0,1])$ we may assume that $U = \cup_{i=1}^k U_j$, where $(U_i, \psi_i), i = 1, ..., k$ are coordinate neighborhoods. Let $\{\chi_i : i = 1, ..., k\}$ be a C^1 partition of unity for U corresponding to $U_i, i = 1, ..., k$. Then we can define, for $m \in U$,

$$f_\alpha(m) := \min_{m' \in U}\{f(m') + \frac{1}{\alpha^2}\sum_{i=1}^{k}\chi_i(m')|\psi_i(m) - \psi_i(m')|^2\}.$$

Applying the above argument to f_α and supposing that $c(t_\alpha) \in U_j$, for some $j \in \{1, \ldots, k\}$, everything goes through with (U, ψ) replaced by (U_j, ψ_j). Q.E.D.

It is interesting to observe that this problem is not 'local' in the sense that the curve $c([0,1])$ is, in general, not contained in one local coordinate system. Thus, the 'lifting' approach in the proof of the fuzzy sum rule does not directly apply. Note how partition of unity is used here to reduce the problem to the situation that involves only one local coordinate system. This approach, combined with the subdifferential representation of the superdifferential, was also used in [16] to derive a version of the mean value inequality with a lower estimate on the function value $f(\bar{m})$ when f is continuous following the model in [15].

On Riemann manifolds, a similar method can also be used to derive a subdifferential criterion for a function to be Lipschitz and an implicit function theorem (see [16, Sections 5.1 and 5.2]).

4.4 Subdifferential of Spectral Functions

Additional properties of the function under consideration may lead to further refinement of the general result. Submanifolds generated by group actions provide a convenient framework for discussing symmetric properties of functions.

We illustrate this point by presenting the recent breakthrough in calculating subdifferentials for spectral functions due to Lewis, Burke and Overton in the framework of nonsmooth functions (the spectral functions) on smooth manifolds (various matrix manifolds).

Let M be the space of complex $N \times N$ matrices. For any $X \in M$ we use $\lambda_1(X), \ldots, \lambda_N(X)$ to denote the N (including repeated) eigenvalues of X in the lexicographical order of real and imaginary parts. We call $\lambda(X) := (\lambda_1(X), \ldots, \lambda_N(X))$ the *eigenvalue mapping.* A spectral function is a function of the form $\phi := f \circ \lambda : M \to \bar{\mathbb{R}}$ where f is invariant under permutation of its variables. The concept of a spectral function encompasses many useful functions related to the eigenvalue mapping such as the *spectral abscissa* $\max\{\mathrm{Re}\ \lambda_n, n = 1, \ldots, N\}$, the *spectral radius* $\max\{|\lambda_n|, n = 1, \ldots, N\}$, the *determinant det* and the *trace tr*. These functions are often intrinsically nonsmooth which makes analyzing their properties difficult. Recently in a series of papers, J. Burke, A. Lewis, and M. Overton made a breakthrough in the analysis of various subdifferentials of the spectral functions (see [6, 7, 19] and their references). In this section we illustrate how to understand their work by using the nonsmooth analysis tools on manifolds established in the previous sections.

As usual we define an inner product on M by

$$\langle X, Y \rangle := tr(X^* Y) = \sum_{n,k=1}^{N} \overline{x_{n,k}} y_{n,k}, \quad X, Y \in M$$

and norm $\|X\| := \sqrt{\langle X, X \rangle}$. We use $GL(N)$ and $O(N)$ to denote the set of all invertible and orthogonal matrices in M, respectively. For $U \in GL(N)$ (resp., $U \in O(N)$) we define a mapping $u : M \to M$ by

$$u(X) = U^{-1} X U \qquad (\text{resp., } \ u(X) = U^* X U).$$

Then u is a diffeomorphism (in fact, a linear invertible transform). Moreover, it is easy to calculate that $u^* : T^*(M) \to T^*(M)$ is defined by $u^*(Y) = UYU^{-1}$ (resp., $u^*(Y) = UYU^*$). Let ϕ be a spectral function on M and $Z = u(X)$ for $U \in GL(N)$ (or $U \in O(N)$). It follows directly from the smooth chain rule of Theorem 1 that

$$u^* \partial \phi(Z) = \partial(\phi \circ u)(X),$$

where $\partial = \partial_F, \partial_L$ or ∂^∞. It is easy to see that $\phi \circ u = \phi$. Thus, we have the following lemma:

Lemma 1. *Let ϕ be a spectral function on M and let $U \in GL(N)$ (resp., $U \in O(N)$). Then*

$$\partial \phi(U^{-1} X U) = U^{-1} \partial \phi(X) U,$$

$$(\text{resp., } \partial \phi(U^* X U) = U^* \partial \phi(X) U),$$

where $\partial = \partial_F, \partial_L$ or ∂^∞.

Next we consider the $GL(N)$ and $O(N)$ orbit of $X \in M$ defined by $GL(N) \cdot X := \{U^{-1}XU : U \in GL(N)\}$ and $O \cdot X := \{U^*XU : U \in O(N)\}$, respectively. It is well known that $GL(N) \cdot X$ and $O(N) \cdot X$ are submanifolds of M and their normal spaces at X are described in the lemma below.

Lemma 2. *Let $X \in M$. Then*

$$N_F(X; O(N) \cdot X) = N_F(X; GL(N) \cdot X) = \{Y \in M : XY^* - Y^*X = 0\}.$$

Lemma 3. *Let ϕ be a spectral function on M. Then $Y \in \partial\phi(X)$ implies that $XY^* = Y^*X$, where $\partial = \partial_F, \partial_L$ or ∂^∞.*

Proof. We need only to prove the case when $\partial = \partial_F$. The rest follows by a limiting process. Observe that by the definition of the Fréchet subdifferential we have

$$\partial_F\phi(X) \subset N_F(X; \phi^{-1}(-\infty, \phi(X)]).$$

Since ϕ is a constant on $O(N) \cdot X$, we have $O(N) \cdot X \subset \phi^{-1}(-\infty, \phi(X)]$. Thus,

$$\partial_F\phi(X) \subset N_F(X; O(N) \cdot X).$$

The lemma follows from the representation of $N_F(O(N) \cdot X; X)$ in Lemma 2. Q.E.D

Now we can deduce the key result that can help us understand the relationship between the subdifferential of a spectral function $\phi = f \circ \lambda$ and that of f. For any $X \in M$, we will use $\operatorname{diag} X$ to denote the diagonal of X as a vector in C^N, i.e., $\operatorname{diag} X = (x_{11}, \ldots, x_{NN})$.

Theorem 5. *Let ϕ be a spectral function on M and let $Y \in \partial\phi(X)$. Then there exists a matrix $U \in O(N)$ such that $R = U^*XU$ is upper triangular with $\operatorname{diag} R = \lambda(X)$ and $S = U^*YU$ is lower triangular and*

$$S \in \partial\phi(R), \tag{6}$$

where $\partial = \partial_F, \partial_L$ or ∂^∞.

Proof. By Lemma 3 $XY^* = Y^*X$. The existence of U as described in the theorem is a direct consequence of the Schur Theorem. The conclusion then follows from Lemma 1. Q.E.D.

Similar results hold when M is the space of Hermitian matrices. In this case S and R are both diagonal matrices. Suppose that $\phi = f \circ \lambda$. Lewis showed in [19] that if S and R are diagonal, then (6) is equivalent to

$$\operatorname{diag} S \in \partial f(\lambda(X)).$$

Burke and Overton showed in [6, 7] that the necessity holds for general S and R. While these relationships are highly nontrivial, we can see that they can be understood as nonsmooth chain rules on the smooth manifolds of matrices.

Acknowledgement. The idea of developing nonsmooth analysis on manifolds came from discussions with E. Sontag. We thank J. M. Borwein, F. H. Clarke and E. Sontag for stimulating conversations and we thank J. S. Treiman for helpful discussions. This research was supported in part by NSF grant #0102496.

References

1. Borwein J, Ioffe A (1996) Proximal analysis in smooth spaces, Set-Valued Analysis 4:1-24
2. Borwein J, Preiss D (1987) A smooth variational principle with applications to subdifferentiability and to differentiability of convex functions, Trans Amer Math Soc 303:517-527
3. Borwein J, Zhu Q (1996) Viscosity solutions and viscosity subderivatives in smooth Banach spaces with applications to metric regularity, SIAM J Control and Optimization 34:1568-1591
4. Borwein J, Zhu Q (1999) A survey of subdifferential calculus with applications, Nonlinear Analysis TMA 38:687-773
5. Brickell F, Clark R (1970) Differentiable Manifolds An Introduction. VN Reinhold Co New York
6. Burke J, Overton M (1992) On the subdifferentiability of functions of a matrix spectrum II Subdifferential formulas. In: Nonsmooth Optimization Methods and Applications (Erice 1991). Gordon and Breach Montreux 19–29
7. Burke J, Overton M (1994), Differential properties of the spectral abscissa and the spectral radius for analytic matrix-valued mappings, Nonlinear Anal 23:467–488.
8. Chryssochoos I, Vinter R (2001) Optimal control problems on manifolds a dynamical programming approach, preprint
9. Clarke F (1973) Necessary Conditions for Nonsmooth Problems in Optimal Control and the Calculus of Variations. PhD Thesis Univ of Washington
10. Clarke F (1983) Optimization and Nonsmooth Analysis. John Wiley & Sons New York
11. Clarke F, Ledyaev Y (1994) Mean value inequalities, Proc Amer Math Soc 122:1075-1083
12. Clarke F, Ledyaev Y (1994) Mean value inequalities in Hilbert space, Trans Amer Math Soc 344:307-324
13. Clarke F, Ledyaev Y, Stern R, Wolenski P (1998) Nonsmooth Analysis and Control Theory. Springer-Verlag New York
14. Ioffe A (1983) On subdifferentiablility spaces, Ann New York Acad Sci 410:107-119
15. Ledyaev Y, Treiman J (2004) Sub- and supergradients of envelopes closures and limits, Mathematicshe Annalen submitted
16. Ledyaev Y, Zhu Q (2004) Nonsmooth analysis on smooth manifolds, Transactions of AMS submitted
17. Ledyaev Y, Zhu Q (2004) Techniques for nonsmooth analysis on smooth manifolds Part II Deformations and Flows, this collection
18. Ledyaev Y, Zhu Q (2004) Multidirectional mean value inequalities and weak monotonicity, Journal of London Mathematical Society submitted
19. Lewis A (1999) Nonsmooth analysis of eigenvalues, Mathematical Programming 84:1-24
20. Matsushima Y (1972) Differential Manifolds. Marcel Dekker New York
21. Michel P, Penot J (1985) Calcul sous-différential pour des fonctions Lipschitziennes et non-Lipschiziennes, CR Acad Sci Paris Ser I Math 298:269-272
22. Mordukhovich B (1976) Maximum principle in problems of time optimal control with nonsmooth constraints, J Appl Math Mech 40:960-969

23. Mordukhovich B (1984) Nonsmooth analysis with nonconvex generalized differentials and adjoint mappings, Dokl Akad Nauk BSSR 28:976-979
24. Mordukhovich B (1988) Approximation Methods in Problems of Optimization and Control. Nauka Moscow
25. Pshenichnyi B (1969) Necessary conditions for an extremum. Nauka Moscow
26. Rockafellar R (1970) Convex Analysis. Princeton University Press Princeton
27. Rockafellar R, Wets J (1998) Variational Analysis. Springer Berlin
28. Sternberg S (1964) Lectures on differential geometry. Prentice-Hall Englewood Cliffs NJ
29. Sussmann H (1994) A strong version of the maximum principle under weak hypotheses. In: Proceedings of the 33rd IEEE Conference on Decision and Control
30. Sussmann H (1996) A strong maximum principle for systems of differential inclusions. In: Proceedings of the 35th IEEE Conference on Decision and Control
31. Treiman J (1989) Finite dimensional optimality conditions B-gradients, J Optim Theory Appl 62:139-150
32. Treiman J, Zhu Q (2004) Extremal principles on smooth manifolds and applications, in preparation
33. Warga J (1976) Derivate containers, inverse functions, and controllability. In: Russell D (ed) Calculus of Variations and Control Theory. Academic Press New York
34. Warga J (1981) Fat homeomorphisms and unbounded derivate containers, J Math Anal Appl 81:545-560
35. Zagrodny D (1988) Approximate mean value theorem for upper subderivatives, Nonlinear Anal TMA 12:1413-1428
36. Zhu Q (1998) The equivalence of several basic theorems for subdifferentials, Set-Valued Analysis 6:171-185

Techniques for Nonsmooth Analysis on Smooth Manifolds II: Deformations and Flows

Yu. S. Ledyaev[1] and Qiji J. Zhu[2]

[1] Department of Mathematics, Western Michigan University, Kalamazoo, MI 49008 and Steklov Institute of Mathematics, Moscow 117966, Russia. ledyaev@wmich.edu

[2] Department of Mathematics, Western Michigan University, Kalamazoo, MI 49008. zhu@wmich.edu

1 Introduction

Nonsmooth analysis, differential analysis for functions without differentiability, has witnessed a rapid growth in the past several decades stimulated by intrinsic nonsmooth phenomena in control theory, optimization, mathematical economics and many other fields. In the past several years many problems in control theory, matrix analysis and geometry naturally led to an increasing interest in nondifferentiable functions on smooth manifolds. Since a smooth manifold is only locally resembles a Euclidean space and, in general, lacks of a linear structure, new techniques are needed for adequately address these problems. A number of results and techniques for dealing with such problems have emerged recently [6, 13, 15, 25].

This is the continuation of our paper [14] in this collection reporting some useful techniques that we developed in the past several years for studying nonsmooth functions on smooth manifolds. In this second part of the paper we discuss the following two issues. (a) Unlike a Banach space, a smooth manifold in general does not have a linear structure. Thus, many known results in Banach spaces that depend on the linear structure of such spaces cannot be directly extended to smooth manifolds. The extremal principles developed in [12, 17, 19], the nonconvex separation theorems discussed in [2, 27] and the multidirectional mean value theorems derived in [8, 9] are typical examples. For this kind of problem, we often need to look at ways to reformulate the original results in a form that does not depend on the linear structure. This reformulation then can be generalized to the setting of a smooth manifold. (b) Using flows determined by vector fields or multi-vector fields in the form of a differential inclusion in the tangent bundle of a manifold is often valuable in dealing with problems on smooth manifolds (both smooth and nonsmooth). For example, a geodesic in a Riemann manifold and its corresponding vector field is a natural replacement of a line segment and its generating vector in a

M. de Queiroz et al. (Eds.): Optimal Control, Stabilization and Nonsmooth Analysis, LNCIS 301, pp. 299–311, 2004.

Euclidean space. Thus, it is not surprising that flows, vector fields and multi-vector fields become the language for describing the analogue of a line or a convex set on a smooth manifold.

We will continue to use the notation and preliminaries from Part I of this paper [14] along with some additional preliminaries in the next section. In the rest of the paper we discuss the two techniques alluded to above and illustrate how to use them by sample results.

2 Preliminaries

We need some additional notations and preliminaries on Riemann manifolds. Again our main references are [5, 16, 22].

Recall that a mapping $g : T(M) \times T(M) \to \mathbb{R}$ is a C^∞ *Riemann metric* if

(1) for each m, $g_m(v, u)$ is a inner product on $T_m(M)$;
(2) if (U, ψ) is a local coordinate neighborhood around m with local coordinate system $(x^1, ..., x^N)$, then

$$g_{ij}(m) := g_m\left(\frac{\partial}{\partial x^i}, \frac{\partial}{\partial x^j}\right) \in C^\infty(M).$$

One can check that (2) is independent of local coordinate systems. The manifold M together with the Riemann metric g is called a *Riemann manifold.* Since any paracompact manifold admits a positive-definite metric structure, in many cases we may assume that M is a Riemann manifold without significant loss of generality.

Let (M, g) be a Riemann manifold . For each $m \in M$, the Riemann metric induces an isomorphism between $T_m(M)$ and $T^*_m(M)$ by

$$v^* = g_m(v, \cdot) \quad (\langle v^*, u\rangle = g_m(v, u), \ \forall u \in T_m(M)).$$

Then we define norms on $T_m(M)$ and $T^*_m(M)$ by

$$\|v^*\|^2 = \|v\|^2 = g_m(v, v).$$

The following generalized Cauchy inequality is crucial: for any $v^* \in T^*_m(M)$ and $u \in T_m(M)$,

$$\langle v^*, u\rangle \le \|v^*\|\|u\|.$$

Let $r : [0, 1] \to M$ be a C^1 curve. The length of r is

$$l(r) = \int_0^1 \|r'(s)\|ds.$$

Let $m_1, m_2 \in M$. Denote the collection of all C^1 curves joining m_1 and m_2 by $C(m_1, m_2)$. Then the distance between m_1 and m_2 is defined by

$$\rho(m_1, m_2) := \inf\{l(r) : r \in C(m_1, m_2)\}.$$

The distance between a point $m \in M$ and a set $S \subset M$ is defined by $\rho(m, S) := \inf\{\rho(m, m') : m' \in S\}$. Finally, we recall the definition of an absolutely continuous mapping $c : [a, b] \to M$. A mapping c is called *absolutely continuous* provided the function $\varphi(c(t))$ is absolutely continuous for any smooth function $\varphi \in C^\infty(M)$. We call such a mapping a *curve* and leave to the reader to prove that for any curve c there exists a mapping $\dot{c}$ such that $\dot{c}(t) \in T_{c(t)}(M)$ for almost all (a.a.) $t \in [a, b]$ and, for any $\varphi \in C^\infty(M)$,

$$\frac{d(\varphi \circ c)(t)}{dt} = \dot{c}(t)(\varphi) \quad \text{a.a. } t \in [a, b].$$

When M is a Riemann manifold we have the following relationship between the normal cone and the subdifferential of the distance function to a set:

Proposition 1. [13] (Normal Cone Representation on Riemann Manifolds). *Let M be a Riemann manifold and let S be a closed subset of M. Then, for any $s \in S$,*

$$N_F(s; S) = \operatorname{cone} \partial_F \rho(s; S).$$

To use flows on a smooth manifold we need to define the concept of a differential inclusion on the tangent bundle. Consider a multifunction F defined on the manifold M whose values are compact convex subsets of the corresponding tangent space, namely,

$$F(m) \subset T_m(M), \quad \forall\, m \in M$$

We say F is *upper semicontinuous at m* provided that $m_i \to m$ and $F(m_i) \ni v_i \to v$ implies that $v \in F(m)$.

We always assume that F is an upper semicontinuous convex compact valued multifunction. The differential inclusion

$$\dot{c}(t) \in F(c(t)), \quad c(0) = m_0, \tag{1}$$

is a well known object (see [1, 11]).

Definition 1. *An absolutely continuous function $c : [0, T] \to M$ is called a solution of the differential inclusion if the inclusion (1) holds for almost all (a.a.) $t \in [0, T]$.*

Let (U, ψ) be a local coordinate neighborhood around m_0. Then it is not hard to see that an absolute continuous function c is a local solution to (1) if and only if $\psi \circ c$ is a local solution to the differential inclusion

$$\frac{d(\psi \circ c)(t)}{dt} \in \psi_* F(\psi^{-1}((\psi \circ c)(t))), \quad \psi \circ c(0) = \psi(m_0). \tag{2}$$

Existence theorems of local solutions to a differential inclusion on manifolds follow directly from corresponding results for differential inclusions in Euclidean spaces. The following is an example:

Theorem 1. *Let F be an upper semicontinuous multifunction with compact and convex values. Then, for any $v \in F(m_0)$, differential inclusion (1) has a local solution c with $c(0) = m_0$ and $c'(0) = v$.*

Proof. Let $v(m')$ be a local extension of v and let $u(m')$ be the projection of $v(m')$ on $F(m')$. Then $u(\cdot)$ is a (local) continuous selection of $F(\cdot)$ and $u(m_o) = v$. We need only to take c be the solution of $\dot{c}(t) = u(c(t))$, $c(0) = m_0$. Q.E.D.

3 Using Deformations

A major difference between the study of nonsmooth problems on smooth manifolds and those on a Banach space is that the linear structure of a Banach space is no longer there. We find that the key to overcome this difficulty often lies in finding modified forms of the Banach space results that do not depend on the linear structure.

We illustrate by providing an extremal principle on a smooth manifold that generalizes the extremal principles of Kruger and Mordukhovich [12, 17, 19]. The original extremal principle dealing with extremal systems of sets relies on shifts of some of the sets. Mordukhovich, Treiman and Zhu [20] generalized this definition for multifunctions and derived corresponding extremal principles that generalized the Kruger-Mordukhovich extremal principle. The essential idea is to define an extremal system through the deformation (i.e., a nonlinear process) of sets rather than the traditional shifting (which depends on the linear structure of the space). It turns out that this is an appropriate form of the extremal principle that can be generalized to a smooth manifold. We now state this result.

Definition 2. *Let M be a smooth manifold and let $D_i, i = 1, 2, ..., I$ be I* metric spaces. *Consider closed-valued multifunctions $S_i : D_i \to X$, $i = 1, 2, ..., I$. We say that $\bar{m}$ is an* extremal point *of the* extremal system *$(S_1, S_2, ..., S_I)$ at $(\bar{s}_1, \bar{s}_2, ..., \bar{s}_I)$ provided that $\bar{m} \in S_1(\bar{s}_1) \cap S_2(\bar{s}_2) \cap ... \cap S_I(\bar{s}_I)$ and there exist a neighborhood U of $\bar{m}$ such that, for any $\varepsilon > 0$, and any neighborhood V of $\bar{m}$ there exist $(s_1, s_2, ..., s_I) \in B_\varepsilon((\bar{s}_1, \bar{s}_2, ..., \bar{s}_I))/\{(\bar{s}_1, \bar{s}_2, ..., \bar{s}_I)\}$, with $V \cap S_i(s_i)) \neq \emptyset, i = 1, 2, ..., I$, and*

$$U \cap S_1(s_1) \cap S_2(s_2) \cap ... \cap S_I(s_I) = \emptyset.$$

To state our extremal principle we need the following condition:

Definition 3. [20] *We say that a closed-valued multifunction S from a metric space Y to the tangent bundle of a smooth manifold M is* normally semicontinuous *at (x, y) provided that,*

$$\text{Limsup}_{x \to \bar{x}, y \to \bar{y}} N_F(x; S(y)) \subset N(\bar{x}; S(\bar{y})),$$

where 'Limsup' stands for the sequential Painlevé-Kuratowski upper limit.

Theorem 2. *Let M be a smooth manifold and let $D_i, i = 1, 2, ..., I$ be metric spaces. Consider closed-valued multifunctions $S_i : D_i \to M$, $i = 1, 2, ..., I$. Suppose that $\bar{m}$ is an extremal point of the extremal system $(S_1, S_2, ..., S_I)$ at $(\bar{s}_1, \bar{s}_2, ..., \bar{s}_I)$. Suppose that $S_i, i = 1, 2, ..., I$ are normal upper semicontinuous at $(\bar{s}_i, \bar{m}), i = 1, 2, ..., I$, respectively. Then there exist $x_i^* \in N(S_i(\bar{s}_i), \bar{x}), i = 1, 2, ..., I$, not all zero, such that*

$$\sum_{i=1}^{I} x_i^* = 0.$$

Sketch of the Proof. Let (U, ψ) be a local coordinate system around $\bar{m}$. Without loss of generality we may assume U is closed and $\bar{m}$ is in the interior of U. Then $\psi(\bar{m})$ is an extremal point of the extremal system $(\psi(S_1 \cap U), \ldots, \psi(S_I \cap U))$. Noting that the normal semicontinuity of S_i is preserved under the mapping ψ, we can apply the extremal principle in [20] to this extremal system (in a Euclidean space). Now we can derive the desired extremal principle on the smooth manifold by lifting the conclusion back with Theorem 1 in [14]. Q.E.D.

As pointed out in [20] this generalization leads to new applications even in Banach spaces. A similar idea can also be used to derive a generalization (see [27]) of a nonconvex separation theorem due to Borwein and Jofre [2] which is useful in analyzing the second welfare theorem in economics.

4 Using Flows on a Smooth Manifold

As pointed out before, a geodesic on a Riemann manifold and its tangent vector field play a role similar to a line in a Euclidean space and its direction vector. Arguably, a class of vector fields corresponding to geodesics will be a reasonable substitute for a convex set. We can put this in the more general framework of differential inclusions and its solution set. In fact, it is shown in [15], in the Banach space setting, that a powerful multidirectional mean value inequality due to Clarke and Ledyaev [8] can be restated in the form of a weak monotonicity result for lower semicontinuous functions with respect to the solution of a differential inclusion. The multidirectional mean value inequality in [8] is stated in reference to a convex set and, therefore, does not have a direct generalization on a smooth manifold. Thus, the weak monotonicity reformulation in [15] provides a good candidate for us to generalize to the setting of a smooth manifold. In this section, we showcase how to use flows on a smooth manifold, in the form of solution sets to a differential inclusion, by implementing the generalization of the weak monotonicity result mentioned above. We also discuss some of the applications for this result.

4.1 Weak Monotonicity

Recall the definition of weak monotonicity:

Definition 4. *A function* $\varphi : M \to \bar{\mathbb{R}}$ *is called* weak monotone decreasing with respect to solutions of (1) *provided for any initial point* $m_0 \in M$ *there exist a* $\tau > 0$ *and a solution* c *of (1) on* $[0, \tau)$ *such that*

$$\varphi(c(t)) \le \varphi(c(0)), \quad \forall t \in [0, \tau). \tag{3}$$

We characterize the weak monotonicity in terms of the lower Hamiltonian

$$h(m, p) := \inf_{v \in F(m)} \langle p, v \rangle. \tag{4}$$

We will also need the upper Hamiltonian

$$H(m, p) := \sup_{v \in F(m)} \langle p, v \rangle. \tag{5}$$

The following assumption is needed:

Assumption (H): F is bounded in the sense that for any function $\varphi \in C^1(M)$,

$$H(m, d\varphi(m)) < \infty, \forall\, m \in M.$$

Theorem 3. *Let* F *satisfy Assumption* (H) *and let* $\varphi : M \to \bar{\mathbb{R}}$ *be a lower semicontinuous function. Then the following are equivalent:*

(i) φ *is weakly monotone decreasing with respect to the solutions of (1).*
(ii) $h(m, p) \le 0$, *for any* $m \in M$ *and* $p \in \partial_F \varphi(m)$.
(iii) $h(m, p) \le 0$, *for any* $m \in M$ *and* $p \in \partial_L \varphi(m) \cup \partial^\infty \varphi(m)$.

Proof. (i) implies (ii) Suppose that φ is weakly monotone decreasing with respect to the solutions of (1). Let $p \in \partial_F \varphi(m)$ and let $g \in C^1(M)$ be such that $\varphi - g$ attains a local minimum at m and $dg(m) = p$. Consider a solution c of (1) with $c(0) = m$ satisfying $\varphi(c(t)) \le \varphi(c(0))$ for all $t \in [0, \tau)$. Then

$$\int_0^t dg(c(r)) \cdot \dot{c}(r) dr = g(c(t)) - g(c(0)) \le 0.$$

This implies that

$$\int_0^t h(c(r), dg(c(r)) dr \le 0.$$

Note that the function $r \to h(c(r), dg(c(r)))$ is lower semicontinuous due to the upper semicontinuity of F. Thus,

$$h(m, p) = h(m, dg(m))) \le \liminf_{t \to 0+} \frac{1}{t} \int_0^t h(c(r), dg(c(r)) dr \le 0.$$

(ii) implies (iii) Note that h is lower semicontinuous and positive homogeneous in p. This follows directly from definitions of the limiting and singular subdifferentials.

(iii) implies (i) Let us fix $m_0 \in \text{dom } \varphi$. Then there exists a local coordinate neighborhood U of m_0 and a local coordinate system (U, ψ) with local coordinates $(x^1, ..., x^N)$. Let B denote the closed unit ball in $\mathbb{R}^N$ centered at the origin. Without loss of generality we can assume that for some positive r

$$U_1 := \text{cl } \psi^{-1}(x_0 + 3rB) \subset U$$

where $x_0 = \psi(m_0)$. Note that the sets U_1 and

$$U_2 := \text{cl } \psi^{-1}(x_0 + rB)$$

are compact.

Remark 1. Since φ is lower semicontinuous we can assume that it is bounded from below on U_1 and, moreover, shifting φ by a constant if necessary, we can assume that it is positive on U_1.

Consider solutions of the differential inclusion

$$\dot{c}(t) \in F(c(t)), \quad c(0) = m_0. \tag{6}$$

We show below that under Assumption (H) there exists constant $\tau > 0$ such all solutions of (6) exist on the interval $[0, \tau]$ and stay in U.

To show it we define a multifunction

$$\tilde{F}(x) := \psi_* F(\psi^{-1}(x))$$

It is easy to see that $\tilde{F}$ is convex-valued and upper semicontinuous. The following lemma demonstrates that $\tilde{F}$ is bounded.

Lemma 1. *Assume the multifunction F is upper semicontinuous and that Assumption* (H) *hold. Then there exists a constant C_1 such that for any $x \in x_0 + 2rB$, $p \in \mathbb{R}^N$ and $\tilde{v} \in \tilde{F}(x)$*

$$\langle \tilde{v}, p \rangle \leq C_1 \|p\| \tag{7}$$

Proof. Let $\tilde{v} \in \tilde{F}(x)$. Then there exists $v \in F(\psi^{-1}(x))$ such that

$$\tilde{v} = \psi_* v$$

This implies that

$$\langle \tilde{v}, p \rangle = \langle v, \psi^* p \rangle = \left\langle \left(\sum_{n=1}^{N} p_n dx^n \right), v \right\rangle = \sum_{n=1}^{N} p_n \langle dx^n, v \rangle \leq$$

$$\sum_{n=1}^{N} |p_n| H(\psi^{-1}(x), dx^n) \leq \|p\| \sum_{n=1}^{N} \max_{m \in U_1} H(m, dx^n(m)).$$

But this means (7). Q.E.D.

It follows from this lemma that if $\tau := r/C_1$ then all solutions of the differential inclusion

$$\dot{x}(t) \in \tilde{F}(x(t)), \quad x(0) = x_0 \tag{8}$$

exist on the interval $[0, \tau]$ and satisfy

$$x(t) \in x_0 + rB, \quad \forall\, t \in [0, \tau].$$

Also we have that any solution $x(t)$ of the following differential inclusion

$$\dot{x}(t) \in \text{ co } \tilde{F}(x(t) + \varepsilon B) \tag{9}$$

stays in $x_0 + rB$ on the interval $[0, \tau]$ for any $\varepsilon \in [0, r)$.

Now we note that any solution $c(t)$ of the (6) is a lifting of some solution $x(t)$ of (8) which implies that $c(t)$ stays in U_2 on the interval $[0, \tau]$.

For $\alpha > 0$, consider the following function φ_α which is analogous to the one defined before:

$$\varphi_\alpha(m) := \min_{m' \in U_1} (\varphi(m') + \frac{1}{2\alpha^2} |\psi(m') - \psi(m)|^2).$$

We fix an arbitrary small positive α satisfying

$$\alpha < r/ \left(1 + \sqrt{2\varphi(m_0)}\right).$$

In view of Remark 1, this implies that for any $m \in U_2$ the minimizer m' in the definition of $\varphi_\alpha(m)$ will be an interior point of U_1.

We construct approximate solutions of (1) as follows. Consider a uniform partition $\pi := \{t_k\}_{k=0}^K$ of the interval $[0, \tau]$ where $t_{k+1} - t_k = \delta$, $\delta := \tau/K$. We can assume that

$$\delta < \min\{\alpha/C_1, 2\alpha^2/C_1^2\tau\}.$$

We define the approximate solution $c_\pi(t)$ of the differential inclusion (1) on $[0, \tau]$ recursively as a lifting of the absolutely continuous arc $x_\pi : [0, \tau] \to U_2$,

$$c_\pi(t) = \psi^{-1}(x_\pi(t)), \tag{10}$$

where x_π is a solution of the differential inclusion

$$\dot{x}_\pi(t) \in \tilde{F}(x_\pi(t) + \Delta(\alpha)B) \tag{11}$$

and the function $\Delta(\alpha)$ is defined as follows:

$$\Delta(\alpha) := \left(\sqrt{2(\varphi(m_0) + 1)} + 1\right) \alpha. \tag{12}$$

We determine $c_\pi(t)$ as follows. Assume that an approximate solution $c_\pi(t)$ has been defined on $[0, t_k]$ and satisfies

$$c_\pi(t) \in U_2, \quad \text{and } \varphi_\alpha(c_\pi(t)) \le \varphi(m_o) + C_1^2 \delta t / 2\alpha^2 \tag{13}$$

on $[0, t_k]$.

We extend c_π to the interval $[t_k, t_{k+1}]$. Let m_k denote a point such that the function

$$m' \to \varphi(m') + \frac{1}{2\alpha^2}|\psi(m') - \psi(c_\pi(t_k))|^2$$

attains a local minimum. In view of Remark 1, this implies that m_k is an interior point of U_1 and

$$p_k := \sum_{n=1}^{N} \frac{1}{\alpha^2}(x^n(c_\pi(t_k)) - x^n(m_k))dx^n(m_k) \in \partial_F \varphi(m_k).$$

Combining condition (iii) and [13, Corollary 4.17] we have

$$h(m_k, p_k) \leq 0.$$

Next we find $v_k \in F(m_k)$ such that

$$\langle p_k, v_k \rangle = h(m_k, p_k) \leq 0. \tag{14}$$

Let $\tilde{v}_k := \psi_{*m_k} v_k$ and define

$$x_\pi(t) := x_\pi(t_k) + (t - t_k)\tilde{v}_k, \quad t \in [t_k, t_{k+1}] \tag{15}$$

Due to Lemma 1 we have that

$$\|\dot{x}_\pi(t)\| \leq C_1, \quad \forall\, t \in [0, t_{k+1}]$$

and

$$x_\pi(t) \in x_0 + rB, \quad \forall\, t \in [0, t_{k+1}].$$

This implies that the first relation in (13) holds on $[0, t_{k+1}]$.

Note that due to (13) and the choice of δ, we have

$$\begin{aligned} |x_\pi(t_k) - \psi(m_k)| &= |\psi(c_\pi(t_k)) - \psi(m_k)| \\ &\leq \sqrt{2\varphi(c_\pi(t_k))}\,\alpha \\ &\leq \sqrt{2(\varphi(m_0) + 1)}\,\alpha. \end{aligned} \tag{16}$$

Now we estimate the increment of the function φ_α along this trajectory for $t \in [t_k, t_{k+1}]$. We use (10), (15), and (7) for this purpose.

$$\begin{aligned} \varphi_\alpha(c_\pi(t)) - \varphi_\alpha(c_\pi(t_k)) &\leq \frac{1}{2\alpha^2}[|\psi(m_k) - \psi(c_\pi(t))|^2 - |\psi(m_k) - \psi(c_\pi(t_k))|^2] \\ &= \frac{1}{2\alpha^2}[2\langle \psi(c_\pi(t)) - \psi(c_\pi(t_k)), \psi(c_\pi(t_k)) - \psi(m_k)\rangle \\ &+ \quad |\psi(c_\pi(t)) - \psi(c_\pi(t_k))|^2] \\ &\leq \frac{1}{2\alpha^2}[2\langle \tilde{v}_k, \psi(c_\pi(t_k)) - \psi(m_k)\rangle(t - t_k) \\ &+ C_1^2(t - t_k)^2] \\ &\leq \frac{1}{2\alpha^2}[2\alpha^2\langle v_k, p_k\rangle(t - t_k) + C_1^2(t - t_k)\delta]. \end{aligned} \tag{17}$$

Due to (14) we obtain that

$$\varphi_\alpha(c_\pi(t)) - \varphi_\alpha(c_\pi(t_k)) \le C_1^2(t - t_k)\delta/2\alpha^2. \tag{18}$$

Using the second relation in (13) for $t = t_k$ we obtain from (18) that this relation is also valid for all $t \in [0, t_{k+1}]$.

To show that $x_\pi(t)$ is a solution of (11) on $[t_k, t_{k+1}]$ we use (16), (13) and the estimate on the choice of δ. Thus, we proved that there exists an arc $c_\pi(t)$ satisfying (10), (15) and (13) on $[0, \tau]$.

Now, by choosing a sequence of partitions π_i with $\delta_i \to 0$ we can assume without loss of generality that x_{π_i} converges uniformly to some arc x which is a solution of the differential inclusion

$$\dot{x}(t) \in \text{ co } \tilde{F}(x(t) + \Delta(\alpha)B)$$

and

$$\varphi_\alpha(c(t)) \le \varphi(m_0),$$

where $c(t) = \psi^{-1}(x(t))$. Then by choosing a sequence of α_i and arcs x_i such that on $[0, \tau]$, $c_i(t) = \psi^{-1}(x_i(t))$ satisfies

$$\varphi_{\alpha_i}(x_i(t)) \le \varphi(m_0).$$

Again without loss of generality we can assume that sequence x_i converges uniformly to some arc x which is a solution of the differential inclusion (8). The corresponding lifting $c(t) = \psi^{-1}(x(t))$ is a solution of (1) satisfying (3). The Theorem is now proved. Q.E.D.

The method discussed here can also be used to derive a strong monotonicity result. We refer to [13] for details.

4.2 Weak Invariance and Linear Growth

Weak monotonicity is closely related to weak invariance and weak linear growth.

Definition 5. *A set $S \subset M$ is called* weak invariant with respect to solutions of (1) *provided for* any *initial point $m_0 \in M$ there* exist *a $\tau > 0$ and a solution c of (1) on $[0, \tau)$ such that*

$$c(t) \in S, \quad \forall t \in [0, \tau). \tag{19}$$

Clearly, a set S is weakly invariant if its indicator function δ_S is weakly monotone decreasing and a function φ is weakly monotone decreasing if all of its level sets

$$\{m \in M : \varphi(m) \le \varphi(m_0)\}$$

are weakly invariant.

For a closed set S, letting $\varphi = \delta_S$, we deduce characterizations of weak invariance as a corollary of the weak monotonicity characterization.

Theorem 4. *Let F satisfy Assumption* (H) *and let S be a closed subset of M. Then the following are equivalent:*

(i) *S is weakly invariant with respect to the solutions of (1).*
(ii) *$h(m,p) \leq 0$, for any $m \in M$ and $p \in N_F(m;S)$.*
(iii) *$h(m,p) \leq 0$, for any $m \in M$ and $p \in N_L(m;S)$.*

Now we recall the definition of weak linear growth.

Definition 6. *Let $\varphi : M \to \bar{\mathbb{R}}$ be a lower semicontinuous function. We say that φ has a* weak linear growth rate r with respect to solutions of (1) *provided for any initial point $m_0 \in M$ there exist $\tau > 0$ and a solution c of (1) on $[0,\tau)$ satisfying $c(0) = m_0$ such that*

$$\varphi(c(t)) \leq \varphi(c(0)) + rt, \quad \forall t \in [0,\tau). \tag{20}$$

It is not hard to check that φ having a weak linear growth rate r with respect to the solution of differential inclusion (1) is equivalent to $\Phi(m,t) = \varphi(m) - rt$ being weak monotone with respect to

$$\frac{d}{dt}(c(t),t) \in F(c(t)) \times \{1\}.$$

Thus, we can derive the following theorem on the weak linear growth from the weak monotonicity characterization:

Theorem 5. *Let F satisfy Assumption* (H) *and let $\varphi : M \to \bar{\mathbb{R}}$ be a lower semicontinuous function. Then the following are equivalent:*

(i) *φ has a weak linear growth rate r with respect to the solutions of (1).*
(ii) *$h(m,p) \leq r$, for any $m \in M$ and $p \in \partial_F\varphi(m)$.*
(iii) *$h(m,p) \leq r$, for any $m \in M$ and $p \in \partial_L\varphi(m) \cup \partial^\infty\varphi(m)$.*

4.3 Implicit Function Theorem

We illustrate the application of the weak growth theorem by using it to prove an implicit function theorem for a general lower semicontinuous function on a smooth manifold.

Let M be a manifold and let P be a parametric set. Consider a function $f : M \times P \to \bar{\mathbb{R}}$. We use $G(p)$ to denote the implicit multifunction determined by $f(m,p) \leq 0$, i.e.,

$$G(p) := \{m \in M : f(m,p) \leq 0\}.$$

In this section $\partial_F f(m,p)$ signifies the Fréchet subdifferential with respect to variable m. Now we can state our implicit function theorem.

Theorem 6. *Let M be a smooth manifold, let Y be a metric space and let U be an open subset of $M \times P$. Consider $f : U \times Y \to \bar{\mathbb{R}}$ that satisfies the following conditions:*

(i) *there exists* $(\bar{m}, \bar{y}) \in U$ *such that*

$$f(\bar{m}, \bar{y}) \leq 0,$$

(ii) *the function* $y \to f(\bar{m}, y)$ *is upper semicontinuous at* $\bar{y}$*;*
(iii) *for any* y *near* $\bar{y}$*, function* $m \to f(m, y)$ *is lower semicontinuous;*
(iv) *there exists a multifunction* $F : M \rightrightarrows T(M)$ *and* $T > 0$ *satisfying Assumption (H) such that, for any* $(m, y) \in U$*, the solution of*

$$\dot{c}(t) \in F(c(t)), c(0) = m$$

exists on $[0, T]$*. Moreover, assume there exists* $\sigma > 0$ *such that, for any* $(m, y) \in U$ *with* $f(m, y) > 0$*,* $p \in \partial_F f(m, y)$ *implies*

$$h_F(m, p) \leq -\sigma.$$

Then there exist open sets $W \subset M$ *and* $V \subset Y$ *containing* $\bar{m}$ *and* $\bar{y}$ *respectively such that, for any* $y \in V$*,* $W \cap G(y) \neq \emptyset$*.*

Proof. Since $y \to f(\bar{m}, y)$ is upper semicontinuous there exists a neighborhood V of $\bar{y}$ such that, for any $y \in V$, $f(\bar{m}, y) \leq \sigma T$. Let W be a neighborhood of $\bar{m}$ that contains all the possible reachable points of the solution of the differential inclusion on interval $[0, T]$ with initial condition $c(0) = \bar{m}$. Then we must have $W \cap G(y) \neq \emptyset$. In fact, assuming the contrary, we have that $f(m, y) > 0$ for all $m \in W$. It follows from condition (iv) and weak linear growth theorem that there exists a solution c of the differential inclusion $\dot{c}(t) \in F(c(t))$, $c(0) = \bar{m}$ with $c(T) \in W$ satisfying

$$f(c(T), y) \leq f(\bar{m}, y) - \sigma T \leq 0,$$

a contradiction. Q.E.D.

When M is a Riemann manifold we can replace condition (iv) by $p \in \partial_F f(m, p)$ implies $\|p\| \geq \sigma$, for this implies condition (iv) with $F(m)$ being the unit ball in $T_m(M)$.

Acknowledgement. The idea of developing nonsmooth analysis on manifolds came from discussions with E. Sontag. We thank J. M. Borwein, F. H. Clarke and E. Sontag for stimulating conversations and we thank J. S. Treiman for helpful discussions. This research was supported in part by NSF grant #0102496.

References

1. Aubin J, Cellina A (1984) Differential Inclusions. Springer–Verlag Berlin
2. Borwein J, Jofré A (1998) A non-convex separation property in Banach spaces, Journal of Operational Research and Applied Mathematics 48:169-180

3. Borwein J, Zhu Q (1996) Viscosity solutions and viscosity subderivatives in smooth Banach spaces with applications to metric regularity, SIAM J Control and Optimization 34:1568-1591
4. Borwein J, Zhu Q (1999) A survey of subdifferential calculus with applications, Nonlinear Analysis TMA 38:687-773
5. Brickell F, Clark R (1970) Differentiable Manifolds An Introduction. VN Reinhold Co New York
6. Chryssochoos I, Vinter R (2001) Optimal control problems on manifolds a dynamical programming approach, preprint
7. Clarke F (1983) Optimization and Nonsmooth Analysis. John Wiley & Sons New York
8. Clarke F, Ledyaev Y (1994) Mean value inequalities, Proc Amer Math Soc 122:1075-1083
9. Clarke F, Ledyaev Y (1994) Mean value inequalities in Hilbert space, Trans Amer Math Soc 344:307-324
10. Clarke F, Ledyaev Y, Stern R, Wolenski P (1998) Nonsmooth Analysis and Control Theory. Springer-Verlag New York
11. Deimling K (1992) Multivalued Differential Equations. de Gruyter Berlin
12. Kruger A, Mordukhovich B (1980) Extremal points and Euler equations in nonsmooth optimization, Dokl Akad Nauk BSSR 24:684-687
13. Ledyaev Y, Zhu Q (2004) Nonsmooth analysis on smooth manifolds, Transactions of the AMS submitted
14. Ledyaev Y, Zhu Q (2004) Techniques for nonsmooth analysis on smooth manifolds Part I Local Problems, this collection
15. Ledyaev Y, Zhu Q (2004) Multidirectional mean value inequalities and weak monotonicity, Journal of London Mathematical Society submitted
16. Matsushima Y (1972) Differential Manifolds. Marcel Dekker New York
17. Mordukhovich B (1976) Maximum principle in problems of time optimal control with nonsmooth constraints, J Appl Math Mech 40:960-969
18. Mordukhovich B (1988) Approximation Methods in Problems of Optimization and Control. Nauka Moscow
19. Mordukhovich B (1994) Generalized differential calculus for nonsmooth and set-valued mappings, J Math Anal Appl 183:250–288
20. Mordukhovich B, Treiman J, Zhu Q (2004) An extended extremal principle with applications to multi-objective optimization, SIAM J Optimization to appear
21. Rockafellar R, Wets J (1998) Variational Analysis. Springer Berlin
22. Sternberg S (1964) Lectures on Differential Geometry. Prentice-Hall Englewood Cliffs NJ
23. Sussmann H (1994) A strong version of the maximum principle under weak hypotheses. In: Proceedings of the 33rd IEEE Conference on Decision and Control 1950-1956
24. Sussmann H (1996) A strong maximum principle for systems of differential inclusions. In: Proceedings of the 35th IEEE Conference on Decision and Control 1809-1814
25. Treiman J, Zhu Q (2004), Extremal principles on smooth manifolds and applications, in preparation
26. Zhu Q (1998) The equivalence of several basic theorems for subdifferentials, Set-Valued Analysis 6:171-185
27. Zhu Q (2004) Nonconvex separation theorem for multifunctions subdifferential calculus and applications, Set-Valued Analysis to appear

Stationary Hamilton-Jacobi Equations for Convex Control Problems: Uniqueness and Duality of Solutions

Rafal Goebel

Center for Control Engineering and Computation, University of California, Santa Barbara, CA 93106-9650. rafal@ece.ucsb.edu

1 Introduction

This note summarizes some of the results of the author obtained in [7], with examples coming also from [6] and [8]. We focus on Hamilton-Jacobi and convex duality characterizations of the value function $V : \mathbb{R}^n \mapsto \overline{\mathbb{R}}$ associated with a convex generalized problem of Bolza on an infinite time interval:

$$V(\xi) = \inf \left\{ \int_0^{+\infty} L(x(t), \dot{x}(t))\, dt \mid x(0) = \xi \right\}. \tag{1}$$

The minimization is carried out over all locally absolutely continuous arcs $x : [0, +\infty) \mapsto \mathbb{R}^n$ (that is, $\dot{x}(\cdot)$ is integrable on bounded intervals), subject to the initial condition $x(0) = \xi$. The Lagrangian $L : \mathbb{R}^{2n} \mapsto \overline{\mathbb{R}}$ is allowed to be nonsmooth and infinite-valued; the key assumption on it is the full convexity:

$$L(x, v) \quad \text{is convex in } (x, v).$$

Such problems of Bolza can model control problems with explicit linear dynamics, convex running costs, and control constraints; see Section 3 for references and examples.

Results on uniqueness of solutions to stationary Hamilton-Jacobi equations applicable to some convex problems can be found in [1], [2], [4]; special cases of duality of convex value functions are described in [6], [11]. Close connection between these seemingly unrelated issues has not been previously observed. We state it in Theorem 1, and obtain new uniqueness and duality results in Theorem 2.

In order for the value function to be well-defined, we assume that $L \geq 0$, $L(0,0) = 0$. Detailed technical assumptions are stated in terms of the Hamiltonian $H : \mathbb{R}^{2n} \mapsto \mathbb{R}$ associated with L by

$$H(x, y) = \sup_{v \in \mathbb{R}^n} \{ y \cdot v - L(x, v) \}, \tag{2}$$

M. de Queiroz et al. (Eds.): Optimal Control, Stabilization and Nonsmooth Analysis, LNCIS 301, pp. 313–322, 2004.

Main Assumption. *The function $H : \mathbb{R}^{2n} \mapsto \mathbb{R}$ is everywhere finite, such that $x \mapsto H(x,y)$ is concave for any fixed y, and $y \mapsto H(x,y)$ is convex for any fixed x. Moreover, there exists a constant α and a finite convex function ϕ with $\phi \geq 0$, $\phi(0) = 0$, such that*

$$-\phi(x) - \alpha|x||y| \leq H(x,y) \leq \phi(y) + \alpha|y||x| \tag{3}$$

for all $(x,y) \in \mathbb{R}^{2n}$.

In particular, the growth conditions above imply that $H(0,0) = 0$ and that $(0,0)$ is a saddle point of H, that is $H(x,0) \leq H(0,0) \leq H(0,y)$ for any $x \in \mathbb{R}^n$ and $y \in \mathbb{R}^n$. (These properties correspond to $L \geq 0$ and $L(0,0) = 0$.) However, we do not require that the saddle point be unique.

The convex (and proper and lsc) value function (1) satisfies a stationary Hamilton-Jacobi equation in the following sense:

$$H(x, -\partial V(x)) = 0 \ \text{ for all } x \in \mathbb{R}^n. \tag{4}$$

Here and in what follows, given a convex function $f : \mathbb{R}^n \mapsto \overline{\mathbb{R}}$, ∂f denotes its subdifferential in the sense of convex analysis:

$$\partial f(x) = \{y \mid f(x') \geq f(x) + y \cdot (x' - x) \text{ for all } x' \in \mathbb{R}^n\},$$

and the equation $H(x, -\partial f(x)) = 0$ means $H(x,-y) = 0$ for all $y \in \partial f(x)$. (For finite convex functions this concept agrees with the notion of a viscosity solution in [1] and that of a generalized solution as described in [3].) We are interested in whether V is the unique proper, lsc, and convex solution to (4), subject to appropriate boundary conditions.

The uniqueness issue turns out to be closely related to whether V is the convex conjugate of an appropriate "dual value function". We recall that the convex conjugate $f^* : \mathbb{R}^n \mapsto \overline{\mathbb{R}}$ of a function $f : \mathbb{R}^n \mapsto \overline{\mathbb{R}}$ is defined as

$$f^*(y) = \sup_{x \in \mathbb{R}^n} \{y \cdot x - f(x)\}.$$

Following [11] and [14], we consider $W : \mathbb{R}^n \mapsto \overline{\mathbb{R}}$ given by

$$W(\eta) = \inf \left\{ \int_0^{+\infty} M(y(t), \dot{y}(t))\, dt \mid y(0) = \eta \right\}. \tag{5}$$

The dual Lagrangian $M : \mathbb{R}^{2n} \mapsto [0, +\infty)$ is the convex conjugate of L with switched variables: $M(y,w) = L^*(w,y)$. Equivalently, M can be constructed directly from the H. Just as L corresponds to H through a formula reciprocal to (2), i.e.,

$$L(x,v) = \sup_{y \in \mathbb{R}^n} \{v \cdot y - H(x,y)\},$$

M corresponds to a Hamiltonian $H(y,x) = -H(x,y)$. (Note that H is indeed a concave-convex function satisfying our assumptions.) That is, we have

$$M(y, w) = \sup_{x \in \mathbb{R}^n} \left\{ w \cdot x - \overline{H}(y, x) \right\} = \sup_{x \in \mathbb{R}^n} \left\{ w \cdot x + H(x, y) \right\}.$$

The 'dual' Hamilton-Jacobi equation, the counterpart to (4) written in terms of $\overline{H}$, translates to $H(-\partial W(y), y) = 0$ for all $y \in \mathbb{R}^n$. The duality relationship of our interest is

$$V(\xi) = \sup_{\eta \in \mathbb{R}^n} \left\{ -\xi \cdot \eta - W(\eta) \right\}, \qquad W(\eta) = \sup_{\xi \in \mathbb{R}^n} \left\{ -\eta \cdot \xi - V(\xi) \right\}, \tag{6}$$

that is, $V = W^*(-\cdot)$, $W = V^*(-\cdot)$. (These equations are equivalent to one another.) The connection between these formulas and the uniqueness of solutions to (4) and to the dual Hamilton-Jacobi equation is suggested by the following example:

Example 1. (One-Dimensional Problems, [6]) Given $\alpha < 0$, $\beta > 0$, consider the following convex function on the real line, and its conjugate:

$$\phi_{a,b}(z) = \begin{cases} +\infty & \text{if} \quad z \in (-\infty, -1), \\ az & \text{if} \quad z \in [-1, 0], \\ bz & \text{if} \quad z \in [0, 1], \\ +\infty & \text{if} \quad z \in (1, +\infty); \end{cases} \qquad \phi^*_{a,b}(u) = \begin{cases} -u & \text{if} \quad u \in (-\infty, -a], \\ 0 & \text{if} \quad u \in [a, b], \\ u & \text{if} \quad z \in [b, +\infty). \end{cases}$$

Given numbers $a, c \leq 0$ and $b, d \geq 0$, let

$$H(x, y) = -\phi^*_{a,b}(x) + \phi^*_{c,d}(y),$$

corresponding to $L(x, v) = \phi^*_{a,b}(x) + \phi_{c,d}(v)$ and $M(y, w) = \phi^*_{c,d}(y) + \phi_{a,b}(w)$. The set where the Hamiltonian equals 0 is shown on the leftmost graph; each point of $[a, b] \times [c, d]$ is a saddle point of H. If $(0, 0)$ is in the interior of this set, there are infinitely many convex functions f satisfying $H(x, -\partial f(x)) = 0$. (For a convex f, $-\partial f$ is a 'nonincreasing function'. Examples below are provided by $-\partial V$ on the middle graph, and $-\partial W^*(-\cdot)$ on the rightmost one.) Boundary conditions $0 \in \partial f(0)$ and $f(0) = 0$ do not change that fact, nor does the restriction to differentiable functions. The same observation holds for the dual Hamilton-Jacobi equation. However, if one of a, d, and one of b, c equals zero (i.e., when $(0, 0)$ is the 'northwest' or 'southeast' corner of the saddle set), V is the unique solution to (4) satisfying $0 \in \partial V(0)$, and a similar uniqueness statement holds for the dual equation.

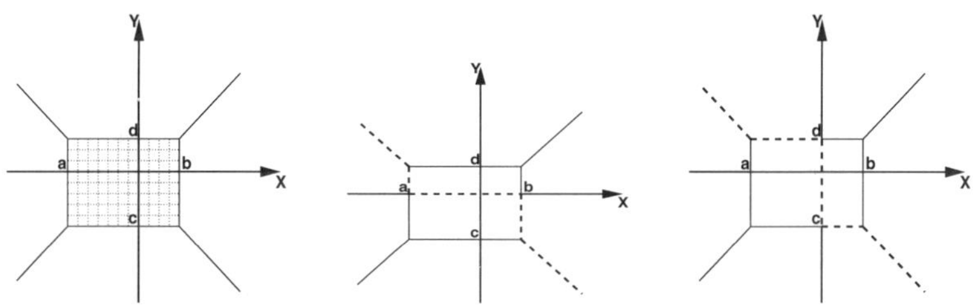

We now turn our attention to general Hamiltonians on $\mathbb{R}^2$, but with structure similar to the one discussed above. We suppose that $[a,b]\times[c,d]$ is the set of saddle points of H, $H(x,0)<0$ for all $x\notin[a,b]$, and $H(0,y)>0$ for all $y\notin[c,d]$. The following "duality-like" relationships hold: for $x\leq a$,

$$V(x)=\sup_y\{-xy-W(y)\}+ad,$$

while for $x\geq b$,

$$V(x)=\sup_y\{-xy-W(y)\}+bc.$$

Similar formulas for $W(y)$ are true. If $ad=bc=0$, then (6) holds. Note that the condition guaranteeing conjugacy agrees with the one on the saddle set, alluded to in the discussion of uniqueness.

Duality and Hamilton-Jacobi description of convex value functions for finite horizon problems was studied in [14]. Theorem 5.1 of that paper, adjusted for our purposes, states the following. Consider a proper, lsc, and convex function $\psi:\mathbb{R}^n\mapsto\mathbb{R}$, and define value functions $V_\phi, W_\psi:[0,+\infty)\times\mathbb{R}^n$ by

$$\begin{aligned} V_\psi(\tau,\xi)&=\inf\left\{\int_0^\tau L(x(t),\dot{x}(t))\,dt+\psi(x(\tau))\mid x(0)=\xi\right\},\\ W_{\psi^*}(\tau,\eta)&=\inf\left\{\int_0^\tau M(y(t),\dot{y}(t))\,dt+\psi^*(-y(\tau))\mid y(0)=\eta\right\}. \end{aligned}\tag{7}$$

For any $\tau\geq 0$, the functions $V_\psi(\tau,\cdot)$, $W_{\psi^*}(\tau,\cdot)$ are proper, lsc, and convex. Moreover, they are conjugate to each other (up to a minus sign), that is

$$W_{\psi^*}(\tau,\eta)=\sup_{\xi\in\mathbb{R}^n}\{-\eta\cdot\xi-V_\psi(\tau,\xi)\},\quad V_\psi(\tau,\xi)=\sup_{\eta\in\mathbb{R}^n}\{-\xi\cdot\eta-W_{\psi^*}(\tau,\eta)\}.$$

Letting $\tau\to+\infty$ in the equations above should lead to (6). However, while choosing a function ψ so that $V_\psi(\tau,\cdot)$ converges to V is not difficult, it is less obvious (and not always true) that W_{ψ^*} will simultaneously converge to W.

As the functions in question may be infinite-valued, and since we need a notion of convergence with respect to which conjugacy is well-behaved, we rely on the convergence of functions in the epigraphical sense, epi-convergence for short. A sequence $k_i:\mathbb{R}^n\to\overline{\mathbb{R}}$, $i=1,2,...$ epi-converges to k if, for every point $x\in\mathbb{R}^n$, $\liminf_{i\to\infty}k_i(x_i)\geq k(x)$ for every sequence $x_i\to x$, while $\limsup_{i\to\infty}k_i(x_i)\leq k(x)$ for some sequence $x_i\to x$. Consult [13] for details.

2 Main Results

Our main result, Theorem 1, shows the equivalence of the uniqueness of solutions to Hamilton-Jacobi equations, the duality relationship between value functions, and of the possibility of simultaneous approximation of infinite-horizon problems by a pair of dual finite-horizon ones. Such approximation will be the key to Theorem 2, giving sufficient conditions for the equivalent statements below to be true. (They may fail in general; recall Example 1.)

Theorem 1. *The following statements are equivalent:*

(a) The value function V is the unique proper, lsc, and convex function satisfying

$$H(x, -\partial f(x)) = 0 \text{ for all } x \in \mathbb{R}^n, \quad f \geq 0, \quad f(0) = 0. \tag{8}$$

(b) The dual value function W is the unique proper, lsc, and convex function satisfying

$$H(-\partial g(y), y) = 0 \text{ for all } y \in \mathbb{R}^n, \quad g \geq 0, \quad g(0) = 0. \tag{9}$$

(c) The conjugacy relationship $V(\xi) = W^*(-\xi)$ holds.
(d) There exists a proper, lsc, and convex function $\psi : \mathbb{R}^n \mapsto \mathbb{R}$ for which

$$\text{epi} - \lim_{\tau \to +\infty} V_\psi(\tau, \cdot) = V, \qquad \text{epi} - \lim_{\tau \to +\infty} W_{\psi^*}(\tau, \cdot) = W. \tag{10}$$

To prove this equivalence, we need a preliminary result.

Lemma 1. *For any proper, lsc, and convex function $f : \mathbb{R}^n \mapsto \overline{\mathbb{R}}$, the following statements are equivalent:*

(a) For all $\xi \in \mathbb{R}^n$, all $\tau \geq 0$,

$$f(\xi) = \inf \left\{ \int_0^\tau L(x(t), \dot{x}(t))\, dt + f(x(\tau)) \mid x(0) = \xi \right\}. \tag{11}$$

(b) $H(x, -\partial f(x)) = 0$ for all $x \in \mathbb{R}^n$.

Proof. Theorem 2.5 of [14], after an appropriate change of variables, states that the value function $U : [0, +\infty) \times \mathbb{R}^n \mapsto \mathbb{R}$ defined by

$$U(\tau, \xi) = \inf \left\{ \int_0^\tau L(x(t), \dot{x}(t))\, dt + f(x(\tau)) \mid x(0) = \xi \right\}$$

satisfies, for all $\tau > 0$, $\xi \in \mathbb{R}^n$, the following non-stationary Hamilton-Jacobi equation: $\rho + H(x, -\eta) = 0$ for any $(\rho, \eta) \in \partial^g U(\tau, \xi)$. Here, $\partial^g U$ denotes the generalized subdifferential of U; see [13] for details. If (a) holds, then $U(\tau, \xi) = f(\xi)$ for all $\tau \geq 0$, all $\xi \in \mathbb{R}^n$, and the generalized subdifferential $\partial^g U(\tau, \xi)$ is $0 \times \partial f(\xi)$. Consequently, we obtain (b).

Now suppose that (b) holds. Consider the function $u : \mathbb{R} \times \mathbb{R}^n \mapsto \overline{\mathbb{R}}$ defined as $u(t, x) = f(x)$ if $t \geq 0$, $u(t, x) = +\infty$ if $t < 0$. Then u is proper, lsc, meets the boundary condition $u(0, x) = f(x)$, and satisfies $\rho + H(x, -\eta) = 0$ for all $(\rho, \eta) \in \partial^g u(t, x) = 0 \times \partial f(x)$ if $t > 0$ while $\rho + H(x, -\eta) \leq 0$ for all $(\rho, \eta) \in \partial^g u(0, x) = (-\infty, 0] \times \partial f(x)$. The last equality follows from the fact that $u(t, x) = f(x) + \delta_{(-\infty, 0]}(t)$ and from the calculus rules of the generalized subdifferential (10.9 in [13]). Note that in fact $(t, x) \mapsto f(x)$ and $(t, x) \mapsto \delta_{(-\infty, 0]}(t)$ are convex functions of (t, x). A uniqueness results of Galbraith [5] states that u must consequently be the value function defined by the right hand side of the equation in (a). Thus, (b) implies (a).

This in particular concludes that the value function V satisfies the Hamilton-Jacobi equation (4). (The Optimality Principle states that $f = V$ satisfies (11).) Dually, the value function W satisfies $H(-\partial W(y), y) = 0$ for all $y \in \mathbb{R}^n$. We are now ready to outline the proof of Theorem 1.

Proof.

- *(a) implies (c):* The value function W solves (9). This is equivalent to $W^*(-\cdot)$ solving (8). Indeed, for any convex function f, $b \in \partial f(a)$ is equivalent to $a \in \partial f^*(b)$ (see 11.3 in [13]) and this translates the dual Hamilton-Jacobi equation for W to (4) for $W^*(-\cdot)$. The same subdifferential relation and the definition of f^* can be used to show the equivalence of boundary conditions: $f \geq 0$ and $f(0) = 0$ is equivalent to $f^* \geq 0$ and $f^*(0) = 0$. With $W^*(-\cdot)$ solving (9), and V being the unique such function by assumption, we must have $V = W^*(-\cdot)$.
- *(b) implies (c):* The proof is symmetric to the one above.
- *(c) implies (a):* Suppose that f solves (8). Then $f \geq V$. Indeed, by Lemma 1 we have, for any $\tau \geq 0$,

$$f(\xi) = \inf\left\{\int_0^\tau L(x(t), \dot{x}(t))\, dt + f(x(\tau)) \mid x(0) = \xi\right\}$$
$$\geq \inf\left\{\int_0^\tau L(x(t), \dot{x}(t))\, dt \mid x(0) = \xi\right\},$$

  and this can be used to conclude the desired inequality. On the other hand, the function g given by $g(y) = f^*(-y)$ solves (9); see the arguments in the proof of *(a) implies (c)*. Then $g \geq W$. The last inequality dualizes to $g^* \leq W^*$, which, thanks to (c), means that $f \leq V$. Thus $f = V$, and the value function is the unique solution to (8).
- *(c) implies (a):* The proof is symmetric to the one above.
- *(d) implies (c):* Conjugacy of convex functions is continuous with respect to epi-convergence, that is, for proper, lsc, and convex functions k_i, $i = 1, 2, ...$ and k, $\text{epi} - \lim_{i \to \infty} k_i = k$ is equivalent to $\text{epi} - \lim_{i \to \infty} k_i^* = k^*$. Consequently, $\text{epi} - \lim_{\tau \to +\infty} V_\psi(\tau, \cdot) = V$, implies $\text{epi} - \lim_{\tau \to +\infty} V_\psi^*(\tau, \cdot) = V^*$. Here, for a fixed $\tau \geq 0$, $V_\psi^*(\tau, \cdot)$ denotes the convex conjugate of $V_\psi(\tau, \cdot)$. Conjugacy of finite time horizon value functions implies that $V_\psi^*(\tau, \cdot) = W_{\psi^*}(\tau, -\cdot)$. Thus $\text{epi} - \lim_{\tau \to +\infty} W_{\psi^*}(\tau, -\cdot) = V^*$. But by the second part of (d), the limit above is $W(-\cdot)$. Consequently, $V^* = W(-\cdot)$.
- *(c) implies (d):* Set $\psi = V$. The Principle of Optimality and the definition of V_ψ imply that $V_\psi(\tau, \cdot) = V$ for all $\tau > 0$. Moreover, by the assumption that $\psi^* = W(-\cdot)$, the same reasoning in the dual setting implies $W_{\psi^*}(\tau, \cdot) = W$. Statements in (d) now trivially follow.

The analysis of the long term behavior of Hamiltonian trajectories (pairs of arcs satisfying optimality conditions) led to conjugacy of the value functions in [11], and to conjugacy-like formulas in [6] (recall Example 1). Here, we take a different path and rely on finite-horizon approximation, as suggested by (d) of Theorem 1. This yields the following result:

Theorem 2. *Assume that*

(a) There exists a compact and convex set C such that, for any $\xi \in \mathbb{R}^n$, if $V(\xi) < +\infty$ then there exists an optimal trajectory x for $V(\xi)$ satisfying $\lim_{t\to\infty} \operatorname{dist}_C(x(t)) = 0$.
(b) There exists a compact and convex set D such that, for any $\eta \in \mathbb{R}^n$, if $W(\eta) < +\infty$ then there exists an optimal trajectory y for $W(\eta)$ satisfying $\lim_{t\to\infty} \operatorname{dist}_D(y(t)) = 0$.
(c) For all $c \in C$, $d \in D$, $c \cdot d \geq 0$.

Then there exists a proper, lsc, and convex function for which (10) *holds, and consequently, all of the equivalent statements in Theorem 1 hold.*

Proof. Let C be a set as described in (a), and ψ be any proper, lsc, and convex function with $\psi \geq 0$ and such that if $\lim_{i\to\infty} \operatorname{dist}_C(x_i) = 0$ then $\lim_{i\to\infty} \psi(x_i) = 0$. It can be then be shown that the finite horizon value function V_ψ defined in (7) satisfies $\operatorname{epi} - \lim_{\tau\to\infty} V_\psi(\tau, \cdot) = V$. A symmetric conclusion can be made for the dual problem.

It is thus sufficient to demonstrate the existence of a proper, lsc, and convex function ψ, with $\psi \geq 0$, $\psi(0) = 0$, such that:

- if a sequence x_i satisfies $\lim_{i\to\infty} \operatorname{dist}_C(x_i) = 0$ then $\lim_{i\to\infty} \psi(x_i) = 0$,
- if a sequence y_i satisfies $\lim_{i\to\infty} \operatorname{dist}_D(y_i) = 0$ then $\lim_{i\to\infty} \psi^*(-y_i) = 0$.

As convex functions are continuous in the interior of their effective domains, it is sufficient that the function ψ be nonnegative, zero on C, and continuous on a neighborhood of C, while ψ^* be zero on $-D$ and continuous on a neighborhood that set. Pick a constant K such that $C \cup D \subset \operatorname{int} K\mathbb{B}$, with $\mathbb{B}$ being the unit ball in $\mathbb{R}^n$. Define

$$\psi(x) = \sigma_{-D}(x) + \delta_{K\mathbb{B}}(x).$$

Here, $\sigma_{-D}(x) = \sup\{x \cdot d \mid d \in -D\}$ is the support function of $-D$ and $\delta_{K\mathbb{B}}$ is the indicator of $K\mathbb{B}$. We then obtain (11.4 and 11.23 in [13]):

$$\psi^*(y) = \inf_{z \in \mathbb{R}^n} \{\delta_{-D}(y - z) + K\|z\|\}.$$

It can be verified that ψ and ψ^* have the desired properties with respect to C and D; assumption (c) is needed here. This finishes the proof.

If $C = D = \{0\}$, as will be the case for control problems we discuss in Section 3, conditions (a) and (b) in the Theorem above essentially say that "optimality implies stability". We add that each of the following conditions implies that assumptions of Theorem 2 hold:

- There exist compact sets C, D with $C \cdot D \geq 0$ such that $H(x, 0) < 0 < H(0, y)$ for all $x \notin C$, all $y \notin D$.

- On some neighborhood of $(0,0)$, the Hamiltonian H is strictly convex in x and strictly concave in y.
- The Lagrangian L is continuous at $(0,0)$ and $L(x,v)>0$ when $x\neq 0$.
- The Lagrangian L is differentiable at $(0,0)$ and there exists K such that $L(x,v)=0$ implies $\|x\|\leq K$.

3 Linear-Quadratic Regulator and Extensions

To conclude, we briefly illustrate the applicability of our results to problems in the classical optimal control format, using the linear-quadratic regulator as an example. Details on how control problems can be modelled in the Bolza format can be found in [3], conditions on the control problem guaranteeing that the resulting Bolza problem fits our framework are in [9], and for constructions of dual control problems consult [12]. Linear systems notions and facts we use can be found in [10].

For a pair of symmetric and positive definite matrices Q and R, consider the following problem:

$$\begin{array}{ll} \text{minimize} & \dfrac{1}{2}\displaystyle\int_0^{+\infty} z(t)\cdot Qz(t)+u(t)\cdot Ru(t)\,dt \\ \text{s.t.} & \begin{cases} \dot{x}(t)=Ax(t)+Bu(t), \\ z(t)=Cx(t), \end{cases} \quad x(0)=\xi. \end{array} \tag{12}$$

If (A,B) is stabilizable and (A,C) is detectable, the value function for (12) is given by $\frac{1}{2}\xi\cdot P\xi$, where P is the unique symmetric positive semidefinite solution to the Riccati equation $A^TP+PA+C^TQC-PBR^{-1}B^TP=0$. However, even under the mentioned assumptions, the value function need not be the unique solution to (8), with the Hamiltonian associated to (12) being

$$H(x,y)=y\cdot Ax-\frac{1}{2}x\cdot C^TQCx+\frac{1}{2}yBR^{-1}B^Ty. \tag{13}$$

Indeed, consider (12) with

$$Q=R=1,\quad A=\begin{bmatrix}-1 & 0\\ 0 & 0\end{bmatrix},\quad B=\begin{bmatrix}0\\ 1\end{bmatrix},\quad C=\begin{bmatrix}0 & 1\end{bmatrix}.$$

Then $H(x,y)=-x_1y_1-\frac{1}{2}x_2^2+\frac{1}{2}y_2^2$, and concentrating on separable solutions $f(x)=f_1(x_1)+f_2(x_2)$ yields a unique convex f_2 but four different f_1's (only one of them being quadratic) for which (8) holds. Indeed, we can take $f_1(x_1)=0$, $f_1(x_1)=\delta_0(x_1)$, $f_1(x_1)=\delta_{(-\infty,0]}(x_1)$, or $f_1(x_1)=\delta_{[0,+\infty)}(x_1)$.

Theorem 2 will yield conditions under which the solution to (8), with H given by (13), is unique, without requiring that it be quadratic. First we describe the control problem dual to (12):

$$\begin{aligned} &\text{minimize } \frac{1}{2}\int_0^{+\infty} w(t)\cdot R^{-1}w(t) + v(t)\cdot Q^{-1}v(t)\,dt \\ &\text{s.t.} \qquad \begin{cases} \dot{y}(t) = -A^T y(t) - C^T v(t), \\ w(t) = B^T x(t), \end{cases} \quad y(0) = \eta. \end{aligned} \tag{14}$$

Detectability of (A, C) implies that all optimal trajectories in (12) (in fact all trajectories with finite cost) converge to 0. The same conclusion holds for (14) if $(-A^T, B^T)$ is detectable – equivalently, if $(-A, B)$ is stabilizable.

Corollary 1. *Suppose that (A, C) is detectable and $(-A, B)$ is stabilizable. Then the equivalent statements in Theorem 1 hold for the pair of problems* (12), (14). *In particular the value function for the control problem* (12) *is the unique proper, lsc, and convex solution to* (8) *with the Hamiltonian* (13).

Linear systems techniques, capable of treating the linear-quadratic regulator (12), are no longer adequate when constraints and nonquadratic penalties are introduced. Our results remain applicable, as long as the convex structure of the problem remains.

Example 2. (Constrained LQR, [7]) Let U be a compact convex set such that $0 \in \operatorname{int} U$, and consider (12) subject to

$$u(t) \in U \quad \text{for almost all } t \in [0, +\infty). \tag{15}$$

The value function V is convex, but is not quadratic, and may not be finite everywhere. As in the unconstrained case, if (A, C) is detectable, all optimal trajectories converge to 0.

The dual problem to (12), (15), as suggested in [12], does not have a quadratic cost, but is unconstrained – the cost in (14) is replaced by

$$\int_0^{+\infty} \rho_{U,R}(w(t)) + \frac{1}{2}v(t)\cdot Q^{-1}v(t)\,dt.$$

Above, $\rho_{U,R}(w) = \sup_{u\in U}\{w\cdot u - \frac{1}{2}u\cdot Ru\}$; this function is convex, finite everywhere, and differentiable with Lipschitz continuous gradient. As $0 \in \operatorname{int} U$, $\rho_{U,R}$ exhibits favorable growth properties (and is in particular quadratic and positive definite around 0) guaranteeing, together with detectability of $(-A^T, B^T)$, that optimal trajectories approach 0. Consequently, under the assumptions of Corollary 1, all of the statements in Theorem 1, with

$$H(x, y) = -y\cdot Ax - \frac{1}{2}x\cdot C^T QCx + \rho_{U,R}(B^T y),$$

hold for the value function W for the dual problem just described, and the value function V of the constrained linear-quadratic regulator.

We point out that under stronger assumptions of observability of (A, C) and controllability of (A, B), both value functions V and W can be shown to be differentiable (in the case of V, differentiable where finite) and strictly convex. Thus, they satisfy corresponding Hamilton-Jacobi equations in the classical sense. Further application of duality leads immediately to optimality conditions, for example: the pair $(x(\cdot), u(\cdot))$ is optimal for the constrained linear-quadratic problem (12), (15) if and only if $u(t)$ maximizes $-u \cdot B^T \nabla V(x(t)) - \frac{1}{2} u \cdot Ru$ over $u \in U$.

References

1. Bardi M, Capuzzo-Dolcetta I (1997) Optimal control and viscosity solutions of Hamilton-Jacobi-Bellman equations. Birkhauser Boston
2. Cannarsa P, Da Prato G (1989) Nonlinear optimal control with infinite horizon for distributed parameter systems and stationary Hamilton-Jacobi equations, SIAM J Control Opt 27:861–875
3. Clarke F (1983) Optimization and Nonsmooth Analysis. Wiley New York
4. Di Blasio G (1991) Optimal control with infinite horizon for distributed parameter systems with constrained controls, SIAM J Control Opt 29:909–925
5. Galbraith G (2001) Extended Hamilton-Jacobi characterization of value functions in optimal control, SIAM J Control Optim 39:281–305
6. Goebel R (2002) Planar generalized Hamiltonian systems with large saddle sets, J Nonlinear Convex Anal 3:365-380
7. Goebel R (2003) Duality and uniqueness of convex solutions to stationary Hamilton-Jacobi equations, submitted
8. Goebel R (2003) Value function for constrained linear-quadratic regulator – convex duality approach, submitted
9. Goebel R, Rockafellar R (2002) Generalized conjugacy in Hamilton-Jacobi theory for fully convex Lagrangians, Journal of Convex Analysis 9:463-473
10. Kwakernaak H, Sivan R (1972) Linear Optimal Control Systems. Wiley-Interscience New York
11. Rockafellar R (1973) Saddle points of Hamiltonian systems in convex problems of Lagrange, Journal of Optimization Theory and Applications 12:367-390
12. Rockafellar R (1987) Linear-quadratic programming and optimal control, SIAM J Control Optim 25:781–814
13. Rockafellar R, Wets R (1998) Variational Analysis. Springer New York
14. Rockafellar R, Wolenski P (2000) Convexity in Hamilton-Jacobi theory 1 Dynamics and duality, SIAM J Control Optim 39:1323–1350

Smooth and Nonsmooth Duality for Free Time Problems

Chadi Nour

Institut Girard Desargues, Université Lyon I, 21 avenue Claude Bernard, 69622 Villeurbanne Cedex, France. chadi@igd.univ-lyon1.fr

Summary. The main result of this paper contains a representation of the minimum cost of a free time control problem in terms of the upper envelope of generalized semisolutions of the Hamilton-Jacobi equation. A corollary generalizes a similar result due to Vinter using smooth subsolutions.

Keywords: free time problems, duality, Hamilton-Jacobi equations, viscosity solutions, proximal analysis, nonsmooth analysis

1 Introduction

The following optimal control problem is considered in Vinter [15]:

$$(Q) \begin{cases} \text{Minimize } \ell(T, x(T)), \\ T \in [0,1], \\ \dot{x}(t) \in F(t, x(t)) \;\; a.e.\; t \in [0,T], \\ x(0) = x_0, \\ (t, x(t)) \in A \subset [0,1] \times \mathbb{R}^n \;\; \forall t \in [0,T], \\ (T, x(T)) \in C \subset [0,1] \times \mathbb{R}^n, \end{cases}$$

where the given data is a point x_0, the function $\ell : \mathbb{R} \times \mathbb{R}^n \to \mathbb{R} \cup \{+\infty\}$, the multivalued function $F : \mathbb{R} \times \mathbb{R}^n \rightrightarrows \mathbb{R}^n$, and the sets A and C. Vinter in addition formulated a convex optimization problem (W), associated with (Q), namely, the minimization of a linear functional under linear constraints of equality type on the set $\mathcal{W}$ of generalized flows, a weak*-convex compact subset of a space of Radon measures also associated with problem (Q). Based on the apparatus of convex analysis and, in particular, on convex duality, he established a very close interconnection between problems (Q) and (W). He proved that the set $\mathcal{W}$ is the convex closure of the set of admissible arcs of the original problem (Q), and also that both problems are solvable and that, moreover, their values coincide. This makes it possible to prove a necessary

M. de Queiroz et al. (Eds.): Optimal Control, Stabilization and Nonsmooth Analysis, LNCIS 301, pp. 323–331, 2004.

and sufficient condition for optimality for problem (Q) related to well-known sufficient conditions, referred to as verification theorems, in dynamic optimization (see [4] and [6]). Simultaneously, Vinter gives a "smooth duality" for the problem (Q); that is, the value of problem (Q) is represented in terms of the upper envelope of smooth subsolutions of the Hamilton-Jacobi equation. This so-called "convex duality" method was first introduced by Vinter and Lewis [17], [18]. For more information about the possibility of approaching control problems via duality theory in abstract spaces, see [7], [8], [11], [12], [15], [17] and [18].

We remark that the problem (Q) treated by Vinter is an optimal control problem with finite horizon ($T \in [0,1]$), and he has affirmed [16] that his generalized flows approach does not extend to free time problems with infinite horizon ($T \in [0,+\infty[$) and do not lead to an upper envelope characterization of the minimum cost, in term of smooth solutions of the *autonomous* Hamilton-Jacobi inequality.

In this article, we consider the following free time problem:

$$(P) \begin{cases} \text{Minimize } T + \ell(x(T)), \\ T \in [0,+\infty[, \\ \dot{x}(t) \in F(x(t)) \;\; a.e.\ t \in [0,+\infty[, \\ x(0) = x_0, \\ x(t) \in A \;\; \forall t \in [0,T], \\ x(T) \in C. \end{cases}$$

Our main result is a "nonsmooth duality" for the problem (P); that is, a representation of the minimum cost of (P) in terms of the upper envelope of generalized semisolutions of the Hamilton-Jacobi equation. This type of duality is well studied in the literature with several techniques for fixed time problems (see for example [1], [2], [4, Chapter 4], [9], [10], [13], [14] and [19]). We use the *proximal subdifferential* to define our generalized semisolutions. This concept of solution appeared in Clarke and Ledyaev [3], where the various concepts were also unified. Using our nonsmooth duality we extend Vinter's smooth duality for free time problems with infinite horizon. Let us enter into the details.

We assume in the problem (P) that the set A is closed, that C is compact, and that the extended-valued function $\ell : \mathbb{R}^n \to \mathbb{R} \cup \{+\infty\}$ is lower semicontinuous and bounded below by a constant ω. As for the multivalued function F, we assume that it takes nonempty compact convex values, has closed graph, and satisfies a linear growth condition: for some positive constants γ and c, and for all $x \in \mathbb{R}^n$,

$$v \in F(x) \Longrightarrow \|v\| \leq \gamma\|x\| + c.$$

Finally, we assume that (P) is nontrivial in the sense that there is at least one admissible trajectory for which the cost is finite.

We associate with F the following function h, called the *(lower) Hamiltonian*:

$$h(x,p) := \min\{\langle p, v\rangle : v \in F(x)\}.$$

The *augmented Hamiltonian* $\bar{h}$ is defined by

$$\bar{h}(x,\theta,\zeta) := \theta + h(x,\zeta).$$

Given a lower semicontinuous function $f : \mathbb{R}^n \to \mathbb{R} \cup \{+\infty\}$ and a point $x \in \mathrm{dom} f := \{x' \in \mathbb{R}^n : f(x') < +\infty\}$, we say that $\xi \in \mathbb{R}^n$ is a *proximal subgradient* of f at x if and only if there exists $\sigma \geq 0$ such that

$$f(y) - f(x) + \sigma\|y - x\|^2 \geq \langle \xi, y - x\rangle,$$

for all y in a neighborhood of x. The set (which could be empty) of all proximal subgradients of f at x is denoted by $\partial_P f(x)$, and is referred to as the proximal subdifferential. The Proximal Density Theorem asserts that $\partial_P f(x) \neq \emptyset$ for all x in a dense subset of dom$f$. We also define the *limiting subdifferential* of f at x by

$$\partial_L f(x) := \{\lim \xi_i : \xi_i \in \partial_P f(x_i), x_i \to x \text{ and } f(x_i) \to f(x)\}.$$

We refer the reader to [4] for full account of proximal and nonsmooth analysis. Now we define Ψ to be the set of all locally Lipschitz functions ψ on $\mathbb{R}^n$ that satisfy the proximal Hamilton-Jacobi inequality

$$\bar{h}(x, \partial_L \psi(x)) \geq 0 \ \ \forall x \in A$$

as well as the boundary condition

$$\psi(x) \leq \ell(x) \ \ \forall x \in C.$$

The following nonsmooth duality is the main result:

Theorem 1.

$$\min(P) = \sup_{\psi \in \Psi} \psi(x_0).$$

This leads directly to the following optimality conditions.

Corollary 1. *Let $(\bar{T}, \bar{x}(\cdot))$ be an admissible trajectory for (P). Then $(\bar{T}, \bar{x}(\cdot))$ is a minimizer for (P) iff there exists a sequence of functions $\{\psi_i\}$ in Ψ such that*

$$\lim_{i \to +\infty} \psi_i(x_0) = \bar{T} + \ell(\bar{x}(\bar{T})).$$

Our theorem, whose proof is self-contained modulo some basic facts from proximal analysis, is new with respect to its very mild regularity hypotheses on F (which need not even be continuous), as well as the presence of a unilateral state constraint. Moreover, the fact that locally Lipschitz functions and limiting subgradients figure in our duality also gives easy access to smooth duality of the type found by Vinter. We extend his result by obtaining a duality in which feature only smooth solutions of an *autonomous* Hamilton-Jacobi inequality.

We note that using our methods we can also prove Vinter's duality presented in [15] and extend it for fixed time problems, but due to space restriction we only treat here the free time with infinite horizon case and we only sketch the proofs. For complete details, see [5].

This article is organized as follows. In the next section we sketch the proof of the above theorem. Section 3 is devoted to the generalization of Vinter's smooth duality.

2 Proof of Theorem 1

First we note that under our hypotheses on F, any trajectory can be extend indefinitely both forward and backward, so all trajectories can be considered as being defined on $]-\infty, +\infty[$. By the compactness property of trajectories and since ℓ is bounded below, it is easy to prove that the problem (P) admits a solution. For all $k \in \mathbb{N}^*$, we consider the function ℓ_k defined by

$$\ell_k(x) := \inf_{y \in \mathbb{R}^n} \{\ell(y) + k\|x - y\|^2\}. \tag{1}$$

The sequence $(\ell_k)_k$ is the quadratic inf-convolution sequence of ℓ. The following lemma gives some properties of ℓ_k:

Lemma 1. *For all* $k \in \mathbb{N}^*$, *we have:*

1. $\ell_k(\cdot) \leq \ell(\cdot)$ *and the set of minimizing points* y *in (1) is nonempty.*
2. ℓ_k *is locally Lipschitz and bounded below by* ω.
3. *For all* $x \in \mathbb{R}^n$,
$$\lim_{k \to +\infty} \ell_k(x) = \ell(x).$$

We also consider a locally Lipschitz approximation for the multifunction F. By [4, Proposition 4.4.4] there exists a sequence of locally Lipschitz multifunctions $\{F_k\}$ also satisfying the hypotheses of F such that:

- For each $k \in \mathbb{N}$, for every $x \in \mathbb{R}^n$,
$$F(x) \subseteq F_{k+1}(x) \subseteq F_k(x) \subseteq \overline{co}\, F(x + 3^{-k+1}B).$$
- $\bigcap_{k \geq 1} F_k(x) = F(x)\ \forall x \in \mathbb{R}^n$.

A standard method of approximating the terminally constrained problem (P) by a problem free of such constraints involves the imposition of a penalty term, the inf-convolution technique, and the preceding approximation of F. We consider for all $k \geq 1$ the following optimal control problem:

$$(P_k)\begin{cases}\text{Minimize } T+\ell_k(x(T))+kd_C(x(T))+k\int_0^T d_A(x(t))\,dt,\\ T\geq 0,\\ \dot{x}(t)\in F_k(x(t)) \;\; a.e.\ t\in[0,+\infty[,\\ x(0)=x_0.\end{cases}$$

Lemma 2. *There exists a sequence* λ_n *strictly increasing in* $\mathbb{N}^*$ *such that:*

$$\lim_{n\to+\infty}\min(P_{\lambda_n})=\min(P)$$

We continue the proof and study the following problem (P_{λ_n}):

$$\begin{cases}\text{Minimize } T+\hat{\ell}_{\lambda_n}(z(T)),\\ T\geq 0,\\ \dot{z}(t)\in \hat{F}_{\lambda_n}(z(t)) \;\; a.e.\ t\in[0,+\infty[,\\ z(0)=(0,x_0),\end{cases}$$

where $\hat{F}_{\lambda_n}$ is the augmented locally Lipschitz multivalued function defined by $\hat{F}_{\lambda_n}(y,x):=\{\lambda_n d_A(x)\}\times F_{\lambda_n}(x)$, $\forall (y,x)\in\mathbb{R}\times\mathbb{R}^n$ and $\hat{\ell}_{\lambda_n}$ is the locally Lipschitz function defined by $\hat{\ell}_{\lambda_n}(y,x)=\ell_{\lambda_n}(x)+\lambda_n d_C(x)+|y|$, $\forall (y,x)\in\mathbb{R}\times\mathbb{R}^n$.

Let $\hat{V}_{\lambda_n}:\mathbb{R}\times\mathbb{R}\times\mathbb{R}^n\to\mathbb{R}$ be the value function of the problem (P_{λ_n}); that is, for every $(\tau,\beta,\alpha)\in\mathbb{R}\times\mathbb{R}\times\mathbb{R}^n$, $\hat{V}_{\lambda_n}(\tau,\beta,\alpha)$ is the minimum of the following problem:

$$\begin{cases}\text{Minimize } T+\hat{\ell}_{\lambda_n}(z(T)),\\ T\geq \tau,\\ \dot{z}(t)\in \hat{F}_{\lambda_n}(z(t)) \;\; a.e.\ t\in[\tau,+\infty[,\\ z(\tau)=(\beta,\alpha).\end{cases}$$

Lemma 3. *The value function* $\hat{V}_{\lambda_n}$ *satisfies the following:*

1. $\hat{V}_{\lambda_n}$ *is locally Lipschitz on* $\mathbb{R}\times\mathbb{R}\times\mathbb{R}^n$.
2. $\hat{V}_{\lambda_n}(\tau,\beta,\alpha)\leq\tau+\hat{\ell}_{\lambda_n}(\beta,\alpha),\;\;\forall(\tau,\beta,\alpha)\in\mathbb{R}\times\mathbb{R}\times\mathbb{R}^n$.
3. $\forall(\tau,\beta,\alpha)\in\mathbb{R}\times[0,+\infty[\times\mathbb{R}^n$ *we have*

$$\hat{V}_{\lambda_n}(\tau,\beta,\alpha)=\tau+\hat{V}_{\lambda_n}(0,0,\alpha)+\beta.$$

4. $\forall(t,y,x)\in\mathbb{R}\times\mathbb{R}\times\mathbb{R}^n,\;\;\forall(\theta,\xi,\zeta)\in\partial_P\hat{V}_{\lambda_n}(t,y,x)$ *we have*

$$\theta+\lambda_n d_A(x)\xi+h_{\lambda_n}(x,\zeta)\geq 0.^1$$

[1]This Hamilton-Jacobi inequality follows since the system $(\hat{V}_{\lambda_n},\hat{F}_{\lambda_n})$ is *strongly increasing* on $\mathbb{R}\times\mathbb{R}\times\mathbb{R}^n$ (the function $\hat{V}_{\lambda_n}(\cdot,z(\cdot))$ being increasing on $[a,b]$ whenever z is a trajectory of $\hat{F}_{\lambda_n}$ on some interval $[a,b]$), using [4, Proposition 4.6.5] which gives a proximal characterization for the strong increase property. We note that h_{λ_n} is the lower Hamiltonian corresponding to F_{λ_n}.

Now let $\psi_{\lambda_n} : \mathbb{R}^n \to \mathbb{R}$ be the function defined by

$$\psi_{\lambda_n}(x) := \hat{V}_{\lambda_n}(0,0,x), \ \ \forall x \in \mathbb{R}^n.$$

Using Lemma 3 and the definition of ∂_L we get:

Lemma 4. $\psi_{\lambda_n} \in \Psi$.

We continue the proof and remark that

$$\psi_{\lambda_n}(x_0) = \hat{V}_{\lambda_n}(0,0,x_0) = \min(P_{\lambda_n}),$$

so

$$\sup_{\psi \in \Psi} \psi(x_0) \geq \psi_{\lambda_n}(x_0) = \min(P_{\lambda_n}).$$

Therefore

$$\min(P) = \lim_{n \to +\infty} \min(P_{\lambda_n}) \leq \sup_{\psi \in \Psi} \psi(x_0).$$

Now we show the reverse inequality by considering $\psi \in \Psi$ making the *temporary hypothesis* that F is locally Lipschitz. Then by reasoning by contradiction and using the definition of ∂_L we have the following lemma.

Lemma 5. *For all open and bounded subset $S \subset \mathbb{R}^n$, for all $\varepsilon > 0$, there exists a neighborhood U of A such that*

$$1 + \bar{h}(x, \partial_P \psi(x)) \geq -\varepsilon \ \ \forall x \in S \cap U.$$

Let $(\bar{T}, \bar{x}(\cdot))$ be a solution of the problem (P). By Gronwall's lemma (see [4, Proposition 4.1.4]) there exists $\rho > 0$ such that $\bar{x}(t) \in B(0;\rho)$, $\forall t \in [0, \bar{T}]$. We apply the preceding lemma for $S = B(0;\rho)$ and for $\varepsilon > 0$ to get the existence of a neighborhood U_ε of A such that

$$1 + \bar{h}(x, \partial_P \psi(x)) \geq -\varepsilon \ \ \forall x \in S \cap U_\varepsilon.$$

Then by [4, Proposition 4.6.5] we get that

$$\psi(x_0) \leq \varepsilon \bar{T} + \psi(\bar{x}(\bar{T})) \leq \varepsilon \bar{T} + \ell(\bar{x}(\bar{T})) = \varepsilon \bar{T} + \min(P)$$

hence by taking $\varepsilon \to 0$ we get

$$\psi(x_0) \leq \min(P)$$

and therefore

$$\min(P) \geq \sup_{\psi \in \Psi} \psi(x_0).$$

To remove the need for the locally Lipschitz hypothesis on F it is sufficient to use the following lemma (which follows by reasoning by contradiction and using a convergence property of ∂_L from [4, Exercise 1.11.21]), and then continue as in the Lipschitz case.

Lemma 6. *For all $n \in \mathbb{N}$ there exists $k_n \geq n$ such that*

$$1 + \bar{h}_{k_n}(x, \partial_L \psi(x)) \geq \frac{-1}{n} \quad \forall x \in A \cap \bar{B}(0;\rho).$$

The proof of Theorem 1 is achieved. □

3 Smooth Duality

An important application of our main result is the smooth duality studied by Vinter in [15]. Using Theorem 1 we show the following result that extends the Vinter's smooth duality for our problem (P).

Corollary 2.

$$\min(P) = \sup_{\varphi \in \Phi} \varphi(x_0)$$

where Φ is the set of all functions $\varphi : \mathbb{R}^n \to \mathbb{R}$ that satisfy:

- $\varphi \in C^1(\mathbb{R}^n, \mathbb{R})$,
- $1 + \langle \varphi'(x), v \rangle \geq 0,\ \forall x \in A,\ \forall v \in F(x)$,
- $\varphi(x) \leq \ell_0(x)\ \forall x \in C$.

Proof. Since for all $\varphi \in \Phi$ we have $\partial_L \varphi(t, x) = \{\varphi'(t, x)\}$, we get that $\Phi \subset \Psi$. Then by Theorem 1 we have

$$\min(P) = \sup_{\psi \in \Psi} \psi(0, x_0) \geq \sup_{\varphi \in \Phi} \varphi(0, x_0).$$

For the reverse inequality, let $\psi \in \Psi$. Using the fact that if ψ is differentiable at $\alpha \in \mathbb{R}^n$ then $\psi'(\alpha) \in \partial_L \psi(\alpha)$, we have the following lemma:

Lemma 7. *Let $\alpha \in A$ be such that ψ is differentiable at α. Then*

$$1 + \langle \psi'(\alpha), v \rangle \geq 0,\ \forall v \in F(\alpha).$$

Since ψ is locally Lipschitz, Rademacher's theorem implies that ψ is differentiable a.e. $\alpha \in \mathbb{R}^n$. Using the sequence F_k and the penalization term

$$k \int_0^T d_A(x(t))\, dt$$

(as in Lemma 2), we can assume that F is Lipschitz and $A = \mathbb{R}^n$. Then by Lemma 7 and by a standard mollification technique (i.e., convolution with mollifier sequence), we have the following lemma.

Lemma 8. *There exists a sequence $\delta_i \to 0$ such that for all $\varepsilon > 0$ there exist $i_0 \in \mathbb{N}$ and a sequence of functions $(\psi_\varepsilon^i)_i$ which satisfy: for $i \geq i_0$ we have $\psi_\varepsilon^i \in \Phi$ and*

$$\psi_\varepsilon^i(x_0) \geq \frac{\psi(x_0) - M|\delta_i| - \varepsilon}{1 + |\delta_i|},$$

where $M := \max_{x \in C} -\psi(x)$.

Clearly the preceding lemma gives the desired inequality. □

It is clear that Corollary 2 leads to a version of the necessary and sufficient conditions of Corollary 1 in which only smooth semisolutions are used.

A well-known and special case of the present framework involves the minimal time function associated to the target C and under the state constraint A:

$$T_C^A(\alpha) := \begin{cases} \operatorname{Inf} T \geq 0, \\ \dot{x}(t) \in F(x(t)) \;\; a.e. \; t \in [0, +\infty[, \\ x(0) = x_0, \\ x(t) \in A \;\; \forall t \in [0, T], \\ x(T) \in C. \end{cases}$$

By Corollary 2 ($\ell_0 = 0$) we have the following characterization of T_C^A, which appears to be new at a technical level:

$$T_C^A(\alpha) = \sup_{\varphi \in \Phi} \varphi(\alpha)$$

where Φ is the set of all functions $\varphi : \mathbb{R}^n \to \mathbb{R}$ that satisfy:

- $\varphi \in C^1(\mathbb{R}^n, \mathbb{R})$,
- $1 + \langle \varphi'(x), v \rangle \geq 0$, $\forall x \in A \; \forall v \in F(x)$,
- $\varphi(x) \leq 0 \; \forall x \in C$.

Acknowledgement. I would like to thank Professor Francis Clarke for his comments, suggestions and encouragement. I am also grateful to the referees who read the paper with great care and made interesting remarks. This work was supported by the National Council for Scientific Research in Lebanon.

References

1. Bardi M, Capuzzo-Dolcetta I (1997) Optimal control and viscosity solutions of Hamilton-Jacobi-Bellman equations. Birkhäuser Boston
2. Barles G (1994) Solutions de viscosité des équations de Hamilton-Jacobi. Springer-Verlag Paris
3. Clarke F, Ledyaev Y (1994) Mean value inequalities in Hilbert space, Trans Amer Math Soc 344:307-324
4. Clarke F, Ledyaev Y, Stern R, Wolenski P (1998) Nonsmooth Analysis and Control Theory (Graduate Texts in Mathematics 178). Springer-Verlag New York
5. Clarke F, Nour C (2003) Nonconvex duality in optimal control, submitted
6. Clarke F, Vinter R (1983) Local optimality conditions and Lipschitizian solutions to the Hamilton-Jacobi equation, SIAM J Control Optim 21:856-870
7. Fleming W (1989) Generalized solutions and convex duality in optimal control. In: Partial Differential Equations and the Calculus of Variations Essays in Honor of Ennio De Giorgi (Progress in Nonlinear Differential Equations and their Applications). Birkhäuser Boston
8. Fleming W, Vermes D (1989) Convex duality approach to the optimal control of diffusions, SIAM J Control Optim 27:1136-1155

9. Frankowska H (1989) Optimal trajectories associated with solution of the Hamilton-Jacobi equation, Appl Math Optim 19:291-311
10. Frankowska H (1993) Lower semicontinuous solutions of the Hamilton-Jacobi equation, SIAM J Control Optim 31:257-272
11. Hang Z (1992) Convex duality and generalized solutions in optimal control problem for stopped processes the deterministic model, SIAM J Control Optim 30:465-476
12. Lions P (1983) Optimal control of diffusion processes and Hamilton-Jacobi-Bellman equations I The dynamic programming principle and applications, Comm Partial Differential Equations 8:1101-1174
13. Soravia P (1993) Discontinuous viscosity solutions to Dirichlet problems for Hamilton-Jacobi equations with convex Hamiltonians, Comm Partial Differential Equations 18:1493-1514
14. Soravia P (1999) Optimality principles and representation formulas for viscosity solutions of Hamilton-Jacobi equations II equations of control problems with state constraints, Differential Integral Equations 12:275-293
15. Vinter R (1993) Convex duality and nonlinear optimal control, SIAM J Control Optim 31:518-538
16. Vinter R, Personal Communication
17. Vinter R, Lewis R (1978) The equivalence of strong and weak formulations for certain problems in optimal control, SIAM J Control Optim 16:546-570
18. Vinter R, Lewis R (1978) A necessary and sufficient condition for optimality of dynamic programming type making no a priori assumptions on the controls, SIAM J Control Optim 16:571-583
19. Wolenski P, Zhuang Y (1998) Proximal analysis and the minimal time function, SIAM J Control Optim 36:1048-1072.

One Sided Lipschitz Multifunctions and Applications

Tzanko Donchev

Department of Mathematics, University of Architecture and Civil Engineering, 1 "Hr. Smirnenski" str., 1046 Sofia, Bulgaria. tdd51us@yahoo.com

Summary. We present an overview of the ordinary differential inclusions with one sided Lipschitz right-hand sides. Some applications and illustrative examples are included.

1 Preliminaries

Consider the differential inclusion:

$$\dot{x}(t) \in F(t, x(t)), \; x(0) = x_0, \; t \in I := [0, 1], \tag{1}$$

where $x \in E$, F is multi-map from $I \times E$ into E with nonempty compact values and I is a (semi)closed real interval. Suppose first that $E \equiv \mathbb{R}^n$ (or even that E is a Hilbert space).

Let A and B be closed bounded sets. The *Hausdorff distance* is defined by

$$D_H(A, B) := \max\{ex(A, B), ex(B, A)\},$$

where

$$ex(B, A) := \sup_{b \in B} \inf_{a \in A} |a - b|.$$

The multi-map F is said to be *upper semicontinuous (USC)* at the point x provided for any $\varepsilon > 0$ there exists $\delta > 0$ such that $ex(F(y), F(x)) < \varepsilon$ when $|y - x| < \delta$. It is called *lower semicontinuous (LSC)* at x provided for any $f \in F(x)$ and any $x_i \to x$ there exists $f_i \in F(x_i)$ with $f_i \to f$. Also, F is called *continuous* provided it is continuous with respect to the Hausdorff distance. The map $F : I \times E \rightrightarrows E$ is said to be *almost USC (LSC, continuous)* provided for every $\varepsilon > 0$ there exists a compact set $I_\varepsilon \subset I$ with Lebesgue measure $meas(I_\varepsilon) > 1 - \varepsilon$ such that F restricted to $I_\varepsilon \times E$ is USC (LSC, continuous).

To prove the existence of solutions one commonly needs some compactness (in case E is a Hilbert space) and that $F(\cdot, \cdot)$ is almost upper semi

M. de Queiroz et al. (Eds.): Optimal Control, Stabilization and Nonsmooth Analysis, LNCIS 301, pp. 333–341, 2004.

continuous (USC) or almost lower semicontinuous (LSC). If, however, we want to prove (for example) the relaxation theorem, then we have to impose some other assumptions. Commonly one uses the (local) Lipschitz condition, saying that $D_H(F(t,x),F(t,y)) \leq L|x-y|$. It can be replaced by $D_H(\overline{\text{co}}\ F(t,x),\overline{\text{co}}\ F(t,y)) \leq L|x-y|$ (in this case $L>0$). Here we want to relax this condition. Consider the differential equation

$$\dot{x}(t) \in f(t,x),\ x(0) = x_0. \tag{2}$$

One can use some dissipative type assumption (in order to prove the continuous dependence of the solution set on the initial conditions, on parameters etc.) as it is done in the books devoted to differential equations [4, 19]. The most popular is the uniformly one sided Lipschitz (UOSL) one. If $\langle\cdot,\cdot\rangle$ is the scalar product, then the UOSL condition is:

$$\langle x-y, f_x - f_y\rangle \leq L|x-y|^2,\ \forall f_x \in f(t,x),\ \forall f_y \in f(t,y),\ \forall\ x,y \in E.$$

Here L may be non-positive. It is well known that under this condition the solution (if exists) is unique and that $f(t,\cdot)$ is single valued almost everywhere in the sense of Baire category (and, when E is finite dimensional, also in the sense of Lebesgue measure). Our objective is to extend this condition to 'truly' differential inclusions.

First we define the *upper Hamiltonian*

$$H_F(t,x,p) = \max_{v\in F(t,x)} \langle p,v\rangle = \sigma(p,F(t,x)) \text{ - the support function}$$

and also the *lower Hamiltonian*

$$h_F(t,x,p) = \min_{v\in F(t,x)} \langle p,v\rangle = -\sigma(-p,F(t,x)).$$

In terms of Hamiltonians the UOSL condition becomes

$$H_F(t,x,x-y) - h_F(t,y,x-y) \leq L|x-y|^2,\ \forall\ x,y \in E.$$

The Lipschitz condition (for $\overline{\text{co}}\ F(t,\cdot)$) implies

$$|H_F(t,x,p) - H_F(t,y,p)| \leq L|x-y||p|, \text{ for every vector } p\ \forall\ x,y \in E.$$

Notice that the last condition also has the same form in terms of lower Hamiltonians.

We will say that $F(t,\cdot)$ is OSL provided

$$H_F(t,x,x-y) - H_F(t,y,x-y) \leq L|x-y|^2,\ \forall\ x,y \in E.$$

In terms of lower Hamiltonians, the OSL condition is the same. Evidently this condition is (essentially) weaker than the both conditions above. Furthermore, $F(t,\cdot)$ is OSL $\Longleftrightarrow$ $\overline{\text{co}}\ F(t,\cdot)$ is OSL. The OSL condition was introduced in [6] for the case of differential inclusions with time lag (the latter being out of our scope).

We refer to [5, 18] for all concepts used in this paper but not explicitly discussed.

2 The Results

First we will present a very short proof of the relaxation theorem. (Compare it to the much longer proofs given in [1, 5, 22, 23].)

Theorem 1. *Suppose $F(\cdot,\cdot)$ has nonempty compact values and is bounded on the bounded sets. If $\overline{\text{co}}\ F(\cdot,\cdot)$ is almost continuous and OSL on x, then the differential inclusion*

$$\dot{x}(t) \in \overline{\text{co}}\ F(t, x(t)),\ x(0) = x_0,\ t \in I, \tag{3}$$

admits a nonempty $C(I,E)$ compact solution set. Moreover, the solution set of (1) is nonempty and dense in the solution set of (3).

Remark 1. The proof of the existence of solutions and the fact that the solution set of (3) is $C(I,E)$ compact is contained in [7]. Notice, however, that the solution set of (3) is in fact R_δ. Furthermore if $G_\varepsilon(\cdot,\cdot) \subset \overline{\text{co}}\ F(t,x)$ is almost LSC with nonempty compact values then the differential inclusion

$$\dot{y}(t) \in G_\varepsilon(t, y(t)),\ y(0) = y_0 \tag{4}$$

has a solution (see Theorem 1 of [7]).

Proof. Define $R(t,x) := \overline{ext}\ F(t,x)$. Since $\overline{\text{co}}\ F(\cdot,\cdot)$ is almost continuous, one has that $R(\cdot,\cdot)$ is almost LSC (see Lemma 2.3.7 of [23]). Let $y(\cdot)$ be a solution of (4). Define:

$$G_\varepsilon(t,y) = \overline{\{v \in R(t,x(t)) :\ \langle x(t) - y, \dot{x}(t) - v\rangle < L|x(t) - y|^2 + \varepsilon\}}.$$

Evidently $G_\varepsilon(\cdot,\cdot)$ is almost LSC with nonempty compact values. Let $y(\cdot)$ be a solution of (4), then $\langle x(t) - y(t), \dot{x}(t) - \dot{y}(t)\rangle < L|x(t) - y(t)|^2 + \varepsilon$. One only has to apply Gronwall's inequality. □

Example 1. Consider the following two dimensional differential inclusion:

$$\begin{aligned} &\dot{x}(t) \in \{-1, +1\},\ x(0) = 0. \\ &\dot{y}(t) = -\sqrt[3]{y(t)} + |x(t)|,\ y(0) = 0. \end{aligned} \tag{5}$$

Obviously the right-hand side of (5) is OSL and bounded on the bounded sets. Therefore the solution set of (5) is dense in the solution set of:

$$\begin{aligned} &\dot{x}(t) \in [-1, +1],\ x(0) = 0. \\ &\dot{y}(t) = -\sqrt[3]{y(t)} + |x(t)|,\ y(0) = 0. \end{aligned}$$

If one changes the sign before $\sqrt[3]{\cdot}$, then it is not true, because in this case

$$y(t) \geq \sqrt{(2t/3)^3}.$$

Notice that if $F(\cdot)$ and $G(\cdot)$ are OSL with constants L_F and L_G then $H(x) = F(x)\bigcup G(x)$ is also OSL with a constant $L_H = \max\{L_F, L_G\}$. This leads us to the next example:

Example 2. Define

$$F(t,x) = \overline{\bigcup_{u\in U} f(t,x,u)}$$

where $f(t,\cdot,u)$ is UOSL (but not necessarily single valued) with a constant L_u and $f(t,x,\cdot)$ is continuous. Assume that F is compact valued and

$$L = \sup_{u\in U} L_u < \infty.$$

Then $F(t,\cdot)$ is OSL with a constant L. If U is noncompact, then it is possible for $F(t,\cdot)$ to be discontinuous (or LSC if $f(t,\cdot,u)$ is continuous).

When $F(\cdot,\cdot)$ is almost LSC the following theorem holds true:

Theorem 2. *Let the map*

$$G(t,x) := \bigcap_{\varepsilon>0} \overline{\text{co}}\ F(t, x+\varepsilon B)$$

be USC on x (and with compact values). Then the solution set of (1) is dense in the solution set of

$$\dot{x}(t) \in G(t,x(t)),\ x(0) = x_0. \tag{6}$$

The proof is the same as the proof of Theorem 1. One has only to use the following inequality (proved in [8, 10]):

$$H_G(t,x,x-y) - H_F(t,y,x-y) \le L|x-y|^2.$$

If one relaxes the OSL condition of F to the OSL condition of G the conclusion of Theorem 2 does not hold:

Example 3. Let $E \equiv \mathbb{R}$ and let $A \subset \mathbb{R}$ be open and dense with the Lebesgue measure $meas(A) < \varepsilon$. Set

$$F(x) = \begin{cases} [0,1] & \text{for } x \in A, \\ 0 & \text{for } x \notin A, \end{cases}.$$

Then $F(\cdot)$ is LSC and $G(x) \equiv [0,1]$ is OSL, but the solution set of (1) is not dense in the solution set of (6).

Let $D \subset E$ be a closed set. A *proximal normal* to D at a point $x \in D$ is a vector $\xi \in E$ such that there exists $\alpha > 0$ with $\langle \xi, x' - x\rangle \le \alpha|x'-x|^2$, $\forall x' \in D$. The set of all such vectors is a cone denoted by $N_D^P(x)$ and is called the *proximal normal cone* to D at x (see [3] for instance).

Recall that given the closed set G , the system ((1), G) is said to be *strongly invariant* provided every solution of (1) remains in G, when $x_0 \in G$.

Theorem 3. *Let $F(\cdot,\cdot)$ be almost LSC and OSL. Assume $G(t,\cdot)$ is compact valued and USC when E^* is uniformly convex, or that $\overline{\text{co}}\, F(\cdot,\cdot)$ is almost continuous (in a general Banach space). The system ((6),G) is strongly invariant iff there exists a null set $\mathcal{A}$ such that $F(t,x) \subset T_G(x)$, $\forall\, x \in G$ and $\forall t \in I \setminus \mathcal{A}$.*

In case E is a Hilbert space the following theorem extends Theorem 3.1 of [3]:

Proposition 1. *Under the conditions of Theorem 3 the system ((6),G) is strongly invariant if and only if there exists a null set $\mathcal{A} \subset I$ such that $H_F(t,x,\varsigma) \leq 0$, $\forall\, \varsigma \in N_G^P(x)$, $\forall\, x \in G$ and $\forall t \in I \setminus \mathcal{A}$.*

The OSL condition is good enough to be used in the discrete approximations of the solution set. We will assume that $F(\cdot)$ is autonomous. Let $I = [0,1]$. Consider the uniform grid $\Delta_N = \{0 < t_1 < \ldots < t_N = 1\}$, where

$$t_i = \frac{i}{N} = ih.$$

The following scheme (called Euler's scheme) is commonly used in the literature:

$$\begin{cases} z(t) = y(t_i) + (t - t_i) f_i, \quad z(0) = x_0, \quad z(t_{i+1}) = \lim\limits_{t \to t_{i+1}} z(t) \\ \quad \text{where} \quad f_i \in F(z(t_i)), \quad i = 0, 1, \ldots, N-1\ , \end{cases} \tag{7}$$

Denote by R_E and R_R the solution sets of (7) and (6) respectively. The following theorem is proved in [10] and partially extends Theorem 3.1 of [21].

Theorem 4. *Let $F(\cdot)$ have closed bounded (but not necessarily compact or convex) values. If $F(\cdot)$ is OSL and bounded on bounded sets, then there exists a constant C such that $D_H(R_E, R_R) \leq Ch^{1/2}$.*

The OSL condition can be introduced when E is a (general) Banach space. When E^* is uniformly convex the OSL condition has the form:

$$H_F(t,x,J(x-y)) - H_F(t,y,J(x-y)) \leq L|x-y|^2 \ \forall\, x,y \in E,$$

where $J(x) = \{v \in E^* : \langle v, x\rangle = |x|^2 \text{ and } |v| = |x|\}$. Notice that Theorems 1 and 2 hold also in this case.

When, however, E is a general Banach space, it is possible for $J(\cdot)$ to be multi-valued and discontinuous. Denote by

$$[x,y]_+ = \lim_{h \to 0^+} h^{-1}\{|x + hy| - |x|\}.$$

The OSL condition is defined as follows:

There exists a constant L such that: for every $x, y \in E$ and a.a. t:

$$\sup_{f_x \in F(t,x)} \ \inf_{f_y \in F(t,y)} [x - y, f_x - f_y]_+ \quad \leq \quad L|x-y|.$$

Notice that F is OSL $\Longrightarrow$ $\overline{\text{co}}\, F$ is OSL, but the converse does not hold.

Theorem 5. *Theorem 1 and Remark 2 hold true also in the case of general Banach space E.*

In this case, the proof is complicated.

Example 4. Consider the following infinite dimensional differential inclusion:

$$\begin{aligned} &\dot{x}(t) \in \{-1, +1\},\ x(0) = 0. \\ &\dot{y}_1(t) = -\sqrt[3]{y_1(t)} + |x(t)|,\ y_1(0) = 0. \\ &\dots \\ &\dot{y}_{n+1} = -\sqrt[3]{y_{n+1}(t)} + \frac{|y_n(t)|}{n^2},\ y_{n+1}(0) = 0. \\ &\dots \end{aligned} \tag{8}$$

The state space is c, i.e. the set of all bounded real-valued sequences

$$\overrightarrow{s} := \{s_i\}_{i=1}^{\infty}$$

such that

$$\lim_{i \to \infty} |s_i| = 0$$

and the norm $|\overrightarrow{s}| = \sup_i |s_i|$. The right-hand side is OSL and continuous. Therefore the solution set of (8) is nonempty and dense in the solution set of the relaxed system. If one changes the sign before $\sqrt[3]{\cdot}$ in infinitely many equations, then the system has no solution. If one changes the sign in one equation (after the first one), then the relaxation theorem is not true.

Finally, we will present some properties of the OSL multi-maps, which are obtained via the differential inclusions theory. Let $F(\cdot)$ be autonomous, i.e., (1) has the form

$$\dot{x}(t) \in F(x),\ x(0) = x_0,\ t \in [0, \infty). \tag{9}$$

We assume that $F(\cdot,\cdot) \equiv \overline{co}F(\cdot,\cdot)$. The following theorem is proved in [11]. Here we present the proof for the reader's convenience, because the latter is not published yet.

Theorem 6. *Let $F(\cdot)$ be continuous or USC, where E^* is uniformly convex. If F is OSL, then the differential inclusion (9) admits a nonempty $C([0,T],E)$ compact solution set for every $T > 0$.*

Proof. The local existence has been proved in [7] when $F(\cdot)$ is continuous and in [8] when $F(\cdot)$ is USC. It is easy to see that every solution $x(\cdot)$ of (9) satisfies

$$|x(t)| \le e^{Lt}\Big(\frac{1 - e^{-Lt}}{L}|F(0)| + |x_0|\Big)$$

when $L \neq 0$ or $|x(t)| \le t|F(0)| + |x_0|$ when $L = 0$. Moreover for $S > 0$ sufficiently small the solution set is $C([0,S],E)$ compact. Hence $D_H(R(t), R(s)) \le MN(s,t)$, where $M = \max\{|F(0)|, |F(x_0)|\}$, $N(t,s) = \max\{e^{Lt}, e^{Ls}\}$ and

$R(t)$ is the reachable set of (9) at the time t. Furthermore $R(t)$ is a nonempty and compact for every $t \in [0, S]$. Suppose the reachable set exists on $[0, T)$ and is nonempty compact for every $t \in [0, T)$. From the inequality above one can conclude that $R(\cdot)$ is Lipschitz continuous on $[0, T)$. Hence

$$\lim_{t \uparrow T} R(t) = R_T$$

exists (with respect to the Hausdorff metric). Furthermore R_T is nonempty compact and hence F is bounded on some neighborhood of R_T. Therefore there exists $\lambda > 0$ such that the result of the (local) existence applies on $[T, T + \lambda)$ for (9) and every $x_0 \in R_T$. Thus the solution set is nonempty $C([0, T + \lambda], E)$ compact. The proof is completed with a simple application of Zorn' lemma. □

Remark 2. Consider the system (9) where the right-hand side $F(\cdot)$ is from Example 4. The state space is l_2, i.e., the set of all real sequences

$$\overrightarrow{s} := \{s_i\}_{i=1}^{\infty}$$

such that

$$\sum_{i=1}^{\infty} |s_i|^2 < \infty.$$

It is easy to see that the right-hand side F is not defined on the whole space, but only for $\overrightarrow{s}$ with

$$\sum_{i=1}^{\infty} |s_i|^{\frac{2}{3}} < \infty.$$

However, the system has a solution on $[0, 1]$ for every such initial condition $\mathbf{s}_0$. In this case $F(\cdot)$ is m-dissipative. We refer to [18] for the theory and applications of the m-dissipative operators. It would be interesting to give a 'truly' multi-valued extension of m-dissipative operators.

The following theorem is proved in [9]:

Theorem 7. *Let $F(\cdot)$ be continuous or USC, where E^* is uniformly convex. If F is OSL with a constant $L < 0$, then $F(\cdot)$ is onto. If $L < 1$ then there exists $x \in E$ such that $x \in F(x)$.*

Recall that we say that F is *coercive* provided

$$\lim_{\|x\| \to \infty} \frac{H_F(x, x)}{\|x\|} = -\infty.$$

When E is Hilbert the following stronger result proved in [14] holds:

Proposition 2. *Let $F(\cdot)$ be OSL ($L = 0$) with closed graph. If F is coercive, then F is surjective.*

Remark 3. The conclusions of Theorem 7 and Proposition 2 hold true also when $F(\cdot)$ is not OSL, but satisfies the opposite inequality, i.e.,

$$\sup_{f_x \in F(t,x)} \inf_{f_y \in F(t,y)} [x-y, f_x - f_y]_+ \geq L|x-y|.$$

Notice that we always required F to be compact valued, which is not true when F is UOSL. However, the case $F(\cdot,\cdot)+G(\cdot,\cdot)$ when the one multifunction is UOSL, while the other is OSL, is not difficult.

Now we present the characterization of the one dimensional OSL multifunction.

Proposition 3. *The USC multifunction* $F : \mathbb{R} \to \mathbb{R}$ *is OSL with a constant L if and only if there exist two monotone nonincreasing real valued (not necessarily continuous) functions* $f(\cdot)$ *and* $g(\cdot)$*, with* $f(x) \leq g(x)$ *such that* $F(x) = Lx + [f(x), g(x)]$.

We refer to the papers [12, 13, 20], where many other examples (and applications) of OSL multi-functions are presented. Some applications of OSL multi-maps to singularly perturbed differential inclusions are considered in [15, 16, 17, 24]. In [8, 9] are proved most of the results presented in the paper.

Acknowledgement. This work has been funded in part by the National Academy of Sciences under the Collaboration in Basic Science and Engineering Program, supported by Contract No. INT-0002341 from the National Science Foundation. The contents of this publication do not necessarily reflect the views or policies of the National Academy of Sciences or the National Science Foundation, nor does mention of trade names, commercial products or organizations imply endorsement by the National Academy of Sciences or the National Science Foundation.

References

1. Aubin J-P, Cellina A (1984) Differential Inclusions. Springer Berlin
2. Clarke F, Ledyaev Y, Stern R, Wolenski P (1995) Qualitative properties of trajectories of control systems a survey, J Dynam Control Systems 1:1–48
3. Clarke F, Ledyaev Y, Stern R, Wolenski P (1998) Nonsmooth Analysis and Control Theory. Springer Berlin
4. Deimling K (1977) Ordinary Differential Equations in Banach Spaces (Lect Notes Math 596). Springer Berlin
5. Deimling K (1992) Multivalued Differential Equations. De Grujter Berlin
6. Donchev T (1991) Functional differential inclusions with monotone right-hand side, Nonlinear Analysis 16:543-552
7. Donchev T (1996) Qualitative properties of a class differential inclusions, Glasnik Matematički 31:269-276
8. Donchev T (1999) Semicontinuous differential inclusions, Rend Sem Mat Univ Padova 101:147-160

9. Donchev T (2002) Properties of one-sided Lipschitz multivalued maps, Nonlinear Analysis 49:13-20
10. Donchev T (2002) Euler approximation of nonconvex differential inclusions, An St Univ Ovidius Constanta Seria Mathematica to appear
11. Donchev T (2003) Surjectivity and fixed points of relaxed dissipative multifunctions differential inclusions approach, submitted
12. Donchev T, Farkhi E (1998) Stability and Euler approximation of one-sided Lipschitz differential inclusions, SIAM J Control Optim 36:780-796.
13. Donchev T, Farkhi E (1999) Approximation of one-sided Lipschitz differential inclusions with discontinuous right-hand sides. In Ioffe A, Reich S, Shafrir I (eds) Calculus of Variations and Differential Equations (Chapman & Hall/CRC Research Notes Mathematics Series 410). CRC Press Boca Raton FL 101–118
14. Donchev T, Georgiev P (2003) Relaxed submonotone mappings, Abstract Appl Analysis 2003:19-31
15. Dontchev A, Slavov I (1991) Upper semicontinuity of solutions of singularly perturbed differential inclusions. In Sebastian H-J, Tammer K (eds) System Modeling and Optimization (Lecture Notes in Control and Information Sciences 143). Springer Berlin 273–280
16. Donchev T, Slavov I (1999) Averaging method for one sided Lipschitz differential inclusions with generalized solutions, SIAM J Control Optim 37:1600-1613
17. Grammel G (1996) Singularly perturbed differential inclusions: an averaging approach, Set-Valued Analysis 4:361-374
18. Hu S, Papageorgiou N (1997, 2000) Handbook of Multivalued Analysis vol I-II. Kluwer Dordrecht
19. Lakshmikantham V, Leela S (1981) Nonlinear Differential Equations in Abstract Spaces. Pergamon Oxford
20. Lempio F, Veliov V (1998) Discrete approximations of differential inclusions, Bayreuter Marhematische Schriften 54:149–232
21. Mordukhovich B (1995) Discrete approximations and refined Euler-Lagrange conditions for differential inclusions, SIAM J Control Optim 33:882-915
22. Pianigiani G (1975) On the fundamental theory of multivalued differential equations, J Differential Equations 25:30-38
23. Tolstonogov A (2000) Differential Inclusions in a Banach Space. Kluwer Dordrecht
24. Veliov V (1994) Differential inclusions with stable subinclusions, Nonlinear Analysis TMA 23:1027–1038

A Characterization of Strong Invariance for Perturbed Dissipative Systems

Tzanko Donchev[1], Vinicio Rios[2], and Peter Wolenski[2]

[1] Department of Mathematics, University of Architecture and Civil Engineering, 1 "Hr. Smirnenski" str., 1046 Sofia, Bulgaria. tdd51us@yahoo.com

[2] Department of Mathematics, Louisiana State University and A. and M. College, Baton Rouge, LA 70803-4918. [rios,wolenski]@math.lsu.edu

Summary. This note studies the strong invariance property of a discontinuous differential inclusion in which the right hand side is the sum of an almost upper semicontinuous dissipative and an almost lower semicontinuous one-sided Lipschitz multifunction. A characterization is obtained in terms of a Hamilton-Jacobi inequality.

1 Introduction and Assumptions

We extend a characterization of strong invariance that was recently proved in [16] to a nonautonomous system with weaker data assumptions. Consider the control system modelled as a differential inclusion

$$\begin{cases} \dot{x}(t) \in F(t, x(t)) & \text{a.e. } t \in [t_0, t_1) \\ x(t_0) = x_0, \end{cases} \tag{1}$$

where the given multifunction $F : \mathcal{I} \times \mathbb{R}^n \to 2^{\mathbb{R}^n}$ has compact convex values and $\mathcal{I} \subseteq \mathbb{R}$ is an open interval with $[t_0, t_1) \subseteq \mathcal{I}$. The focus in this paper is on the case where the multifunction F has the form

$$F(t, x) = D(t, x) + G(t, x), \tag{2}$$

where D is upper semicontinuous and dissipative in x and G is lower semicontinuous and one-sided Lipschitz; precise assumptions will be given below.

We briefly recall the main concepts of invariant systems theory for the system (1) using the language and notation from [6, 7]. In addition to F, a closed set $S \subseteq \mathbb{R}^n$ is also given. The pair (S, F) is *weakly invariant* (respectively, *strongly invariant*) if for every $[t_0, t_1) \in \mathcal{I}$ and every $x_0 \in S$, at least one (respectively, every) solution $x(\cdot)$ of (1) satisfies $x(t) \in S$ for all $t \in [t_0, t_1)$. We refer the reader to [5, 13, 15] for the history of invariance theory, which is elsewhere called viability theory [1, 2].

M. de Queiroz et al. (Eds.): Optimal Control, Stabilization and Nonsmooth Analysis, LNCIS 301, pp. 343–349, 2004.

There are two main approaches for characterizing the notions of invariance, one involving tangential conditions and the other using normal cones and Hamilton-Jacobi inequalities. Data assumptions for the characterization of weak invariance are very general, whereas for strong invariance, simple examples show that additional assumptions are needed. A Lipschitz assumption has usually been invoked for this purpose in the literature [6, 13]. Our main result shows that strong invariance can be characterized under more general assumptions, but is given only in the "normal" framework via Hamilton-Jacobi inequalities.

The Hamiltonian is prominently featured in our results, but the assumptions are most easily formulated in these terms as well. Recall that the minimized and maximized Hamiltonians associated to a multifunction F are given, respectively, by

$$h_F(t,x,p) = \inf\{\langle p, v\rangle : v \in F(t,x)\}, \quad \text{and}$$
$$H_F(t,x,p) = \sup\{\langle p, v\rangle : v \in F(t,x)\}.$$

We shall only consider global data assumptions in order to illustrate the main ideas clearly; local versions can be proven by similar arguments, but require more technical bookkeeping, and will be published elsewhere. The convex valued multifunction F satisfies the *Lipschitz property* (in x, uniformly in $t \in \mathcal{I}$) provided there exists a constant $k > 0$ so that

$$\left|H_F(t,x,p) - H_F(t,y,p)\right| \leq k|p||x-y| \quad \forall x,\, y,\, p \in \mathbb{R}^n, \forall t \in \mathcal{I}. \tag{3}$$

A modification of the Lipschitz condition is the following: F is *one-sided Lipschitz* (in x, uniformly in $t \in \mathcal{I}$) provided there exists a constant $k > 0$ so that

$$H_F(t,x,x-y) - H_F(t,y,x-y) \leq k|x-y|^2 \quad \forall x,\, y \in \mathbb{R}^n,\, \forall t \in \mathcal{I}. \tag{4}$$

The one-sided Lipschitz notion was introduced in [10]; see also [12]. It is obvious that (4) is weaker than (3), and is strictly weaker, since for example in dimension one, $F(x) = \sqrt[3]{-x}$ satisfies (4) but not (3).

We now consider a special class of one-sided Lipschitz maps (2). One can show $h_F = h_D + h_G$, where h_D, h_G denote the minimized Hamiltonians for D, G respectively. Similarly, $H_F = H_D + H_G$. The maps $D(\cdot,\cdot)$ and $G(\cdot,\cdot)$ will be assumed to have the following properties.

(H1) $D(\cdot,\cdot)$ and $G(\cdot,\cdot)$ have nonempty, compact, and convex values.
(H2) $D(\cdot,\cdot)$ is almost USC and dissipative in x.
(H3) $G(\cdot,\cdot)$ is almost LSC and one-sided Lipschitz in x.

Recall that D is *upper semicontinuous (USC)* (called outer continuous in [17]) in x at a point (t,x) if for every $\varepsilon > 0$, there exists $\delta > 0$ such that $D(t,x+\delta B) \subset D(t,x) + \varepsilon B$, where $B \subset \mathbb{R}^n$ denotes the unit ball.

The dissipative requirement is that

$$H_D(t,y,y-x) - h_D(t,x,y-x) \le 0 \quad \forall x,\, y \in \mathbb{R}^n, \text{ a.e. } t \in \mathcal{I}. \tag{5}$$

Properties of dissipative maps (which means $-D$ is monotone in the language of [17]) have been well studied; see [2, 3, 17], and for physical applications, [8, 9]. The condition that $L : \mathbb{R}^n \to 2^{\mathbb{R}^n}$ is *lower semicontinuous (LSC)* (called inner continuous in [17]) at x means that for all $x \in \mathbb{R}^n$, all $y \in L(x)$, and every sequence $x_i \to x$, there exists $y_i \in L(x_i)$ such that $y_i \to y$. The multifunction $G(\cdot,\cdot)$ is said to be *almost LSC* provided there exists a sequence $\{I_k\}_{k=1}^{\infty}$ of pairwise disjoint compact sets such that G restricted to $I_k \times \mathbb{R}^n$ is LSC for every k and the set $\bigcup_{k=1}^{\infty} I_k \subset \mathcal{I}$ is of full measure. Almost USC maps are defined analogously.

Note that under (H1)-(H3), the sum $F = D + G$ is one-sided Lipschitz with constant k equal to the one-sided Lipschitz constant for G. However, F is neither necessarily almost USC nor almost LSC.

2 Strong Invariance

A closed subset $S \subseteq \mathbb{R}^n$ is given. Clarke [4] showed that strong invariance is equivalent to

$$F(x) \subseteq T_C^S(x) \quad \forall x \in S, \tag{6}$$

where $T_C^S(x)$ is the Clarke tangent cone (see [6]). Krastanov [14] gave an infinitesimal characterization of normal-type, by showing strong invariance is equivalent to

$$H_F(x,\zeta) \le 0 \quad x \in S,\, \zeta \in N_S^P(x), \tag{7}$$

where $N_S^P(x)$ denotes the proximal normal cone:

$$N_S^P(x) = \{\zeta \in \mathbb{R}^n :\ \exists \sigma > 0 \text{ such that } \langle \zeta, y-x\rangle \le \sigma |x-y|^2\ \forall y \in S\}.$$

See [5] for a Hilbert space version. These results assume the data F is autonomous and Lipschitz. Donchev [12] was the first to extend these characterizations beyond the autonomous Lipschitz case to "almost continuous" one-sided Lipschitz multifunctions. Rios and Wolenski [16] proved an autonomous normal-type characterization that allows for a discontinuous component. It is interesting to note that the latter result does not include a tangential-type equivalence, and neither will the main result of the present paper. In fact, (6) is no longer necessary for strong invariance in the the case of purely dissipative maps, as can be seen by considering

$$F(x) = \begin{cases} \{-\frac{x}{|x|}\} & \text{if} \quad x \neq 0 \\ [-1,1] & \text{if} \quad x = 0, \end{cases}$$

and $S = \{0\}$. It is easy to see (S,F) is strongly invariant and (6) fails.

3 The Main Result

We assume $F = D+G$ and (H1)-(H3) are satisfied. We remark that since F is not necessarily semicontinuous, the existence of solutions to (1) is not assured by the usual theory. However, in our case, existence follows since there exists an almost continuous (Caratheodory) selection $g(\cdot,\cdot)$ of $G(\cdot,\cdot)$ (see [2, 8]). Then $D+g$ is almost upper semicontinuous and has a solution, and obviously any solution to $D+g$ is a solution to $D+G$. The following statement is the constrained time-independent version of the weak invariant normal-type result proved in [12], which we will later invoke:

Theorem 1. *Suppose F is almost upper semicontinuous and $S \subset \mathbb{R}^n$ is closed. Then (S,F) is weakly invariant if and only if there exists a null set $A \subset \mathcal{I}$ such that*

$$h(t,x,\zeta) \leq 0, \tag{8}$$

for all $t \in \mathcal{I} \setminus A$, $x \in S$, and $\zeta \in N_S^P(x)$.

Remark 1. The condition (8) holds precisely at all points of density of I_n, $n \geq 1$, where F is USC in $I_n \times \mathbb{R}^n$. Recall that a point $t \in \mathbb{R}$ is a *point of density* of a measurable set $\mathcal{M} \subset \mathbb{R}$ provided

$$\lim_{\delta \to 0} \frac{\mu([t-\delta, t+\delta] \bigcap \mathcal{M})}{2\delta} = 1,$$

(where μ denotes the Lebesgue measure in $\mathbb{R}$), and that the set $\tilde{\mathcal{M}}$ of all points of density of $\mathcal{M}$ has full measure in $\mathcal{M}$.

The following theorem is the main contribution of this paper.

Theorem 2. *The system $(S, D+G)$ is strongly invariant if and only if there exists a set $I \subset \mathcal{I}$ of full measure such that*

$$h_D(t,x,\xi) + H_G(t,x,\xi) \leq 0, \ \forall (t,x) \in I \times S, \ \forall \zeta \in N_S^P(x). \tag{9}$$

Proof. Suppose the system $(S, D+G)$ is strongly invariant. Without loss of generality we can assume that there is collection of pairwise disjoint compact sets $\{J_k\}_{k=1}^{\infty}$ such that $\bigcup_{k=1}^{\infty} J_k \subset \mathcal{I}$ has full measure, and $D(\cdot,\cdot)$ and $G(\cdot,\cdot)$ are USC and LSC respectively in each $J_k \times \mathbb{R}^n$. Let $I_k \subset J_k$ be the set consisting of all points of density of J_k, and consider $I = \bigcup_{k=1}^{\infty} I_k$. Now fix $t_0 \in I$, $x_0 \in S$, and $\zeta \in N_S^P(x_0)$, and choose $v \in G(t_0, x_0)$ so that

$$H_G(t_0, x_0, \zeta) = \langle v, \zeta \rangle.$$

Michael's Theorem assures the existence of a selection g of G that is continuous on every $J_k \times \mathbb{R}^n$, and such that $g(t_0, x_0) = v$, i.e., $g(\cdot,\cdot)$ is Caratheodory. This means $g : \mathcal{I} \times \mathbb{R}^n \to \mathbb{R}^n$ is measurable on $\mathcal{I}$ for all fixed x and continuous in x for almost all $t \in \mathcal{I}$. The multifunction $D+g$ given by $(D+g)(t,x) :=$

$\{d + g(t,x) : d \in D(t,x)\}$ is then USC in each J_k. By the assumption of strong invariance of $D + G$, it readily follows that $D + g$ is weakly invariant, and therefore Theorem 1 implies

$$h_D(t_0, x_0, \zeta) + h_g(t_0, x_0, \zeta) \le 0, \tag{10}$$

where $h_g(t,x,\zeta) := \langle g(t,x), \zeta\rangle$. By the choice of $v \in G(t_0, x_0)$, one has $h_g(t_0, x_0, \zeta) = H_G(t_0, x_0, \zeta)$, and so (9) follows immediately from (10). This proves the necessity.

For the sufficiency, assume that (9) holds, and let $[t_0, t_1) \subseteq \mathcal{I}$ and $x(\cdot)$ be a trajectory of (1) with $x(t_0) = x_0 \in S$. The multifunctions $\{\dot{x}(t) - d : d \in D(t, x(t))\}$ and $G(t, x(t))$ are both measurable in t on $[t_0, t_1)$, and therefore their intersection is also measurable (see [8]). Furthermore, since $x(\cdot)$ is a solution of (1), this intersection is nonempty for almost all $t \in [t_0, t_1)$. Consequently ([8]) there exists a measurable selection

$$v(t) \in \left[\{\dot{x}(t) - d : d \in D(t, x(t))\} \bigcap G(t, x(t))\right] \quad \text{a.e. } t \in [t_0, t_1).$$

For $\delta > 0$, define

$$G_\delta(t,x) := \text{cl}\ \{g \in G(t,x) : \langle g - v(t), x - x(t)\rangle < k|x - x(t)|^2 + \delta\}, \tag{11}$$

where $k > 0$ is the constant in the one-sided Lipschitz assumption on G. It is clear that $G_\delta(t,x)$ is convex and compact, and it is nonempty since G is assumed to be one-sided Lipschitz and $v(t) \in G(t, x(t))$. Moreover, the almost lower semicontinuity of G implies that of $G_\delta(\cdot,\cdot)$. Let $\{J_k\}_{k=1}^\infty$ be a collection of pairwise disjoint compact sets such that $\bigcup_{k=1}^\infty J_k \subset [t_0, t_1]$ has full measure, and $G_\delta(\cdot,\cdot)$ and $D(\cdot,\cdot)$ are LSC and USC respectively in each $J_k \times \mathbb{R}^n$. Using again Michael's theorem we find a selection g of G_δ that is continuous in each $J_k \times \mathbb{R}^n$. Let

$$\tilde{I} = \bigcup_{k=1}^{\infty} (I_k \bigcap I),$$

where $I_k \subset J_k$ is the set consisting of all points of density of J_k. Note that $\tilde{I} \subset I$ has full measure in $[t_0, t_1)$. The fact that $g(\cdot,\cdot) \in G_\delta(\cdot,\cdot) \subset G(\cdot,\cdot)$ and the assumption (9) yield that for all $t \in \tilde{I}$, all $x \in S$, and all $\zeta \in N_S^P(x)$, we have

$$\begin{aligned} h_D(t,x,\zeta) + h_g(t,x,\zeta) &\le h_D(t,x,\zeta) + H_{G_\delta}(t,x,\zeta) \\ &\le h_D(t,x,\zeta) + H_G(t,x,\zeta) \le 0. \end{aligned}$$

This implies the weak invariance of $(S, D + g)$ (see Theorem 1), and so there exists an absolutely continuous arc $y_\delta(\cdot)$ satisfying

$$\dot{y}_\delta(t) - g(t, y_\delta(t)) =: d(t) \in D(t, y_\delta(t)),\ \ y_\delta(t_0) = x_0$$

and $y_\delta(t) \in S$ for all $t \in [t_0, t_1)$. Recall that for almost all $t \in [t_0, t_1)$,

$$v(t) \in G(t, x(t)),$$
$$\dot{x}(t) - v(t) \in D(t, x(t)), \text{ and}$$
$$g(t, y_\delta(t)) \in G_\delta(t, y_\delta(t)).$$

By definition (11) and the dissipativity of $D(t, \cdot)$, one has

$$\begin{aligned}\frac{1}{2}\frac{d}{dt}|y_\delta(t) - x(t)|^2 &= \langle \dot{y_\delta}(t) - \dot{x}(t), y_\delta(t) - x(t)\rangle \\ &= \langle g(t, y_\delta(t)) - v(t), y_\delta(t) - x(t)\rangle \\ &+ \langle d(t) - (\dot{x}(t) - v(t)), y_\delta(t) - x(t)\rangle \\ &\le k|y_\delta(t) - x(t)|^2 + \delta.\end{aligned}$$

Gronwall's inequality implies $|y_\delta(t) - x(t)|^2 \le 2\delta e^{2|k|(t-t_0)}(t - t_0)$. Since $\delta > 0$ is arbitrary, it follows that $y_\delta(t) \to x(t)$ as $\delta \downarrow 0$ for all $t \in [t_0, t_1)$. Since $y_\delta(t) \in S$ for all $t \in [t_0, t_1)$ and S is closed, we conclude $x(t) \in S$. Hence $(S, D + G)$ is strongly invariant as claimed. □

The following corollary exploits the mixed-type Hamiltonian condition (9) to reveal a property of trajectories of dissipative and one-sided Lipshitz maps. Namely, strong invariance is preserved under the sum if the pieces are separately invariant.

Corollary. *Under assumptions of Theorem 2, if (S, D) and (S, G) are strongly invariant, then $(S, D + G)$ is strongly invariant.*

Proof. By setting first $D(\cdot, \cdot) \equiv 0$ and then $G(\cdot, \cdot) \equiv 0$, we obtain from Theorem 2 that there is a set of full measure $I \subset \mathcal{I}$ (common for both G and D) such that $H_G(t, x, \zeta) \le 0$ and $h_D(t, x, \zeta) \le 0$, for all $t \in I$, $x \in S$ and $\zeta \in N_S^P(x)$. Adding these two inequalities, and again applying the criterion (9), the result follows immediately. □

Acknowledgement. This work has been funded in part by the National Academy of Sciences under the Collaboration in Basic Science and Engineering Program, supported by Contract No. INT-0002341 from the National Science Foundation. The contents of this publication do not necessarily reflect the views or policies of the National Academy of Sciences or the National Science Foundation, nor does mention of trade names, commercial products or organizations imply endorsement by the National Academy of Sciences or the National Science Foundation.

References

1. Aubin J-P (1991) Viability Theory. Birkhäuser Boston
2. Aubin J-P, Cellina A (1984) Differential Inclusions. Springer Berlin
3. Brezis H (1970) On a characterization of flow-invariant sets, Commun Pure Appl Math 23:261-263

4. Clarke F (1975) Generalized gradients and applications, Trans of the American Math Society 205:247-262
5. Clarke F, Ledyaev Y, Radulescu M (1997) Approximate invariance and differential inclusions in Hilbert spaces, J Dynam Control Syst 3:493–518
6. Clarke F, Ledyaev Y, Stern R, Wolenski P (1995) Qualitative properties of trajectories of control systems: a survey, J Dynam Control Systems 1:1–48
7. Clarke F, Ledyaev Y, Stern R, Wolenski P (1998) Nonsmooth Analysis and Control Theory. Springer New York
8. Deimling K (1992) Multivalued Differential Equations. De Grujter Berlin
9. Deimling K, Szilágyi P (1993) Periodic solutions of dry friction problems, Z Angew Math Phys 44:53-60
10. Donchev T (1991) Functional differential inclusions with monotone right-hand side, Nonlinear Analysis 16:543-552
11. Donchev T (1999) Semicontinuous differential inclusions, Rend Sem Mat Univ di Padova 101:147–160
12. Donchev T (2002) Properties of the reachable set of control systems, Systems Control Letters 46:379-386
13. Frankowska H, Plaskacz S, Rzezuchowski T (1995) Measurable viability theorems and the Hamilton-Jacobi-Bellman equation, Journal of Differential Equations 116:265-305
14. Krastanov M (1995) Forward invariant sets, homogeneity and small-time local controllability. In: Jakubczyk B et al (eds) Nonlinear Control and Differential Inclusions (Banach Center Publ 32). Polish Acad Sci Warsaw 287–300
15. Redheffer R, Walter W (1975) Flow-invariant sets and differential inequalities in normed spaces, Applicable Analysis 5:149-161
16. Rios V, Wolenski P (2003) A characterization of strongly invariant systems for a class of non-Lipschitz multifunctions. In: Proc 42nd IEEE CDC to appear
17. Rockafellar R, Wets R (2000) Variational Analysis. Springer New York

Generalized Differentiation for Moving Objects

Boris S. Mordukhovich[1] and Bingwu Wang[2]

[1] Department of Mathematics, Wayne State University, Detroit, MI 48202, USA. boris@math.wayne.edu
[2] Department of Mathematics, Eastern Michigan University, Ypsilanti, MI 48071, USA. bwang@emunix.emich.edu

1 Introduction

This work is devoted to the study of the analysis of moving sets/mappings in the form $\{\Omega_t\}_{t\in T}$ or $\{F_t\}_{t\in T}$. Such objects arise commonly both in applications and theories including parameterized optimization, non-autonomous control, mathematical economics, etc. As the principal part of nonsmooth variational analysis, the generalized differentiation theory for the non-moving objects has been developed greatly in the past few decades, which greatly advanced the related applications and studies. Being a natural extension and much required by applications, the study of the corresponding moving objects, however, has been quite limited. In this paper, we summarize and generalize some of the recently developed concepts, namely, extended normal cone, extended coderivative and extended subdifferential, which are natural generalizations of the corresponding normal cone, coderivative and subdifferential for non-moving objects, respectively. We then develop complete calculus rules for such extended constructions. Similarly to the case of non-moving objects, the developed calculus of extended constructions requires certain normal compactness conditions in infinite-dimensional spaces. As the second goal of the paper, we propose new sequential normal compactness for the moving objects and then establish calculus rules for the extended normal compactness that are crucial to applications. We illustrate that most results for the non-moving situations could be generalized to the moving case. The main results are developed in Asplund spaces, and our main tool is a fuzzy intersection rule based on the extremal principle.

The paper is organized as follows. In Section 2 we present necessary notations and definitions of generalized differential constructions as well as generalized normal compactness. Some preliminary discussions of these constructions and conditions are also collected here. In Section 3 we present calculus rules for the extended normal cones, subdifferentials and coderivatives. In Section 4 we establish calculus of extended normal compactness. In the concluding Section 5 we propose a new kind of derivative for moving objects and sketch the

M. de Queiroz et al. (Eds.): Optimal Control, Stabilization and Nonsmooth Analysis, LNCIS 301, pp. 351–361, 2004.

ways of developing general calculus rules in Banach spaces using this concept. Restricted by the length of paper, most details of the proofs are omitted. The readers are referred to [12] for details and more discussions.

2 Definitions and Preliminaries

The notations in this paper are standard (cf. [6], [14]). Let X be a Banach space. By X^* we mean its topological dual and $I\!B(X)$ its closed unit ball. Let Ω be a nonempty subset of X with $\bar{x} \in \Omega$. The *Fréchet ε-normal cone* of Ω at $\bar{x}$ is defined as

$$\widehat{N}_\varepsilon(\bar{x};\Omega) := \Big\{ x^* \in X^* \Big| \limsup_{x \xrightarrow{\Omega} \bar{x}} \frac{\langle x^*, x - \bar{x}\rangle}{\|x - \bar{x}\|} \le \varepsilon \Big\}.$$

When $\varepsilon = 0$, the above set reduces to a cone which is called *Fréchet normal cone* and is denoted by $\widehat{N}(\bar{x};\Omega)$ for simplicity.

Throughout the paper by T we mean an index set, which is a metric space. Let $\bar{t} \in T$ and $\{\Omega_t\}_{t\in T}$ be a collection of subsets in X with $\bar{x} \in \Omega_{\bar{t}}$. Then the *extended normal cone* $\check{N}(\bar{x};\Omega_{\bar{t}})$ is defined by

$$\check{N}(\bar{x};\Omega_{\bar{t}}) := \mathop{\mathrm{Limsup}}_{\substack{\varepsilon\downarrow 0,\, t\to\bar{t},\\ x\to\bar{x} \text{ with } x\in\Omega_t}} \widehat{N}_\varepsilon(x;\Omega_t),$$

where Limsup is the so-called *sequential Painlevé-Kuratowski upper limit* (see, for example, [6]). It is clear that when T is a singleton or Ω_t is the same set for all $t \in T$, the extended normal cone reduces to the basic (limiting) normal cone to a single set.

Let $F\colon X \rightrightarrows Y$ be a set-valued mapping between Banach spaces. Then the *Fréchet ε-coderivative* $\widehat{D}^*_\varepsilon F(\bar{x},\bar{y})\colon Y^* \rightrightarrows X^*$ of F at $(\bar{x},\bar{y}) \in \operatorname{gph} F$ is given by

$$\widehat{D}^*_\varepsilon F(\bar{x},\bar{y})(y^*) := \{x^* \in X^* \mid (x^*, -y^*) \in \widehat{N}_\varepsilon((\bar{x},\bar{y}); \operatorname{gph} F)\} \quad \forall y^* \in Y^*.$$

Let $\{F_t\}_{t\in T}$ be a collection of set-valued mappings from X to Y. Then the *extended coderivative* $\check{D}^* F_{\bar{t}}(\bar{x},\bar{y})\colon Y^* \rightrightarrows X^*$ at $(\bar{x},\bar{y}) \in \operatorname{gph} F_{\bar{t}}$ is defined as

$$\check{D}^* F_{\bar{t}}(\bar{x},\bar{y})(\bar{y}^*) := \mathop{\mathrm{Limsup}}_{\substack{\varepsilon\downarrow 0,\, y^* \xrightarrow{w^*} \bar{y}^*,\, t\to\bar{t}\\ (x,y)\to(\bar{x},\bar{y}) \text{ with } (x,y)\in \operatorname{gph} F_t}} \widehat{D}^*_\varepsilon F_t(x,y)(y^*) \quad \forall \bar{y}^* \in Y^*.$$

It is clear that we have the relation

$$\check{D}^* F_{\bar{t}}(\bar{x},\bar{y})(y^*) = \{x^* \in X^* \mid (x^*, -y^*) \in \check{N}((\bar{x},\bar{y}); \operatorname{gph} F_{\bar{t}})\}$$

for each $y^* \in Y^*$. We may also define the *extended mixed coderivative* $\check{D}^*_M F_{\bar{t}}(\bar{x},\bar{y})$ at $(\bar{x},\bar{y}) \in \operatorname{gph} F_{\bar{t}}$ as

$$\breve{D}^*_M F_{\bar{t}}(\bar{x},\bar{y})(\bar{y}^*) := \mathop{\mathrm{Limsup}}_{\substack{\varepsilon\downarrow 0,\, y^*\to\bar{y}^*,\, t\to\bar{t} \\ (x,y)\to(\bar{x},\bar{y}) \text{ with } (x,y)\in \mathrm{gph}\, F_t}} \widehat{D}^*_\varepsilon F_t(x,y)(y^*) \quad \forall \bar{y}^* \in Y^*.$$

Let $\varphi\colon X \to I\!R \cup \{\infty\}$ be an extended-real-valued function. The set-valued mapping $E_\varphi\colon X \rightrightarrows I\!R$ is defined so that $\mathrm{gph}\, E_\varphi = \mathrm{epi}\, \varphi$. Let $\{\varphi_t\}_{t\in T}$ be a collection of functions from X to $I\!R\cup\{\infty\}$. Then the *extended subdifferential* $\breve{\partial}\varphi_{\bar{t}}(\bar{x})$ and *extended singular subdifferential* $\breve{\partial}^\infty\varphi_{\bar{t}}(\bar{x})$ at $(\bar{t},\bar{x})$ are given as

$$\breve{\partial}\varphi_{\bar{t}}(\bar{x}) := \breve{D}^* E_{\varphi_{\bar{t}}}(\bar{x},\varphi_{\bar{t}}(\bar{x}))(1),$$
$$\breve{\partial}^\infty\varphi_{\bar{t}}(\bar{x}) := \breve{D}^* E_{\varphi_{\bar{t}}}(\bar{x},\varphi_{\bar{t}}(\bar{x}))(0).$$

Let δ_Ω be the indicator function of the set Ω that is defined as $\delta_\Omega(x) = 0$ for $x \in \Omega$ and ∞ otherwise. Then the following relation holds for $\bar{x} \in \Omega_{\bar{t}}$:

$$\breve{\partial}\delta_{\Omega_{\bar{t}}}(\bar{x}) = \breve{\partial}^\infty\delta_{\Omega_{\bar{t}}}(\bar{x}) = \breve{N}(\bar{x};\Omega_{\bar{t}}).$$

Now we turn to the *extended sequential normal compactness.*

Definition 1. *Let $I := \{1,\ldots,p\}$ be an index set and X_i $(i\in I)$ be Banach spaces with $\Omega_t \subset X_1\times\cdots\times X_p$ for all $t\in T$. Suppose $\bar{x}\in\Omega_{\bar{t}}$ and $I_1\subset I$, $I_2 := I\backslash I_1$.*

(i) $\{\Omega_t\}_{t\in T}$ *is* extendedly partially sequentially normally compact (ePSNC) *at $(\bar{t},\bar{x})$ with respect to $\{X_i \mid i\in I_1\}$ if, for any $\varepsilon_k\downarrow 0$, $t_k\to\bar{t}$, $x_k\to\bar{x}$ with $x_k\in\Omega_{t_k}$ and*

$$x^*_k = (x^*_{1,k},\ldots,x^*_{p,k}) \in \widehat{N}_{\varepsilon_k}(x_k;\Omega_{t_k}),$$

we have

$$\left[x^*_{i,k} \xrightarrow{w^*} 0 \quad \forall i\in I_1, \quad \|x^*_{i,k}\|\to 0 \quad \forall i\in I_2\right] \Longrightarrow \left[\|x^*_{i,k}\|\to 0 \quad \forall i\in I_1\right].$$

In particular, we say that $\{\Omega_t\}_{t\in T}$ *is* extendedly sequentially normally compact (eSNC) *at $(\bar{t},\bar{x})$ provided $I_1 = I$ (In this case, no product structure is needed).*

(ii) $\{\Omega_t\}_{t\in T}$ *is* strongly extendedly partially sequentially normally compact (strong ePSNC) *at $(\bar{t},\bar{x})$ with respect to $\{X_i\mid i\in I_1\}$ if, for any $\varepsilon_k\downarrow 0$, $t_k\to\bar{t}$, $x_k\to\bar{x}$ with*

$$x_k\in\Omega_{t_k}, \quad x^*_k\in\widehat{N}_{\varepsilon_k}(x_k;\Omega_{t_k}),$$

we have

$$\left[x^*_{i,k}\xrightarrow{w^*} 0 \quad \forall i\in I\right] \Longrightarrow \left[\|x^*_{i,k}\|\to 0 \quad \forall i\in I_1\right].$$

When T reduces to a singleton or Ω_t is a constant set with respect to $t\in T$, the above notions reduce to PSNC, SNC, strong PSNC, respectively (cf. [3], [6], [8], [11], [15]). Similarly to the latter notions, we can also define corresponding concepts for a collection of mappings/functions $\{F_t\}_{t\in T}$ by considering their graphs/epigraphs. In particular, we say that $\{F_t\}_{t\in T}$ is ePSNC at $(\bar{t},\bar{x},\bar{y}) \in \operatorname{gph} F_{\bar{t}}$ provided $\{\operatorname{gph} F_t\}_{t\in T}$ is ePSNC at $(\bar{t},\bar{x},\bar{y})$ with respect to X; and we say that $\{F_t\}_{t\in T}$ is eSNC at $(\bar{t},\bar{x},\bar{y}) \in \operatorname{gph} F_{\bar{t}}$ provided $\{\operatorname{gph} F_t\}_{t\in T}$ is eSNC at $(\bar{t},\bar{x},\bar{y})$. We say that $\{\varphi_t\}_{t\in T}$ is eSNEC at $(\bar{t},\bar{x})$ provided $\{\operatorname{epi}\varphi_t\}_{t\in T}$ is eSNC at $(\bar{t},\bar{x},\varphi_{\bar{t}}(\bar{x}))$.

We proceed to see some general properties of these normal compactness conditions. We say that a collection of vectors $\{x_t\}_{t\in T}$ in X is a *null sequence* at $\bar{t}\in T$ if, for any $t_k \to \bar{t}$, $x_{t_k} \to 0$. Then it is clear that the following holds:

Proposition 1. *Let $\{\Omega_t\}_{t\in T}$ be a collection of subsets in X and $\{x_t\}_{t\in T}$ be any null sequence at $\bar{t}$. Then $\{\Omega_t\}$ and $\{\Omega_t + x_t\}$ share the same normal cone and normal compactness at $\bar{t}$.*

While simple, the preceding proposition points out an important fact about extended normal cones, coderivatives, subdifferentials and compactness: these constructions/conditions are *not essentially related to the graph* $\{(t,x) \mid x \in \Omega_t, t\in T\}$, since the graph could be altered without affecting the underlying constructions/conditions. In particular, it does not make much sense to seek relations between these constructions/conditions for the moving objects and those for their graphs. Some examples and more discussions are given in [12].

Next let us single out some situations when such normal compactness conditions do hold.

Definition 2. *We say that the collection $\{\Omega_t\}_{t\in T}$ with $\bar{x}\in\Omega_{\bar{t}}$ is* uniformly compactly epi-Lipschitzian (CEL) *at $(\bar{t},\bar{x})$ provided there exists a neighborhood U of $\bar{x}$, O of 0, V of $\bar{t}$ and a compact set C, and a $\gamma_t > 0$ for each $t\in V$ such that*

$$\Omega_t \cap U + \gamma O \subset \Omega_t + \gamma C \quad \forall \gamma \in (0,\gamma_t) \quad \forall t \in V.$$

We say that the collection is uniformly epi-Lipschitzian *provided C can be selected as a singleton.*

This is a natural generalization of the CEL and epi-Lipschitzian properties for non-moving objects (see [1], [2], [13], etc.). We can generalize corresponding results to moving situations as in the following two propositions:

Proposition 2. *A collection $\{\Omega_t\}_{t\in T}$ of convex subsets in X is uniformly epi-Lipschitzian at $(\bar{t},\bar{x})$ if, for some neighborhood V of $\bar{t}$, $\cap_{t\in V}\Omega_t$ has nonempty interior.*

Proposition 3. *A collection $\{\Omega_t\}_{t\in T}$ is eSNC at $(\bar{t},\bar{x})$ if it is uniformly CEL at this point.*

For the case of moving set-valued mappings we propose the *uniform local Lipschitz-like* property in contrast to the *local Lipschitz-like (Aubin)* property of non-moving objects. The corresponding relation to normal compactness follows as a proposition.

Definition 3. *A collection* $\{F_t\}_{t\in T}$ *of mappings from* X *to* Y *is* uniformly locally Lipschitz-like *around* $(\bar{t},\bar{x},\bar{y})$ *with* $(\bar{x},\bar{y}) \in \operatorname{gph} F_{\bar{t}}$ *provided there exist* $\ell \geq 0$, *neighborhoods* U *of* $\bar{x}$ *and* V *of* $\bar{y}$ *such that*

$$F_t(x) \cap V \subset F_t(u) + \ell\|x-u\| I\!B(Y) \quad \forall x,u \in U, t \in T.$$

Proposition 4. *A collection* $\{F_t\}_{t\in T}$ *of mappings from* X *to* Y *is ePSNC at* $(\bar{t},\bar{x},\bar{y})$ *if it is uniformly locally Lipschitz-like at this point.*

3 Calculus of Extended Differentiation in Asplund Spaces

In this section we present basic calculus of the extended differentiation constructions from Section 2 for the case Asplund spaces. It turns out that such results can be established following methods similar to those in [10], which are based on the *fuzzy intersection rule* established therein (see also [8], [11]). This rule is derived from the *extremal principle* (see, for instance, [4], [5], [6], [7], [10]) which is convenient in applications. It is proved in [15] that this fuzzy intersection rule actually gives another characterization of Asplund spaces.

We start with the generalization of *qualification conditions.*

Definition 4. *Let* $\{\Omega_{j,t}\}_{t\in T}$ $(j=1,2)$ *be two collections of sets in* $X \times Y$ *with* $(\bar{x},\bar{y}) \in \Omega_{1,\bar{t}} \cap \Omega_{2,\bar{t}}$. *We say* $\{\Omega_{j,t}\}_{t\in T}$ $(j=1,2)$ *satisfy the* extended mixed qualification condition *at* $(\bar{t},\bar{x},\bar{y})$ *with respect to* Y *if for any sequences* $\varepsilon_k \downarrow 0$, $t_k \to \bar{t}$, $(x_{j,k},y_{j,k}) \to (\bar{x},\bar{y})$ *with* $(x_{j,k},y_{j,k}) \in \Omega_{j,t_k}$, *and*

$$(x^*_{j,k},y^*_{j,k}) \xrightarrow{w^*} (x^*_j,y^*_j)$$

with

$$(x^*_{j,k},y^*_{j,k}) \in \widehat{N}_{\varepsilon_k}((x_{j,k},y_{j,k});\Omega_{j,t_k}), \quad j=1,2,$$

one has

$$\left[x^*_{1,k}+x^*_{2,k} \xrightarrow{w^*} 0, \|y^*_{1,k}+y^*_{2,k}\| \to 0\right] \Longrightarrow (x^*_1,y^*_1) = (x^*_2,y^*_2) = 0$$

as $k \to \infty$. *In particular, we say that* $\{\Omega_{j,t}\}_{t\in T}$ $(j=1,2)$ *satisfy the* extended limiting qualification condition *at* $(\bar{t},\bar{x},\bar{y})$ *provided* $X=\{0\}$.

Now we present the pointbased *intersection rule* for extended normals, which is the basis for deriving all other results in this section.

Theorem 1. *Let* $X_1,\ldots,X_p$ *be Asplund spaces, let* $\{\Omega_{j,t}\}_{t\in T}$ $(j=1,2)$ *be two collections of closed sets in* $X_1 \times \cdots X_p$ *with* $\bar{x} \in \Omega_{1,\bar{t}} \cap \Omega_{2,\bar{t}}$, *and let* $I_1, I_2 \subset I := \{1,\ldots,p\}$ *with* $I_1 \cup I_2 = I$. *Assume that:*

(i) One of $\{\Omega_{1,t}\}_{t\in T}$, $\{\Omega_{2,t}\}_{t\in T}$ *is ePSNC with respect to* $\{X_i \mid i \in I_1\}$ *at* $(\bar{t},\bar{x})$ *and the other is strongly ePSNC with respect to* $\{X_i \mid i \in I_2\}$ *at* $(\bar{t},\bar{x})$.

(ii) The collections of sets $\{\Omega_{1,t}\}_{t\in T}$, $\{\Omega_{2,t}\}_{t\in T}$ *satisfy the extended limiting qualification condition at* $(\bar{t},\bar{x})$.

Then the intersection rule holds:

$$\check{N}(\bar{x};\Omega_{1,\bar{t}}\cap\Omega_{2,\bar{t}})\subset\check{N}(\bar{x};\Omega_{1,\bar{t}})+\check{N}(\bar{x};\Omega_{2,\bar{t}}).$$

To proceed, we need generalizations of the inner semicontinuity and inner semicompactness notions to moving settings as follows.

Definition 5. *Let* $S_t\colon X\rightrightarrows Y$ $(t\in T)$ *be a collection of set-valued mappings. Then we say that*

(i) $\{S_t\}_{t\in T}$ *is* extendedly inner semicompact *at* $(\bar{t},\bar{x})$ *provided for any sequence* $(t_k,x_k)\to(\bar{t},\bar{x})$ *with* $S_{t_k}(x_k)\neq\emptyset$ *there is a sequence* $y_k\in S_{t_k}(x_k)$ *that contains a subsequence converging to a point* $\bar{y}\in S_{\bar{t}}(\bar{x})$.
(ii) $\{S_t\}_{t\in T}$ *is* extendedly inner semicontinuous *at* $(\bar{t},\bar{x},\bar{y})$, *where* $\bar{y}\in S_{\bar{t}}(\bar{x})$, *provided for any sequence* $(t_k,x_k)\to(\bar{t},\bar{x})$ *with* $S_{t_k}(x_k)\neq\emptyset$ *there is a sequence* $y_k\in S_{t_k}$ *that contains a subsequence converging to* $\bar{y}$.

Using the intersection rule and the above definitions, we derive the following *sum rules* for extended normals, which seem to be new even for non-moving objects:

Theorem 2. *Let* $\{\Omega_{j,t}\}_{t\in T}$ $(j=1,2)$ *be two collections of closed sets in an Asplund space* X *and* $\bar{x}\in\Omega_{1,\bar{t}}+\Omega_{2,\bar{t}}$. *Assume that the collection of set-valued mappings* $\{S_t\}_{t\in T}$, *where* $S_t\colon X\rightrightarrows X\times X$ *with*

$$S_t(x)=\{(u,v)\mid u+v=x,u\in\Omega_{1,t},v\in\Omega_{2,t}\},$$

is extendedly inner semicompact at $(\bar{t},\bar{x})$. *Then one has*

$$\check{N}(\bar{x};\Omega_{1,\bar{t}}+\Omega_{2,\bar{t}})\subset\bigcup_{(u,v)\in S_{\bar{t}}(\bar{x})}\check{N}(u;\Omega_{1,\bar{t}})\cap\check{N}(v;\Omega_{2,\bar{t}}).$$

Moreover, if $\{S_t\}_{t\in T}$ *is extendedly inner semicontinuous at* $(\bar{t},\bar{x},\bar{u},\bar{v})$ *for some* $(\bar{u},\bar{v})\in S_{\bar{t}}(\bar{x})$, *then*

$$\check{N}(\bar{x};\Omega_{1,\bar{t}}+\Omega_{2,\bar{t}})\subset\check{N}(\bar{u};\Omega_{1,\bar{t}})\cap\check{N}(\bar{v};\Omega_{2,\bar{t}}).$$

The next result gives convenient formulas to compute extended normals to inverse images.

Theorem 3. *Let* X, Y *be Asplund spaces, let* $\{F_t\}_{t\in T}$ *be a collection of set-valued mappings from* X *to* Y *with closed graphs, and let* $\{\Theta_t\}_{t\in T}$ *be a collection of closed subsets of* Y *with* $\bar{x}\in F_{\bar{t}}^{-1}(\Theta_{\bar{t}})$. *Suppose that the collection of set-valued mappings* $x\mapsto F_t(x)\cap\Theta_t$ $(t\in T)$ *is extendedly inner semicompact at* $(\bar{t},\bar{x})$. *Then*

$$\check{N}(\bar{x};F_{\bar{t}}^{-1}(\Theta_{\bar{t}}))\subset\bigcup\{\check{D}^*F_{\bar{t}}(\bar{x},\bar{y})(y^*)\mid y^*\in\check{N}(\bar{y};\Theta_{\bar{t}}),\bar{y}\in F_{\bar{t}}(\bar{x})\cap\Theta_{\bar{t}}\}$$

provided that for each $\bar{y}\in F_{\bar{t}}(\bar{x})\cap\Theta_{\bar{t}}$ *the following assumptions hold:*

(i) *Either* $\{F_t^{-1}\}_{t\in T}$ *is ePSNC at* $(\bar t,\bar y,\bar x)$, *or* $\{\Theta_t\}_{t\in T}$ *is eSNC at* $(\bar t,\bar y)$.
(ii) *One has the qualification condition*

$$\check D^*_M F_{\bar t}^{-1}(\bar y,\bar x)(0)\cap[-\check N(\bar y;\Theta_{\bar t})]=\{0\}.$$

Based on this result, we can derive formulas of extended normals for collections of level sets $\{\Omega_t\}_{t\in T}$ in the inequality and equality forms $\{x\in X \mid \varphi_t(x)\le 0\}$ or $\{x\in X\mid \varphi_t(x)=0\}$, where $\{\varphi_t\}_{t\in T}$ is a collection of extended-real-valued functions; see [12] for more details.

Our next goal is to derive *sum rules* and *chain rules* for moving set-valued mappings. Let $F_{j,t}\colon X\rightrightarrows Y$ $(j=1,2,\ t\in T)$ be two collections of set-valued mappings. Define a new collection of set-valued mappings $S_t\colon X\times Y\rightrightarrows Y\times Y$ by

$$S_t(x,y):=\{(y_1,y_2)\in Y^2\mid y_1\in F_{1,t}(x),y_2\in F_{2,t}(x),y_1+y_2=y\}. \tag{1}$$

Theorem 4. *Let* $F_{j,t}\colon X\rightrightarrows Y$ $(j=1,2,\ t\in T)$ *be two closed-graph collections of set-valued mappings between Asplund spaces* X *and* Y, *and let* $(\bar x,\bar y)\in \operatorname{gph}(F_{1,\bar t}+F_{2,\bar t})$. *Suppose that the collection of mappings* $\{S_t\}$ *in (1) is extendedly inner semicompact at* $(\bar t,\bar x,\bar y)$. *Then*

$$\check D^*(F_{1,\bar t}+F_{2,\bar t})(\bar x,\bar y)(y^*)\subset\bigcup_{(\bar y_1,\bar y_2)\in S_{\bar t}(\bar x,\bar y)}\left[\check D^*F_{1,\bar t}(\bar x,\bar y_1)(y^*)+\check D^*F_{2,\bar t}(\bar x,\bar y_2)(y^*)\right]$$

for all $y^*\in Y^*$ *provided that for every* $(\bar y_1,\bar y_2)\in S_{\bar t}(\bar x,\bar y)$ *the following assumptions hold:*

(i) *Either* $\{F_{1,t}\}$ *is ePSNC at* $(\bar t,\bar x,\bar y_1)$, *or* $\{F_{2,t}\}$ *is ePSNC at* $(\bar t,\bar x,\bar y_2)$.
(ii) *One has the qualification condition*

$$\check D^*_M F_{1,\bar t}(\bar x,\bar y_1)(0)\cap[-\check D^*_M F_{2,\bar t}(\bar x,\bar y_2)(0)]=\{0\}. \tag{2}$$

Considering set-valued mappings associated with epigraphs of extended-real-valued functions, we can establish sum rules for extended subdifferentials and singular subdifferentials of parameter-dependent functions, which are omitted here.

Given $G_t\colon X\rightrightarrows Y$ and $F_t\colon Y\rightrightarrows Z$ $(t\in T)$, define a new collection of set-valued mappings $S_t\colon X\times Z\rightrightarrows Y$ by

$$S_t(x,z):=G_t(x)\cap F_t^{-1}(z)=\{y\in Y\mid y\in G_t(x),z\in F_t(y)\} \tag{3}$$

for all $t\in T$, $x\in X$, and $z\in Z$. We have the *coderivative chain rules.*

Theorem 5. *Let* X, Y, Z *be Asplund spaces, let*

$$\{G_t\colon X\rightrightarrows Y\}_{t\in T},\quad \{F_t\colon Y\rightrightarrows Z\}_{t\in T}$$

be two collections of set-valued mappings with closed graphs for all $t\in T$, *and let* $(\bar x,\bar z)\in\operatorname{gph}(F_{\bar t}\circ G_{\bar t})$. *Assume that* S_t *is extendedly inner semicompact at* $(\bar t,\bar x,\bar z)$ *and that for every* $\bar y\in S_{\bar t}(\bar x,\bar z)$ *the following hold:*

(i) *Either* $\{G_t^{-1}\}_{t\in T}$ *is ePSNC at* $(\bar{t},\bar{y},\bar{x})$, *or* $\{F_t\}_{t\in T}$ *is ePSNC at* $(\bar{t},\bar{y},\bar{z})$.
(ii) *The qualification condition is fulfilled:*

$$\check{D}^*_M G_{\bar{t}}^{-1}(\bar{y},\bar{x})(0)\cap(-\check{D}^*_M F_{\bar{t}}(\bar{y},\bar{z})(0))=\{0\}.$$

The one has the chain rule

$$\check{D}^*(F_{\bar{t}}\circ G_{\bar{t}})(\bar{x},\bar{z})\subset\bigcup_{\bar{y}\in G_{\bar{t}}(\bar{x})\cap F_{\bar{t}}^{-1}(\bar{z})}\check{D}^*G_{\bar{t}}(\bar{x},\bar{y})\circ\check{D}^*F_{\bar{t}}(\bar{y},\bar{z}),$$

where the inclusions hold for all $\bar{z}\in Z^*$ *on both sides.*

4 Calculus of Extended Normal Compactness in Asplund Spaces

As demonstrated in [9], it is crucial to develop calculus of normal compactness in order to apply calculus rules of generalized differentiation to optimization problems. We address the calculus issues for extended normal compactness in this section. It turns out that most of the results in [8], [11] can be generalized to the case of moving sets/mappings.

First let us present the basic intersection rule the proof of which can be obtained similarly to [11] using the method of the preceding section.

Theorem 6. *Let* $X_1,\ldots,X_p$ *be Asplund spaces, let* $\{\Omega_{j,t}\}_{t\in T}$ $(j=1,2)$ *be two collections of closed sets in* $X_1\times\cdots X_p$ *with* $\bar{x}\in\Omega_{1,\bar{t}}\cap\Omega_{2,\bar{t}}$, *and let* $I_1,I_2\subset I:=\{1,\ldots,p\}$ *with* $I_1\cup I_2=I$. *Assume that:*

(1) *For each* $j=1,2$ *the collection* $\{\Omega_{j,t}\}_{t\in T}$ *is ePSNC with respect to* $\{X_i\mid i\in I_j\}$ *at* $(\bar{t},\bar{x})$.
(2) *Either* $\{\Omega_{1,t}\}$ *is strongly ePSNC at* $(\bar{t},\bar{x})$ *with respect to* $\{X_i\mid i\in I_1\backslash I_2\}$, *or* $\{\Omega_{2,t}\}$ *is strongly ePSNC at* $(\bar{t},\bar{x})$ *with respect to* $\{X_i\mid i\in I_2\backslash I_1\}$
(3) *The collections of sets* $\{\Omega_{1,t}\}_{t\in T}$ *and* $\{\Omega_{2,t}\}_{t\in T}$ *satisfy the extended mixed qualification condition at* $(\bar{t},\bar{x})$ *with respect to* $\{X_i\mid i\in(I_1\backslash I_2)\cup(I_2\backslash I_1)\}$.

Then the collection of sets $\{\Omega_{1,t}\cap\Omega_{2,t}\}_{t\in T}$ *is ePSNC at* $(\bar{t},\bar{x})$ *with respect to* $\{X_i\mid i\in I_1\cap I_2\}$.

Similarly to the calculus of generalized differentiation in the previous section, we can derive normal compactness rules for sums of sets, inverse images, and level sets, which are omitted here. Let us just present sum and chain rules for the ePSNC property.

Theorem 7. *Let* $F_{j,t}\colon X\rightrightarrows Y$ $(j=1,2,\ t\in T)$ *be two closed-graph collections of set-valued mappings between Asplund spaces* X *and* Y, *and let* $(\bar{x},\bar{y})\in\operatorname{gph}(F_{1,\bar{t}}+F_{2,\bar{t}})$. *Suppose that the collection of mappings* $\{S_t\}$ *in (1) is extendedly inner semicompact at* $(\bar{t},\bar{x},\bar{y})$ *and impose the following assumptions for every* $(\bar{y}_1,\bar{y}_2)\in S_{\bar{t}}(\bar{x},\bar{y})$:

(i) *Each* $\{F_{j,t}\}$ *is ePSNC at* $(\bar{t}, \bar{x}, \bar{y}_j)$, $j = 1, 2$, *respectively.*
(ii) *The qualification condition (2) holds.*

Then $\{F_{1,t} + F_{2,t}\}_{t \in T}$ *is ePSNC at* $(\bar{t}, \bar{x}, \bar{y})$.

Theorem 8. *Let* X, Y, Z *be Asplund spaces, let*

$$\{G_t : X \rightrightarrows Y\}_{t \in T}, \quad \{F_t : Y \rightrightarrows Z\}_{t \in T}$$

be two closed-graph collections of set-valued mappings, and let $(\bar{x}, \bar{z}) \in \operatorname{gph}(F_{\bar{t}} \circ G_{\bar{t}})$. *Suppose that the collection of mappings* $\{S_t\}_{t \in T}$ *in (3) is extendedly inner semicompact at* $(\bar{t}, \bar{x}, \bar{z})$ *and that for every* $\bar{y} \in S_{\bar{t}}(\bar{x}, \bar{z})$ *the following assumptions hold:*

(i) *Either* $\{G_t\}_{t \in T}$ *is ePSNC at* $(\bar{t}, \bar{y}, \bar{x})$ *and* $\{F_t\}_{t \in T}$ *is ePSNC at* $(\bar{t}, \bar{y}, \bar{z})$, *or* $\{G_t\}_{t \in T}$ *is eSNC at* $(\bar{t}, \bar{y}, \bar{z})$.
(ii) *One has the qualification condition*

$$Ker\, \check{D}^* G_{\bar{t}}(\bar{y}, \bar{x})(0) \cap \check{D}^*_M F_{\bar{t}}(\bar{y}, \bar{z})(0) = \{0\}. \tag{4}$$

Then $\{F_t \circ G_t\}_{t \in T}$ *is ePSNC at* $(\bar{t}, \bar{x}, \bar{z})$.

5 Calculus Rules in General Banach Spaces

In this section we present some calculus results for both generalized differentiation and extended normal compactness in general Banach spaces. Let us start with a new differentiability concept, which is a generalization of strict differentiability to the case of moving mappings $\{f_t\}_{t \in T}$. In fact, it plays a similar role in developing calculus for moving objects as the strict differentiability does for non-moving ones.

Definition 6. *Let* $\{f_t\}_{t \in T}$ *be a collection of single-valued mappings between Banach spaces* X *and* Y. *Then the collection is* extendedly strictly differentiable *at* $(\bar{t}, \bar{x})$ *provided* $\lim_{t \to \bar{t}} f_t(\bar{x}) = f_{\bar{t}}(\bar{x})$ *and there exists a continuous linear operator* $A: X \to Y$ *such that*

$$\lim_{x_1, x_2 \to \bar{x}, x_1 \neq x_2, t \to \bar{t}} \frac{\|f_t(x_1) - f_t(x_2) - A(x_1 - x_2)\|}{\|x_1 - x_2\|} = 0.$$

By the definition, the extended strict differentiability of $\{f_t\}_{t \in T}$ at $(\bar{t}, \bar{x})$ implies the strict differentiability of $f_{\bar{t}}$ at $\bar{x}$ with $A = \nabla f_{\bar{t}}(\bar{x})$. Thus A is unique, and we call it the *extended strict derivative* and denote it by $\check{\nabla} f_{\bar{t}}(\bar{x})$. One can also observe that in this case f_t is Lipschitzian in a common neighborhood of $\bar{x}$ with a uniform Lipschitzian constant for all t close to $\bar{t}$, and that the following continuity-like property holds: $\lim_{t \to \bar{t}, x \to \bar{x}} f_t(x) = f_{\bar{t}}(\bar{x})$. By a direct calculation we can further show that

$$\check{D}^*_M f_{\bar{t}}(\bar{x}) = \check{D}^* f_{\bar{t}}(\bar{x}) = \{[\check{\nabla} f_{\bar{t}}(\bar{x})]^*\}.$$

Using this concept, we can establish sum rules and chain rules, both for extended differential constructions and for extended normal compactness, in the general Banach space setting. Let us just present calculus rules for generalized differentiation; the ones for normal compactness are formulated in a similar way (cf. the preceding section).

Theorem 9. *Let $\{f_t\}_{t\in T}$, $\{G_t\}_{t\in T}$ be collections of functions and set-valued mappings from X to Y, respectively. Assume $\{f_t\}_{t\in T}$ is extendedly strictly differentiable at $(\bar{t},\bar{x})$. Then for all $\bar{y}\in G_{\bar{t}}(\bar{x})$, one has*

$$\check{D}^*(f_{\bar{t}}+G_{\bar{t}})(\bar{x},f_{\bar{t}}(\bar{x})+\bar{y})=[\check{\nabla}f_{\bar{t}}(\bar{x})]^*+\check{D}^*G_{\bar{t}}(\bar{x},\bar{y}).$$

Theorem 10. *Let $\{F_t\}_{t\in T}$ be a collection of set-valued mappings from X to Y, and let $\{g_t\}_{t\in T}$, be a collection of functions from Y to Z with $(\bar{x},\bar{z})\in$ gph $(g_{\bar{t}}\circ F_{\bar{t}})$. Assume that the collection $\{S_t\}_{t\in T}$ of mappings from $X\times Z$ to Y with $S_t(x,z):=F(x)\cap g^{-1}(z)$ is extendedly inner semicompact at $(\bar{t},\bar{x},\bar{z})$, and that for each $\bar{y}\in S_{\bar{t}}(\bar{x},\bar{z})$ the collection $\{g_t\}_{t\in T}$ is extendedly strictly differentiable at $(\bar{t},\bar{y})$. Then for all $z^*\in Z^*$ one has*

$$\check{D}^*(g_{\bar{t}}\circ F_{\bar{t}})(\bar{x},\bar{z})(z^*)\subset\bigcup_{\bar{y}\in F_{\bar{t}}(\bar{x})\cap g_{\bar{t}}^{-1}(\bar{z})}\check{D}^*F_{\bar{t}}(\bar{x},\bar{y})([\check{\nabla}g_{\bar{t}}(\bar{y})]^*z^*).$$

If we assume instead that $\{S_t\}_{t\in T}$ is extendedly inner semicontinuous at $(\bar{t},\bar{x},\bar{z},\bar{y})$ for some $\bar{y}\in S_{\bar{t}}(\bar{x},\bar{z})$ and that $\{g_t\}_{t\in T}$ is extendedly strictly differentiable at $(\bar{t},\bar{y})$, then

$$\check{D}^*(g_{\bar{t}}\circ F_{\bar{t}})(\bar{x},\bar{z})(z^*)\subset\check{D}^*F_{\bar{t}}(\bar{x},\bar{y})([\check{\nabla}g_{\bar{t}}(\bar{y})]^*z^*).$$

Acknowledgement. The research of the first author was partly supported by the National Science Foundation under grants DMS-0072179 and DMS-0304989.

References

1. Borwein J, Lucet Y, Mordukhovich B (2000) Compactly epi-Lipschitzian convex sets and functions in normed spaces, J Convex Anal 7:375–393
2. Borwein J, Strojwas H (1985) Tangential approximations, Nonlinear Anal 9:1347–1366
3. Fabian M, Mordukhovich B (2003) Sequential normal compactness versus topological normal compactness in variational analysis, Nonlinear Anal 54:1057–1067
4. Mordukhovich B (2000) An abstract extremal principle with applications to welfare economics, J Math Anal Appl 251:187–216
5. Mordukhovich B (2001) The extremal principle and its applications to optimization and economics. In: Rubinov A, Glover B (eds) Optimization and Related Topics (Applied Optimization Volumes 47). Kluwer Academic Dordrecht 343–369

6. Mordukhovich B, Shao Y (1996) Nonsmooth sequential analysis in Asplund spaces, Trans Amer Math Soc 348:1235–1280
7. Mordukhovich B, Treiman J, Zhu Q (2004) An extended extremal principle with applications to multiobjective optimization, SIAM J Optim to appear
8. Mordukhovich B, Wang B (2001) Sequential normal compactness in variational analysis, Nonlinear Anal 47:717–728
9. Mordukhovich B, Wang B (2002) Necessary suboptimality and optimality conditions via variational principles, SIAM J Control Optim 41:623–640
10. Mordukhovich B, Wang B (2002) Extensions of generalized differential calculus in Asplund spaces, J Math Anal App 272:164–186
11. Mordukhovich B, Wang B (2003) Calculus of sequential normal compactness in variational analysis, J Math Anal App 282:63–84
12. Mordukhovich B, Wang B (2004) Calculus of extended generalized differentiation and normal compactness, preprint
13. Rockafellar R (1979) Directionally Lipschitzian functions and subdifferential calculus, Proc London Math Soc 39:331–355
14. Rockafellar R, Wets R (1998) Variational Analysis. Springer Berlin
15. Wang B (2004) Extensions of normal compactness conditions with applications, preprint

Lecture Notes in Control and Information Sciences

Edited by M. Thoma and M. Morari
2000–2004 Published Titles:

Vol. 250: Corke, P.; Trevelyan, J. (Eds)
Experimental Robotics VI
552 p. 2000 [1-85233-210-7]

Vol. 251: van der Schaft, A.; Schumacher, J.
An Introduction to Hybrid Dynamical Systems
192 p. 2000 [1-85233-233-6]

Vol. 252: Salapaka, M.V.; Dahleh, M.
Multiple Objective Control Synthesis
192 p. 2000 [1-85233-256-5]

Vol. 253: Elzer, P.F.; Kluwe, R.H.; Boussoffara, B.
Human Error and System Design and Management
240 p. 2000 [1-85233-234-4]

Vol. 254: Hammer, B.
Learning with Recurrent Neural Networks
160 p. 2000 [1-85233-343-X]

Vol. 255: Leonessa, A.; Haddad, W.H.; Chellaboina V.
Hierarchical Nonlinear Switching Control Design with Applications to Propulsion Systems
152 p. 2000 [1-85233-335-9]

Vol. 256: Zerz, E.
Topics in Multidimensional Linear Systems Theory
176 p. 2000 [1-85233-336-7]

Vol. 257: Moallem, M.; Patel, R.V.; Khorasani, K.
Flexible-link Robot Manipulators
176 p. 2001 [1-85233-333-2]

Vol. 258: Isidori, A.; Lamnabhi-Lagarrigue, F.; Respondek, W. (Eds)
Nonlinear Control in the Year 2000 Volume 1
616 p. 2001 [1-85233-363-4]

Vol. 259: Isidori, A.; Lamnabhi-Lagarrigue, F.; Respondek, W. (Eds)
Nonlinear Control in the Year 2000 Volume 2
640 p. 2001 [1-85233-364-2]

Vol. 260: Kugi, A.
Non-linear Control Based on Physical Models
192 p. 2001 [1-85233-329-4]

Vol. 261: Talebi, H.A.; Patel, R.V.; Khorasani, K.
Control of Flexible-link Manipulators Using Neural Networks
168 p. 2001 [1-85233-409-6]

Vol. 262: Dixon, W.; Dawson, D.M.; Zergeroglu, E.; Behal, A.
Nonlinear Control of Wheeled Mobile Robots
216 p. 2001 [1-85233-414-2]

Vol. 263: Galkowski, K.
State-space Realization of Linear 2-D Systems with Extensions to the General nD (n>2) Case
248 p. 2001 [1-85233-410-X]

Vol. 264: Baños, A.; Lamnabhi-Lagarrigue, F.; Montoya, F.J
Advances in the Control of Nonlinear Systems
344 p. 2001 [1-85233-378-2]

Vol. 265: Ichikawa, A.; Katayama, H.
Linear Time Varying Systems and Sampled-data Systems
376 p. 2001 [1-85233-439-8]

Vol. 266: Stramigioli, S.
Modeling and IPC Control of Interactive Mechanical Systems – A Coordinate-free Approach
296 p. 2001 [1-85233-395-2]

Vol. 267: Bacciotti, A.; Rosier, L.
Liapunov Functions and Stability in Control Theory
224 p. 2001 [1-85233-419-3]

Vol. 268: Moheimani, S.O.R. (Ed)
Perspectives in Robust Control
390 p. 2001 [1-85233-452-5]

Vol. 269: Niculescu, S.-I.
Delay Effects on Stability
400 p. 2001 [1-85233-291-316]

Vol. 270: Nicosia, S. et al.
RAMSETE
294 p. 2001 [3-540-42090-8]

Vol. 271: Rus, D.; Singh, S.
Experimental Robotics VII
585 p. 2001 [3-540-42104-1]

Vol. 272: Yang, T.
Impulsive Control Theory
363 p. 2001 [3-540-42296-X]

Vol. 273: Colonius, F.; Grüne, L. (Eds)
Dynamics, Bifurcations, and Control
312 p. 2002 [3-540-42560-9]

Vol. 274: Yu, X.; Xu, J.-X. (Eds)
Variable Structure Systems: Towards the 21st Century
420 p. 2002 [3-540-42965-4]

Vol. 275: Ishii, H.; Francis, B.A.
Limited Data Rate in Control Systems with Networks
171 p. 2002 [3-540-43237-X]

Vol. 276: Bubnicki, Z.
Uncertain Logics, Variables and Systems
142 p. 2002 [3-540-43235-3]

Vol. 277: Sasane, A.
Hankel Norm Approximation for Infinite-Dimensional Systems
150 p. 2002 [3-540-43327-9]

Vol. 278: Chunling D. and Lihua X. (Eds)
H_∞ Control and Filtering of
Two-dimensional Systems
161 p. 2002 [3-540-43329-5]

Vol. 279: Engell, S.; Frehse, G.; Schnieder, E. (Eds)
Modelling, Analysis, and Design of Hybrid Systems
516 p. 2002 [3-540-43812-2]

Vol. 280: Pasik-Duncan, B. (Ed)
Stochastic Theory and Control
564 p. 2002 [3-540-43777-0]

Vol. 281: Zinober A.; Owens D. (Eds)
Nonlinear and Adaptive Control
416 p. 2002 [3-540-43240-X]

Vol. 282: Schröder, J.
Modelling, State Observation and
Diagnosis of Quantised Systems
368 p. 2003 [3-540-44075-5]

Vol. 283: Fielding, Ch. et al. (Eds)
Advanced Techniques for Clearance of
Flight Control Laws
480 p. 2003 [3-540-44054-2]

Vol. 284: Johansson, M.
Piecewise Linear Control Systems
216 p. 2003 [3-540-44124-7]

Vol. 285: Wang, Q.-G.
Decoupling Control
373 p. 2003 [3-540-44128-X]

Vol. 286: Rantzer, A. and Byrnes C.I. (Eds)
Directions in Mathematical Systems
Theory and Optimization
399 p. 2003 [3-540-00065-8]

Vol. 287: Mahmoud, M.M.; Jiang, J. and Zhang, Y.
Active Fault Tolerant Control Systems
239 p. 2003 [3-540-00318-5]

Vol. 288: Taware, A. and Tao, G.
Control of Sandwich Nonlinear Systems
393 p. 2003 [3-540-44115-8]

Vol. 289: Giarré, L. and Bamieh, B.
Multidisciplinary Research in Control
237 p. 2003 [3-540-00917-5]

Vol. 290: Borrelli, F.
Constrained Optimal Control
of Linear and Hybrid Systems
237 p. 2003 [3-540-00257-X]

Vol. 291: Xu, J.-X. and Tan, Y.
Linear and Nonlinear Iterative Learning Control
189 p. 2003 [3-540-40173-3]

Vol. 292: Chen, G. and Yu, X.
Chaos Control
380 p. 2003 [3-540-40405-8]

Vol. 293: Chen, G. and Hill, D.J.
Bifurcation Control
320 p. 2003 [3-540-40341-8]

Vol. 294: Benvenuti, L.; De Santis, A.; Farina, L. (Eds)
Positive Systems: Theory and Applications (POSTA 2003)
414 p. 2003 [3-540-40342-6]

Vol. 295: Kang, W.; Xiao, M.; Borges, C. (Eds)
New Trends in Nonlinear Dynamics and Control,
and their Applications
365 p. 2003 [3-540-10474-0]

Vol. 296: Matsuo, T.; Hasegawa, Y.
Realization Theory of Discrete-Time Dynamical Systems
235 p. 2003 [3-540-40675-1]

Vol. 297: Damm, T.
Rational Matrix Equations in Stochastic Control
219 p. 2004 [3-540-20516-0]

Vol. 298: Choi, Y.; Chung, W.K.
PID Trajectory Tracking Control for Mechanical Systems
127 p. 2004 [3-540-20567-5]

Vol. 299: Tarn, T.-J.; Chen, S.-B.; Zhou, C. (Eds.)
Robotic Welding, Intelligence and Automation
214 p. 2004 [3-540-20804-6]

Vol. 300: Nakamura, M.; Goto, S.; Kyura, N.; Zhang, T.
Mechatronic Servo System Control
Problems in Industries and their Theoretical Solutions
212 p. 2004 [3-540-21096-2]
(forthcoming)

Printing and Binding: Strauss GmbH, Mörlenbach